T0219088

Artificial Intelligence in Industry 4.0 and 5G Technology

Artificial Intelligence in Industry 4.0 and 5G Technology

Edited by

Pandian Vasant
MERLIN Research Centre, Ton Duc Thang University
Vietnam

Elias Munapo
Department of Business Statistics and Operations Research
North West University, Mahikeng
Mmabatho, SA

J. Joshua Thomas
UOW Malaysia, KDU Penang University College
Penang, Malaysia

Gerhard-Wilhelm Weber
Department of Marketing and Economic Engineering
Poznan University of Technology
Poznan, PL

Registered Office
John Wiley & Sons, Inc., 111 River Street, Hoboken, NJ 07030, USA

Editorial Office
111 River Street, Hoboken, NJ 07030, USA

For details of our global editorial offices, customer services, and more information about Wiley products visit us at www.wiley.com.

Wiley also publishes its books in a variety of electronic formats and by print-on-demand. Some content that appears in standard print versions of this book may not be available in other formats.

Library of Congress Cataloging-in-Publication Data Applied for
ISBN: 9781119798767

Cover Design: Wiley
Cover Image: © Fit Ztudio/Shutterstock

Set in 9.5/12.5pt STIXTwoText by Straive, Chennai, India

Contents

List of Contributors

Haidar Abbas
Department of Business
Administration, Salalah College of
Applied Sciences
University of Technology and Applied
Sciences
Sultanate of Oman

Sadia S. Ali
Department of Industrial Engineering
College of Engineering
Jeddah 21589
Saudi Arabia

Nancy M. Arratia-Martinez
Universidad de las Américas Puebla
Department of Business
Administration, Ex-Hacienda Santa
Catarina Mártir S/N
Puebla, 72810
México

Arrieta-M Luisa F
Simon Bolivar University
Barranquilla
Atlántico
Colombia

Paulina A. Avila-Torres
Universidad de las Américas Puebla
Ex-Hacienda Santa Catarina Mártir
Department of Business
Administration
San Andrés Cholula Puebla, C.P. 72810
México

Swapnoj Banerjee
Maulana Abul Kalam Azad University
of Technology
Meghnad Saha Institute of Technology
Department of Computer Science and
Engineering
Kolkata
West Bengal 700150
India

Choo K. Chin
University of Northern Iowa
Bachelor of Arts in Computer
Information System

Joshua E. Chukwuere
North-West University
Department of Information Systems
South Africa

Irraivan Elamvazuthi
Persiaran UTP
Seri Iskandar
Perak

Timothy Ganesan
Member of American Mathematical
Society
University Drive NW
University of Calgary
Alberta
Canada

Alexander N. Gneushev
Moscow Institute of Physics and
Technology
Department of Control and Applied
Mathematics
Moscow Region, 141701
Russia

and

Federal Research Center "Computer
Science and Control" of Russian
Academy of Sciences
Dorodnicyn Computing Center
Moscow
Russia

Alexey D. Grigorev
Moscow Institute of Physics and
Technology
Department of Control and Applied
Mathematics
Moscow Region, 141701
Russia

Soumodipto Halder
Maulana Abul Kalam Azad University
of Technology
Meghnad Saha Institute of Technology
Department of Computer Science and
Engineering
Kolkata
West Bengal 700150
India

Y. Bevish Jinila
Department of Information
Technology
Sathyabama Institute of Science and
Technology
Chennai
India

Cinthia Joy
Sydney International School of
Technology and Commerce
Sydney
Australia

Rajbir Kaur
Government Girls College
Panchkula
Haryana 134001
India

Vladimir A. Kozub
Moscow Institute of Physics and
Technology
Department of Control and Applied
Mathematics
Moscow Region, 141701
Russia

and

AEROCOSMOS Research Institute for
Aerospace Monitoring
Moscow, 105064
Russia

Igor S. Litvinchev
Federal Research Center "Computer
Science and Control" of Russian
Academy of Sciences
Dorodnicyn Computing Center
Moscow, 119333
Russia

Lopez-I Fernando
Autonomous University of Nuevo
León
Department of Mechanical and
Electronic Engineering
San Nicolás de los Garza
Nuevo León
México

Elias Manopo
Department of Statistics and
Operations Research, School of
Economic Sciences
North West University
Mahikeng
South Africa

Ivan A. Matveev
Federal Research Center "Computer
Science and Control" of Russian
Academy of Sciences
Dorodnicyn Computing Center
Moscow, 119333
Russia

Boitumelo Molefe
North-West University
Department of Information Systems
South Africa

Subhash Mondal
Maulana Abul Kalam Azad University
of Technology
Meghnad Saha Institute of Technology
Department of Computer Science and
Engineering
Kolkata, West Bengal 700150
India

Alexander B. Murynin
AEROCOSMOS Research Institute for
Aerospace Monitoring
Moscow, 105064
Russia

and

Federal Research Center "Computer
Science and Control" of Russian
Academy of Sciences
Dorodnicyn Computing Center
Moscow, 119333
Russia

Denis Y. Nartsev
Moscow Institute of Physics and
Technology
Department of Control and Applied
Mathematics
Moscow Region, 141701
Russia

Juan Jesús Tello Rodríguez
Universidad Autonoma de Nuevo
León
Facultad de Ingeniería Mecánica y
Eléctrica (FIME), Posgrado en
Ingeniería en Sistemas (PISIS)
San Nicolás de los Garza
Nuevo León 66451
México

Diganta Sengupta
Maulana Abul Kalam Azad University
of Technology
Meghnad Saha Institute of Technology
Department of Computer Science and
Engineering
Kolkata
West Bengal 700150
India

S. Prayla Shyry
Sathyabama Institute of Science and
Technology
Chennai
India

Isidro Soria-Arguello
Departamento de Ingeniería Química,
Industrial y de Alimentos
Universidad Iberoamericana
Ciudad de Mexico 01219
Mexico

J. Joshua Thomas
Department of Computing
UOW Malaysia
KDU Penang University College
Penang
Malaysia

Rafael Torres-Esobar
Facultad de Ingeniería
Universidad Anáhuac México
Huixquilucan 52786
Mexico

Pandian Vasant
Ton Duc Thang University
Modeling Evolutionary Algorithms
Simulation & Artificial Intelligence
(MERLIN), Faculty of Electrical &
Electronic Engineering
Ho Chi Minh City 700000
Vietnam

Deng H. Xiang
He Ying Metal Industries Sdn Bhd.
Malaysia

Preface

Regarded from our editorial views, this monograph has been wished to become a most valuable reference work of 2021 within the domains of modern information and decision aiding, urban and international, local and regional, ecological and spatiotemporal, industrial and natural operational research, artificial intelligence and creative arts and sciences. Proceeding diligent work from the sides of the authors and of us all, the endeavor of this work succeeded by stimulating, gathering, and compiling the newest research on the present state and inventions of electric and electronic, informational and energetic, green and recoverable, creative and re-creative means and their smart and dynamic employment, with care and responsibility. Our thus given monograph and handbook about scientific research on concerns about natural and human resources and their supply is a remarkable scholarly resource. It analyzes and discusses the efficient utilization of those resources that have a supportive impact on sustainable development and relations of, within and between us humans and our communities, cities and rural countryside, and finally migrations, social peace, and peace among our countries.

For this book's international directions, it has advanced toward a very special resource outlining the remarkable progress obtained worldwide, related to artificial intelligence, operational research, electronic, information and mobility devices and methods, renewable energy, natural resources, etc. The book is on the way to becoming respected on a global level for its broad analytic and practical contents.

"*Artificial Intelligence in Industry 4.0 and 5G Technology*" provides details on cutting-edge methodologies utilized in business and industrial sectors. It gives a holistic background on innovative optimization applications, focusing on main technology sectors such as 5G networks, Industry 4.0, and robotics. It discusses topics such as hyper-heuristics algorithmic enhancements and performance measurement approaches and provides keen insights into the implementation of meta-heuristic strategies to many-objectives optimization real-life problems in business, economics, and finance. With this book, the esteemed readers can learn to solve real-world sustainable optimization problems effectively using the appropriate techniques from emerging fields including artificial intelligence, hybrid evolutionary and swarm intelligence, hyper-heuristics programming, and many-objectives optimization.

"Artificial Intelligence in Industry 4.0 and 5G Technology" is a well-chosen collection of creative research about the methodologies and utilization of deep learning approaches in business, economics and finance, science and engineering, neuroscience, and medicine. While highlighting topics including intelligent optimization and computational modeling, data hybridization, and artificial intelligence, this work is ideally shaped and elaborated for high-tech experts and engineers, IT specialists and big-data analysts, data scientists and engineers, researchers and academicians, philanthropes, and political decision makers who look for contemporary studies on deep learning and its fruitful utilization in upcoming smart and green industries.

Deep Learning is a method that is transforming the world of data and analytics. *Optimization* of this new approach is still unclear; however, there is a need for research on various applications and techniques of deep learning in the areas of artificial intelligence and machine learning. Modern methodologies and tools from neuro-imaging, brain imaging, etc., today generate high-quality neurophysiological data with a resolution quality never reached before. These accelerating dynamics generate promising pathways to improve our comprehension of the nervous system and eventually of deeper learning. Computational issues occurred because of the high complexity of neuronal systems and the big number of constituents with still unknown connections between them. Highly innovative considerations and methods of computational neuroscience lead to more realistic biophysical representations which provide amazing chances for conditional behavior and links among brain regions in economic, professional, and our daily decision-making spheres.

This handbook encompasses three areas called "units:"

Unit 1 Industry 4.0:

Advanced Techniques and Technologies for Energy Saving, Artificial Intelligence in Smart Agriculture and Agroengineering and their AI-optimized Hardware, Biometrics, Big-Data, Cloud Computing, Cybersecurity, Embedded Systems, Fractional Differential Approach on Machine Learning, Graph-based Data Analysis, Grid Computing, Internet of Things (IoT), Intelligent Spindle Frameworks, Knowledge Representation and Reasoning, Smart Manufacturing, Spindle Frameworks, Manufacturing Intelligence and Informatics, Unmanned Arial Vehicles (drones technology).

Unit 2 Artificial Intelligence:

Augmented AI, Adaptive Systems, Bioinformatics, Data Mining, Deep Learning, Evolutionary Computations, Fuzzy Logic, Hybrid and Nonlinear Systems,

Knowledge Representation and Reasoning, Machine Learning, Meta Heuristics, Mathematical Modelling in Artificial Intelligence, Natural Language Processing, Natured-Inspired Algorithms, Robotics Automation, Swarm Intelligence.

Unit 3 5G Technology:

Innovative Smart Cities Design and Applications, Mathematical Optimization in Engineering and Business Applications, Multi-task Learning, Radio Communications Technologies, Remote Access and Control, Robotic Process Automation, Sequential and Image Processing, Smart Communications Systems, Smart Grid, Speech Recognition, Sensors, Virtual Machines, Vehicular Networks, Wireless Sensor Networks, Wearable Technologies.

Subsequently, we provide a short introduction to the fifteen chapters of this work.

In the first chapter *"Dynamic Key-based Biometric End User Authentication Proposal for IoT in Industry 4.0,"* the coauthors *Subhash Mondal, Swapnoj Banerjee, Soumodipto Halder, and Diganta Sengupta* recalls research and growth in IoT, both in the prospects of architecture as well as a voluminous increase in inter-networked devices. This chapter concentrates on securing the data acquisition at the end-user node level by means of biometric authentication. Extended AES algorithm incorporates an S-box, which defines the edge security through a nonlinear behavior. This seems to be a first attempt to combine Minutiae Extraction algorithm, Key Generation algorithm, AES, and fingerprint authentication to generate a single Edge-Framework for fingerprint authentication for the end users.

By intelligence algorithms, the second chapter *"Decision Support Methodology for Scheduling Orders in Addictive Manufacturing"* co-authored by *Juan Jesús Tello Rodriguez and Fernando Lopez Irarragorri* studies additive manufacturing, a family of manufacturing technologies where a 3-dimensional solid is manufactured, shaped from a computerized model by depositing thin layers of material. It is considered one of the most important emerging technologies because of the multiple benefits it brings for businesses, and because it is harmoniously coupled with Industry 4.0 and the digitization of manufacturing. This work addresses a variant of the job-shop re-scheduling problem in additive manufacturing. The new approach is applied to a real-world problem.

In the third chapter called *"Significance in consuming 5G built Artificial Intelligence in smart cities,"* *Y.Bevish Jinila, Cinthia Joy Godly, Joshua Thomas and S.Prayla Shyry* recall that smart cities have a greater potential toward convenient, comfortable, and automated applications that could help humans make things easier. The traditional systems do not have sufficient models. AI has experienced a great "boom" in smart transportation. The advent of 5G has created a greater impact in the field of telecommunication, where the data transfer speed enormously increased. In this chapter, the significance of 5G-built AI in smart cities is presented. The 5G-built AI model for smart cities is also modeled. The proposed model highlights the importance of applying 5G with AI to improve the performance of the system.

The fourth chapter, titled *"Neural network approach to segmentation of economic infrastructure objects on high-resolution satellite images,"* authored by *V.A. Kozub, A.B. Murynin, I.S. Litvinchev, I.A. Matveev, and P. Vasant*, addresses the problem of semantic segmentation of infrastructure objects in high-resolution satellite images, considered as an integral part of the method for constructing digital terrain models. Semantic segmentation involves classes such as buildings, roads, and railways. A set of labeled satellite images is collected, and neural network architecture is selected and trained. In order to reduce the imbalance of classes in the training sample, a probabilistic method of augmentation is developed and applied.

The fifth chapter *"The impact of data security on the Internet of Things"* written by coauthors *Joshua Ebere Chukwuere and Boitumelo Molefe* reminds us that the IoT is making lives easier and more productive; security on the IoT has also increased. The main purpose of this chapter is to determine the impact of data security on IoT. A quantitative methodology was used where data were collected using questionnaires and analyzed using Statistical Package for Social Sciences (SPSS) software. The results show that security threats negatively impact the IoT which affects IoT devices adversely. Lastly, users of IoT should be made aware of these security threats and how they can keep their devices safe.

In the sixth chapter *"Sustainable renewable energy and waste management on weathering corporate pollution,"* the coauthors *Choo Kwong Chin and Deng Huo Xiang* investigate the fourth Industrial Revolution, the man, and the earth. The majority of the environmental problems can be traced to industrialization, particularly since the "great acceleration" in global economic activity. As the fourth Industrial Revolution is pacing fast, innovations are becoming faster, more efficient, and widely accessible than ever before. Emerging technologies, including the IoT, virtual reality, AI, Block-Chain technology, etc. This study in this chapter shows that enterprises responsibly consider the environmental impacts of their activities and undertake actions aimed at preserving the environment and its resources.

The seventh chapter *"Adam adaptive optimization method for neural network models regression in image recognition tasks,"* coauthored by *D. Yu. Nartsev, A. N. Gneushev, and I. A. Matveev* proposes to apply adaptive optimization in neural network regression tasks like estimating image quality or aligning objects in the frame. Two sample tasks are presented. The first one is the evaluation of the degree of blurring in the iris recognition system. Eye images are taken from BATH and CASIA databases and Gaussian blurring generates samples. The second sample task is the alignment of the face in an image. Both tasks are solved by the direct estimation of parameters with neural networks. The resulting accuracy of parameter estimation is acceptable for practical use. The Adam algorithm and its modifications, such as AdamW and Radam, are applied.

In the eighth Chapter *"Application of integer programming in allocating energy resources in rural Africa"*, the author, *Elias Manopo,* presents a solution method for the quadratic assignment problem. In this approach, the quadratic assignment problem is first linearized into a linear binary problem. The linear binary problem is then converted into a convex quadratic problem, which is then solved efficiently by interior point algorithms. In addition to allocating resources, the quadratic assignment problem can also be used in numerical analyses and dartboard constructions, archaeology, and statistical analysis, chemical reactions and processes, economic problem modeling and decision frameworks, hospital layouts, and backboard wiring problems, campus planning models, and many more.

In the ninth chapter *"The Feasibility of Drones as the Next Step in Innovative Solution for Emerging Society,"* the coauthors *Sadia Samar Ali, Rajbir Kaur, and Haidar Abbas* state that organizations are looking for technologies capable of accomplishing multiple tasks, providing economic benefits and an edge over their competitors. Technology-related innovations are taking businesses toward a broad spectrum, related to philanthropy, social welfare, and well-being of all the related stakeholders. Drone technology has arrived to advance human efficiencies. This study examines various aspects of drone technology viz. utilities, complexities involved, and its prospects, especially in the Indian Context. Furthermore, it aims to identify and establish the various factors impacting "Technovation."

The 10th Chapter called *"Designing a distribution network for a Soda Company: Formulation and efficient solution procedure"* by coauthors *Isidro Soria Arguello, Rafael Torres Escobar, Hugo Alexer Pérez-Vicente and Pandian Vasant* aims at professionals and university students interested in the redesign and optimization of primary distribution networks related to consumer products, in a three-tier distribution system. A MILP model is presented to design a distribution network for a soda company. To solve the model efficiently, Lagrangian relaxation technique is used. Four scenarios are analyzed. The solutions obtained are feasible in the original problem. It is guaranteed that the demand of each distribution center will be satisfied by only one macro-center.

In the 11th Chapter named *"Machine learning and MCDM approach to characterize student attrition in higher education,"* the coauthors *Luisa Fernanda Arrieta and Fernando Lopez-Irarragorri* discuss school dropout as a global problem. The authors present a methodology based on multicriteria decision making, and machine learning addresses assess factors related to student attrition in universities. As a real case, data from the "Universidad Simon Bolivar" are developed to demonstrate the application of the proposed methodology. Some interesting findings of factors related to student dropout are discussed.

In the 12th Chapter "*A concise review on recent optimization and deep learning applications in blockchain technology*," the coauthors *Timothy Ganesan, Irraivan Elamvazuthi, Pandian Vasant, and J. Joshua Thomas* investigate blockchain technology as one of the most influential disruptive technologies today. The objective is to provide a concise and recent review of the implementation of optimization and deep learning techniques on blockchain-based systems. Application areas considered are computational optimization frameworks, the IoT, smart grids, supply chain management, and healthcare data systems. Implementation of optimization and deep learning techniques are presented and discussed.

In the 13th Chapter, "*Inventory routing problem with fuzzy demand and deliveries with priority*," the coauthors, *Paulina A. Avila-Torres and Nancy Maribel Arratia Martinez*, consider the inventory routing problem based on a producer and distributor company of gases with three main products. The main customers are industry and hospitals. The inventory level is monitored to establish the amount of product to deliver. The problem studied aims to guarantee a minimum inventory level of the customer prioritizing a group of customers in each route and taking into account demand as fuzzy. A mathematical model based on the vehicle routing problem is presented with the main objective is to minimize the distribution cost. The model is tested with a case of study.

In the 14th Chapter, "*A comparison of defuzzification methods for selecting projects*," the coauthors *Nancy Maribel Arratia Martinez, Paulina A. Avila-Torres, and Fernando López* state that study the problem of project portfolio selection when the resources are subject to budget availability. They model the uncertainty in the total budget to carry out projects and the uncertainty in restricted budgets to different areas using triangular fuzzy numbers. Naturally, different methods result in different project portfolios. For this reason, the authors discuss the results of the application of different methods to compare fuzzy numbers, using SAUGMECON as the solution method.

The 15th Chapter, "*Re-identification-based models for multiple object tracking*" by the coauthors *A.D. Grigorev, A.N. Gneushev, and I.S. Litvinchev* addresses the multiple objects tracking problem. By factorization of the posterior distribution of objects' parameters, it is proven that the original problem is equivalent to the procedure containing two steps. Given that track measurements are not equal in terms of their usefulness for re-identification, the technique of track descriptor pre-filtering is introduced. Known quality assessment methods and an alternative detector-based approach are taken into account. Computational experiments are conducted. The results showed computational efficiency and increased stability.

We, the *editors*, hope that the chosen fields and picked themes within this work will reflect a key selection of worldwide research coping with emerging and complex, sometimes long-lasting problems of *Artificial Intelligence in Industry 4.0 and 5G Technology* and their domains in everyone's life, in professional and daily lives, in natural sciences and development, in city and country planning, in high technology and the arts, in economics and finance, in healthcare and medicine, by approaches, methods and results of operational research and artificial intelligence. We are very thankful to the publishing house of *Wiley* for the distinction of hosting our compendium as a pioneering scientific endeavor. Special gratitude is conveyed to the editors of publishing house *Wiley*, to editorial managers and staff for their continuous advice and help in every respect. We convey our thanks to all the authors for their hard work and willingness to share their novel insights and amazing inventions with our worldwide community. We sincerely hope that our authors' studies will crystalize and initialize collaboration and progress at a global and premium stage, and as a service of humility and excellence to humankind and the entire creation.

Profile of Editors

Pandian Vasant is an Editor-in-Chief of the International Journal of Energy Optimization and Engineering (IJEOE) and a Research Associate at MERLIN Research Centre of Ton Duc Thang University. He holds PhD in Computational Intelligence (UNEM, Costa Rica), MSc (University Malaysia Sabah, Malaysia, Engineering Mathematics), and BSc (Hons, Second Class Upper) in Mathematics (University of Malaya, Malaysia). His research interests include Soft Computing, Hybrid Optimization, Innovative Computing and Applications. He has co-authored numerous research articles in journals, conference proceedings, presentations, special issues guest editor, book chapters (300 publications indexed in Google Scholar) and General Chair of the EAI International Conference on Computer Science and Engineering in Penang, Malaysia (2016) and Bangkok, Thailand (2018). In the years 2009 and 2015, Dr. Pandian Vasant was awarded top reviewer and outstanding reviewer for the journal Applied Soft Computing (Elsevier), respectively. He has 31 years of working experience at various universities. Currently, he is an Editor-in-Chief of International Journal of Energy Optimization & Engineering, Member of AMS (USA), NAVY Research Group (TUO, Czech Republic), MERLIN Research Centre (TDTU, Vietnam), and General Chair of the International Conference on Intelligent Computing and Optimization (https://www.icico.info/). H-Index Google Scholar = 34; i-10-index = 144.
http://www.igi-global.com/ijeoe
E-mail: pvasant@gmail.com

Elias Munapo is a Professor of Operations Research and he holds a BSc. (Hons) in Applied Mathematics (1997), MSc. in Operations Research (2002), and a Ph.D. in Applied Mathematics (2010). All these qualifications are from the National University of Science and Technology (N.U.S.T.) in Zimbabwe. In addition, he has a certificate in outcomes-based assessment in Higher Education and Open distance learning, from the University of South Africa (UNISA) and another certificate in University Education Induction Programme from the University of

KwaZulu-Natal (UKZN). Elias Munapo is a Professional Natural Scientist certified by the South African Council for Natural Scientific Professions (SACNASP) in 2012. Elias Munapo has vast experience in university education and has worked for five (5) institutions of higher learning. The institutions are Zimbabwe Open University (ZOU), Chinhoyi University of Technology (CUT), University of South Africa (UNISA), University of KwaZulu-Natal (UKZN) and North-West University (NWU). Elias Munapo has successfully supervised/co-supervised 10 doctoral students and over 20 master's students to completion. Professor Munapo has published over 100 research articles. Of these publications, one is a book, several are book chapters and conference proceedings and the majority are journal articles. In addition, he has been awarded the North West University Institutional Research Excellence Award (IREA) thrice and is an editor of a couple of journals, has edited several books and is a reviewer of several journals.
Email: emunapo@gmail.com

Elias Munapo - Business Statistics and Operations Research, Northwest University, South Africa
E-mail: elias.munapo@nwu.ac.za
https://commerce.nwu.ac.za/business-statistics-and-operations-research/elias-munapo, https://scholar.google.co.za/citations?user=4Npmpr0AAAAJ&hl=en

J. Joshua Thomas is an Associate Professor in Computer Science at UOW Malaysia KDU Penang University College. He obtained his PhD (Intelligent Systems Techniques) in 2015 from University Sains Malaysia, Penang and Master's degree in 1999 from Madurai Kamaraj University, India. He served as Head and deputy head of the Department of Computing between the years 2012 to 2017. From July to September 2005, he worked as a research assistant at the Artificial Intelligence Lab in University Sains, Malaysia. From March 2008 to March 2010, he worked as a research associate at the same University. His work involves intelligent systems techniques in which he adopts computational algorithm implementation in inter-discipline field areas. His expertise is evident in working with International collaborators in publications, visiting research fellows to share knowledge. Recently, work in Deep Learning, data analytics, especially targeting Graph Convolutional Neural Networks (GCNN) and Graph Recurrent Neural Networks (GRNN), Hyper-Graph Attentions, End-to-end steering learning systems, design algorithms in drug discovery and Quantum machine learning. He is a principal investigator, co-investigator in various grants funding, including internal, national and international levels. He is an editorial board member for the Journal of Energy Optimization and Engineering (IJEOE), Book author, guest editor for Applied Sciences, Computations (MDPI), Mathematics Biosciences and Engineering (MBE), and Computer Modelling in Engineering &

Sciences (CMES). He has authored and edited several books. He has published more than 50 research papers in leading international conference proceedings and peer-reviewed journals He has been a Regular Invitee, Plenary, Keynote speaker, and Workshop Presenter in IAIM2019, LCQAI2021, ICRITCC'21, and IAIM2022. He is a "Visiting Research Fellow" to the Sathyabama Institute of Science and Technology, Chennai, India.

Weblink: https://www.uowmkdu.edu.my/research/our-people/dr-joshua-thomas

Email: joshua.j.thomas@gmail.com

J. Joshua Thomas, UOW Malaysia; KDU Penang University College, Malaysia
E-mail: jjoshua@kdupg.edu.my
https://www.uowmkdu.edu.my/research/our-people/dr-joshua-thomas

Gerhard-Wilhelm Weber is a Professor at Poznan University of Technology, Poznan, Poland, at Faculty of Engineering Management. His research is on OR, Financial Mathematics, Optimization and Control, neuro- and bio-sciences, data mining, education, and development. He is involved in the organization of scientific life internationally. He received his Diploma and Doctorate in mathematics, and economics/business administration, at RWTH Aachen, and his Habilitation at TU Darmstadt. He held Professorships by proxy at the University of Cologne, and TU Chemnitz, Germany. At IAM, Middle East Technical University (METU), Ankara, Turkey, Prof. Weber served as a Professor in the programs of Financial Mathematics and Scientific Computing, and Assistant to the Director, and he has been a member of further graduate schools, institutes and departments of METU. Further, he has affiliations at the Universities of Siegen, Ballarat, Aveiro, North Sumatra, and the Malaysia University of Technology. Furthermore, he is an "Advisor to EURO Conferences" and "IFORS Fellow".

Weblink: https://www.researchgate.net/profile/Gerhard_Wilhelm_Weber
Email: gerhard-wilhelm.weber@put.poznan.pl

Gerhard-Wilhelm Weber- Poznań University of Technology, Poland
E-mail: gerhard-wilhelm.weber@put.poznan.p l
https://www.researchgate.net/profile/Gerhard_Wilhelm_Weber

Acknowledgments

The editors would like to acknowledge the contributions of everyone involved in the development of this book and, in particular, would like to thank all authors for their contributions. Our sincere thanks go to the chapter authors who contributed their valuable time and special thanks go to the people who, in addition to the preparations, actively participated in the evaluation process.

My thanks go to the members of the Editorial Board who have done so much in making the book of high quality, especially to the critical positions of the members of the Editorial Committee, Professor Gerhard Wilhelm Weber, Professor Pandian Vasant, Professor Ugo Fiore for their excellent progress and valuable help.

The editors are confident that readers will find this book extremely useful because, among other things, it gives them an opportunity to learn about the results of the published research. It would be an indication of the enormous role Wiley has played in publishing this information.

Finally, thanks to the editors who would like to thank all the editorial staff Team Wiley, especially Lemore Sarah, for helping the editors and authors prepare the manuscript of this book in a professional and extremely kind manner, and they went a long way to make it this book on the exceptional quality edition of the book.

I would like to thank our Lord and Savior Jesus Christ who gave life and strength to the who worked throughout the book preparation process to complete the book during the time of Covid-19 pandemic.

1

Dynamic Key-based Biometric End-User Authentication Proposal for IoT in Industry 4.0

Subhash Mondal, Swapnoj Banerjee, Soumodipto Halder, and Diganta Sengupta

Maulana Abul Kalam Azad University of Technology, Meghnad Saha Institute of Technology, Department of Computer Science and Engineering, Behind Urbana Complex Near, Ruby General Hospital, Nazirabad Rd, Uchhepota, Kolkata, West Bengal 700150, India

1.1 Introduction

Internet of Things (IoT) has emerged as one of the top research and subsequent application domains lately, accounting for a rise from 4.7 billion connected devices in 2016 [1] to 11.7 billion in 2020 [2], and having a projected upper ceiling expansion of 30 billion in 2025 [2]. The phenomenal rise in IoT has also led to Distributed Denial of Service (DDoS) attacks by cybercriminals [1], which is expected to gain further momentum by 2025 owing to the explosive rise in IoT devices, approximately four devices per person [1]. Hence, the security of the IoT architecture in both the middleware and the edges has attracted huge global research. This chapter concentrates on securing the edge framework of the IoT architecture by generating a dynamic (cipher) key based on fingerprint impressions (images) of the end user. IoT devices are typically grouped into different clusters [3] having gateways (cluster heads). These gateways are further inter-connected via the cloud platform [4, 5]. The devices within the cluster communicate with each other and with the internet through the gateways, which have varied topological placements. Communication among the different devices within the cluster and the gateways within diverse topologies is generally managed through the most popular communication protocols such as the MQ Telemetry Transport (MQTT) [6] and Constrained Application Protocol (CoAP) [7]. The difference between these two is given in ([8]). Furthermore, wireless communication forms the backbone of IoT infrastructure (gateway) wherein the terminal nodes send the message packets to the framework (platform), which subsequently transmits the packets wirelessly [9]. Choice of these communication protocols [6, 7] depends upon the

Artificial Intelligence in Industry 4.0 and 5G Technology, First Edition.
Edited by Pandian Vasant, Elias Munapo, J. Joshua Thomas, and Gerhard-Wilhelm Weber.
© 2022 John Wiley & Sons, Inc. Published 2022 by John Wiley & Sons, Inc.

application, but the protocols all experience certain security vulnerabilities, which are generally addressed by authentication. Moreover, authentication schemes such as Elliptic Curve Cryptography [10], Self-Certified Keys Cryptosystem [11], and Hash Functions [12] are employed at the wireless communication end. We propose an authentication system that acts at the node level itself de-voiding the requirement at the middleware much before the wireless communication. The proposed authentication system is primitively a template-based architecture that uses a modified Advanced Encryption Standard (AES) algorithm for data (image) pre-processing and feature (minutiae) extraction. Biometrics used for authentication varies from fingerprints, irises, and facial images.

As discussed earlier, IoT has witnessed a huge rise in connected devices over the past few years and is poised to increase even further [1]. This increase demands proper data management in terms of storage and processing. Several Big Data architectures have been employed for such data management leading to the use of the cloud infrastructure [13]. In terms of data management, cloud services also aid in high scalability and flexibility. Moreover, cloud-driven IoT aids in real-time processing at minimal cost [14]. The bottleneck in cloud-driven IoT is authentication as the data are available across different topologies and can be accessed by any third party. Latency accounts for a further issue, which occurs while analyzing and retrieving data from the cloud. Secured cloud access with proper authentication is managed by Edge Computing [15], which provides solutions not only in the current IoT landscapes [16, 17] but also in 5G. Our proposal works primarily at the authentication of the edge level and also through the cloud infrastructure. End users provide their fingerprint images, which are stored as biometric templates in the cloud database. The end users have to initially authenticate themselves to connect their devices to the IoT network. The second stage calls for encrypting the biometric template using the AES algorithm followed by template matching at the cloud level.

The rest of the paper is arranged as follows. Section 1.2 provides the related work, while Section 1.3 details the proposed approach. Comparative analysis is given in Section 1.4 and the conclusion in Section 1.5.

1.2 Literature Review

IoT and Wireless Sensor Networks (WSN) generally operate in a hand-shaking mode, thereby raising major authentication issues. Primary investigation reveals four dimensions of authentication (Figure 1.1). Authentication factors such as ownership, knowledge, and biometrics [18] play a key role in authentication mechanisms. Knowledge and biometric factors constitute passwords and fingerprints, respectively. The four authentication techniques have been further elaborated.

Figure 1.1 IoT authentication dimensions.

Physical Unclonable Functions (PUF), defined as physical objects which are physically defined digital fingerprint responses, serve as unique identifiers for a particular input and challenge. A lightweight system for the attestation and authentication of things (AAoT) has been proposed [19], which is based on PUFs and proposes a tamper-proof authentication feature for smart embedded devices. Memory resources are attenuated via random memory filling based on PUF technology. Mutual authentication is achieved by block optimization of AAoT. A theoretical PUF-based authentication protocol has been experimentally validated in [20] resulting in a huge reduction in computational and storage burden. In (Wallrabenstein, 2016), PUF-based algorithms were used for the generation of digital signature, device authentication purposes followed by enrolment and decryption [20].

Blockchain-based authentication proposals have been implemented wherein Li et al. [21] confirm that such proposals can be made secure under the condition that the cooperating group of attacker nodes lower hash rates than the collective control of the honest nodes. Prior to the use of blockchain technology for IoT architecture authentication, it is mandatory that the complete network architecture be decentralized. Decentralization helps in efficient management and authentication of smaller blocks as proposed in [22], and the mechanism is termed as "Bubbles of Trust." The bubbles were deployed as smart contracts, which have identification and trust capabilities among themselves. The process generates a transaction that includes two identifiers – one belonging to the device and the other to the group. A valid transaction results in a bubble creation at the device end, which generated the transaction. Blockchain technology validates the uniqueness of the identifiers. Although the scheme in [22] ensures complete anonymity, the proposal lacks identification assurance. This bottleneck can be addressed by deploying private blockchain architectures, though they suffer from new device or service insertion. Li et al. [21] have provided each device a unique identification number, which is further grafted into the blockchain, thereby enabling the device authentication and violating central authority.

The third proposal in Figure 1.1 is the key-based authentication system, which provides an improvement in the brute force authentication techniques.

Adriano Witkovski et al. [23] have proposed an authentication scheme where they provide single-sign-on in IoT architecture. The proposal generates integration via interaction between the key-based internet scheme and the key-based IoT scheme. The proposal limits the device count in the IoT network to 50 and the message sizes to 4096 bytes, which to some extent limits the deployment capabilities and feasibilities. The proposal in [24] contains an agreement scheme between a symmetric key-based authentication and the session key. The authentication proposal scheme in [25] uses the symmetric key for wireless sensor networks claiming a 52.63% efficiency. The proposal has leverage over the Denial-of-Service (DoS) attacks and user traceability but attracts high computation costs. Shah and Venkatesan [26] proposed a secure vault system, which basically contains equal-sized keys. This vault aids in a multi-key-based mutual authentication process between the IoT device and the IoT server. Every successful communication changes the contents of the vault whose initial content had been shared between the two stakeholders – IoT device and the server. Mohammad Wazid et al. presented a proposal for resource-constrained smart devices [27], which focus on a one-way hash function, symmetric encryption and decryption processes, and bitwise exclusive or (XOR) operation. The scheme has been supported through the popular real-or-random (ROR) model.

The final dimension shown in Figure 1.1 is the biometric-based authentication system. This paper also focuses on this branch of authentication techniques. In this system, primarily the client nodes access the IoT application server located at different geographical locations. The client nodes have certain vulnerabilities as follows:

- user impersonation attack
- offline password guessing attack
- user-specific key theft attack
- insider attack.

All the above attacks can be efficiently addressed via biometric authentication, which has already proven to be reliable in contrast to the traditional password-dependent authentication schemes. IoT security has been addressed by "lightweight biometric multi-factor remote user authentication" in [28], which employed a gateway node for initial user registration, subsequently connecting the sensor node via the smart device. Other such proposals working in a similar architecture include those in [29, 30] but generating high message exchange cost and insufficient security advancement. Huang et al. have combined biometric, smart card, and password techniques for designing the authentication framework but lack a threat identification network [31]. Authentication systems for wireless sensor networks have been proposed in [32] and for cloud-based biometric identification via Cloud-Id in [33] but at the cost of high computational complexity similar to the process proposed in [34]. Other notable proposals include [35] where Rivest–Shamir–Adleman (RSA) and Blowfish algorithms are employed to generate an algorithm, which takes care of secure communication of the message packets over the internet. Punithavathi et al. [36, 37] have proposed a template bases enrolment and authentication framework, which addresses the issues of availability, integrity, and confidentiality via data integrity, data encryption, and session key agreement. Although the proposal in [36, 37] has been accepted to be highly advanced in contrast to traditional biometric techniques, the major challenge exists in the deployment of such a system in the cloud environment.

Securing the IoT devices and corresponding networks is termed IoT Security. Enabling security features at the device end still calls for extensive research and applications/algorithms in this domain are heavily appreciated. In addition to the security features, malware installation on the devices forms another issue of huge concern as it has network infection capabilities. Therefore, IoT security is an area of great importance in the advanced technology domain. Security requirements in IoT infrastructure are alphabetically defined in Table 1.1 and security threats are shown in Table 1.2.

1.3 Proposed Framework

The authentication process is based on a key-based biometric encryption technique in our proposal. The fingerprint image generated by the biometric is taken as the input from the user. These fingerprint images are fed into a cloud-based database during the enrolment process. These fingerprint images generate the dynamic cipher key for the AES encryption process. During the authentication phase, users provide their biometric fingerprint images. This image is pre-processed to enhance the image quality for feature (minutiae) extraction. The extracted minutiae points are used to generate the cipher key for the encryption process. The input fingerprint image and a stored fingerprint image of an enrolled user are encrypted using this cipher key and the resultant cipher texts are matched. Results of the matching determine the authentication of a user. The framework for our proposal is presented in Figure 1.2.

1.3.1 Enrolment Phase

The process flow for the Enrolment phase is presented in Figure 1.3. Algorithm 1.1 presents the algorithm for this phase. During this phase, the biometric sensor acquires the fingerprint image of the user. The fingerprint image captured using the biometric sensor is used as the input data and stored in the cloud-based database as discussed earlier.

Table 1.1 Key terms of IoT security.

Term	Responsibilities
Anonymity	The attacker cannot access the authentic information
Authentication	It helps in the verification of the IoT device communication
Authorization	It is responsible for ensuring the authorization of the various resource devices of the communicating parties
Availability	The IoT devices are available to the users on the requirement
Confidentiality	It helps in preventing unauthorized access and data protection
Data freshness	This adheres to the fact that the dataset transmitted by the IoT device is unique with no repetition
Forward secrecy	If a message is compromised, then this adheres that random session-dependent public keys are generated to avoid compromising the other messages
Integrity	The data acquired from the IoT sensor remain unaltered during transmission and also un-compromised at the warehouse
Non-repudiation	It ensures that the recipient of the data from an IoT sensor or transmitter holds responsibility of earlier received messages and avoids denial
Resilience against sensing device capture attack	If a certain node or part of IoT architecture is compromised, this ensures that the remaining parts of the architecture are protected from breach
Unbreakable service	This is responsible for continuous services addressing the emergency issues like physical theft or energy issues. This typically ensures seamless services in IoT frameworks

Algorithm 1.1 Enrolment Phase of the User

procedure Enrolment (*Buser*)
begin

       $Buser \leftarrow U$: Capture the biometric fingerprint image

       Establish a network connection between cloud server and client

         $Buser \leftarrow BS$: Captured biometric image is sent to the fingerprint image database on the cloud.

       $DB \leftarrow Buser$: Store biometric fingerprint image of the user in the database

       return Message "Enrolment successful"

end

Table 1.2 Security threats to IoT architecture.

Term	Responsibilities
Denial of service attack	In this type of attack, the entire system is under the control of an illegal entity thereby restricting access to the legal entities
Eavesdropping attack	If the wireless communication channel between the IoT device and the network is insecure, then attackers can eavesdrop and gain access to the messages in this type of attack
Gateway node bypassing attack	All the messages pass through a gateway in an IoT network. If an unauthorized node gains access to the network, chances are there that the node bypasses the gateway and gains access to the network in this type of attacks
Impersonation attack	In this type of attack, an illegal entity first gains access to an authentic communicated message and then it plays that message to pretend to be a legal entity
Man-in-the-middle Attack	Before the message goes from one sensor/node to another sensor/node, it is intercepted in the middle of the communication channel by a third party
Offline guessing attack	This claims illegal entry by accessing passwords through brute force process
Parallel session attack	If an IoT protocol has numerous parallel initiations, then there are chances wherein an unauthorized entity may intercept one of the parallel sessions and gain information
Password change attack	Password changes are generally kept in logs. An attacker may initiate password changes successively and get access to the actual password
Stolen smart device attack	If an IoT device is stolen, then the stored information on the device can be accessed and that leads to unauthorized entry into the network. The information can be accessed via power analysis mechanism

1.3.2 Authentication Phase

Enrolled users have to authenticate themselves to get access to the IoT network. This phase comprises multiple steps of pre-processing, key generation, encryption, and matching. Fingerprint image of the user is captured using the biometric sensor. The captured fingerprint image needs to undergo multiple pre-processing steps to enhance the image for minutiae extraction discussed in the following sub-section.

1.3.2.1 Pre-processing

Pre-processing is a vital pre-requisite step for the minutiae extraction process. This includes modification of the pixel intensity of the input image by using

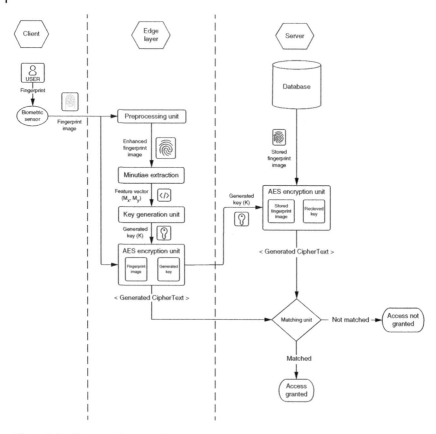

Figure 1.2 Proposed framework.

Figure 1.3 Enrolment phase of a new user.

mathematical operations to generate better output images by enhancing the quality of the input image. Image enhancement techniques have been developed by Hong et al. [38] and were later modified [39]. The decadal growth in technology in this specific domain can be witnessed in [40–44]. Additional stages of Segmentation, Binarization, and thinning of the image are performed for minutiae extraction. Figure 1.4 presents the pre-processing steps.

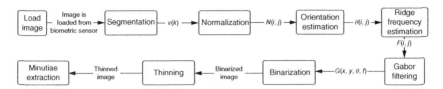

Figure 1.4 High-level diagram for pre-processing.

Step 1: Load Image. This step captures the fingerprint image of the user via the biometric and is subsequently fed to image processing methods for further processing. Further steps are elaborated from Steps 2–5.

Step 2: Segmentation. The focus of this step is to primarily separate the foreground image (Region of Interest – ROI) from the background region. The ROI majorly contains the ridges and the valleys. Segmentation is done to facilitate reliable extraction of minutiae as segmentation helps in noise removal and false minutiae removal using the Variance Threshold technique discussed in [38] and reflected by Eq. (1.1) for a block size of $Z \times Z$.

$$V(k) = \frac{1}{Z^2} \sum_{x=0}^{Z-1} \sum_{y=0}^{Z-1} (I(x,y) - M(k))^2 \tag{1.1}$$

where

$V(k)$ is the variance of the block k,

$I(x, y)$ is the gray-level value at pixel (x, y), and

$M(k)$ is the mean gray-level value for block k.

For cases where $V(k) >$ Global Threshold, those areas become part of the ROI.

Step 3: Normalization. Image normalization refers to the standardization of the intensity values of an image by adjusting the range of gray-level values so that it lies within a desired range of values. Equation (1.2) presents the equation for normalization as presented in [38].

$$A(x,y) = \begin{cases} M_0 + \sqrt{\dfrac{V_0(I(x,y) - M)^2}{V}} & \text{if } I(i,j) > B \\[3mm] M_0 - \sqrt{\dfrac{V_0(I(x,y) - M)^2}{V}}, & \text{otherwise} \end{cases} \tag{1.2}$$

where

$I(x, y) = $ gray-level value at pixel (x, y),

$A(x, y) = $ normalized gray-level value at pixel (x, y),

B and V are estimated mean and variance of $I(x, y)$, respectively, and

B_0 and V_0 denote the desired mean and variance values, respectively.

Step 4: Orientation Estimation. As discussed earlier, the ROI contains the ridges and valleys. The orientation of these ridges in the ROI constitutes the orientation field. The estimate for the orientation is required for Gabor filtering. Therefore, the pixel orientation of (x, y) is achieved using the process as described in [38] and is presented in the following steps.

1. A block of size $Z \times Z$ is centered at pixel (x, y) in the normalized fingerprint image.
2. Sobel operator is used in both the vertical and the horizontal traverses for each pixel in the block, thereby computing the gradient magnitudes of x and y.
3. Equations (1.3) to (1.5) estimate the local orientation. The readers may refer to [38] for details regarding these equations.

$$V_x(x, y) = \sum_{u=x-\frac{Z}{2}}^{x+\frac{Z}{2}} \sum_{v=y-\frac{Z}{2}}^{y+\frac{Z}{2}} 2\partial_x (u, v) \partial_y (u, v) \tag{1.3}$$

$$V_y(x, y) = \sum_{u=x-\frac{Z}{2}}^{x+\frac{Z}{2}} \sum_{v=y-\frac{Z}{2}}^{y+\frac{Z}{2}} \partial_x^2 (u, v) \partial_y^2 (u, v) \tag{1.4}$$

$$\theta(x, y) = \frac{1}{2} \tan^{-1} \frac{V_Y (x, y)}{V_x (x, y)} \tag{1.5}$$

where $\theta(x, y)$ provides the local orientation least square estimate for the block centered at pixel (x, y).

The next step in the image pre-processing (image smoothening) is achieved by the Gaussian filter.

Step 5: Estimation of the Ridge Frequency. As discussed earlier, the local ridge frequency serves as an important feature for Gabor filtering. The frequency characteristic reflects the ridges within the local frequency of the fingerprint image. As discussed earlier, the image is split into $Z \times Z$ blocks and the gray-level values are extracted, and they are then arranged orthogonally to the orientation of the local ridge generating a near sinusoidal wave depicting the fingerprint ridges as the local minimum points. Before determining the ridge spacing, an extra projection smoothing step is done. The noise in the projection is further reduced using a Gaussian low-pass filter. The size of the filter has been restricted to $Z \times Z$. The consecutive minima points present in the projected waveform are calculated using the median pixel count generating the ridge spacing. Equation (1.6) [38] presents the ridge frequency $F(x, y)$.

$$F(x, y) = \frac{1}{S(x, y)} \tag{1.6}$$

Step 6: Gabor Filtering. A 2D-Gabor filter is generated using a Gaussian envelope. It consists of a particular frequency and orientation. The frequency and
orientation-selective features of the Gabor filters make them popular as they
can be adjusted for providing maximum response specifically oriented and specific ridges in the ROI. Gaussian modulation of a cosine waveform generates the
even-symmetric Gabor filter in the spatial domain and is presented in Eqs. (1.7)–
(1.9) [38].

$$G(x, y; \theta, f) = \exp\left\{ -\frac{1}{2}\left[\frac{x_\theta^2}{\sigma_x^2} + \frac{y_\theta^2}{\sigma_y^2} \right] \right\} \cos(2\pi f x_\theta) \qquad (1.7)$$

$$x_\theta = x\cos\theta + y\sin\theta \qquad (1.8)$$

$$y_\theta = -x\sin\theta + y\cos\theta \qquad (1.9)$$

where

θ = Gabor filter orientation,
f = cosine wave frequency,
σ_x, σ_y = standard deviations along the x and y axes for the Gaussian envelope,
x_θ, y_θ = x and y axes of the filter coordinate frame.

Pixel convolution at (x, y) mandated computation of $F(x, y)$ and corresponding
orientation value $O(x, y)$. The fingerprint image was spatially convolved using the
Gabor filter. The generation of the enhanced image E using the Gabor filter G is
shown in Eq. (1.10) from [38]. The features w_x, w_y denote the width and height of
the mask used for Gabor filtering.

$$E(x, y) = \sum_{u=-\frac{w_x}{2}}^{\frac{w_x}{2}} \sum_{v=-\frac{w_y}{2}}^{\frac{w_y}{2}} G(u, v; O(x, y), F(x, y))N(x - u, y - v) \qquad (1.10)$$

Step 7: Binarization. The conversion of a gray-level image to binary images is
required for minutiae extraction algorithms. In a binarized image, there are
typically two levels of pixels: black and white. The black pixels denote the ridges
in contrast to the white pixels, which correspond to the valleys. Hence, the
conversion of the gray-level image into a binary image is termed Binarization.
The process of Binarization sets the global threshold to zero. Then, if

$$\text{Grey level} \begin{cases} > 0 : \text{pixel value} = 1 \\ \leq 0 : \text{pixel value} = 0 \end{cases}$$

Step 8: Thinning. Thinning is the process of making the boundaries in the image
one pixel thick. It is basically a morphological operation. The process involves

P_4	P_3	P_2
P_5	P	P_1
P_6	P_7	P_8

Figure 1.5 Neighborhood of the central pixel P in 3×3 window.

not only the boundary sharpening, but also keeping the connectivity of the ridges intact. The output image from the thinning process is generally termed as the skeleton binary image, which is further used for minutiae extraction. There are many popular algorithms in thinning such as Gradient and Watershed algorithm presented in [45]. A Rule-Based Thinning algorithm was proposed in [46]. This paper concentrates on the standard algorithm/toolkit from MATLAB as presented in [47].

1.3.2.2 Minutiae Extraction and False Minutiae Removal

Crossing-Number-Based Minutiae Extraction We have used the Crossing Number (CN) method for minutiae extraction, which involves the thinned skeletal images generated in the previous section. A 3×3 window has been used for extracting the ridge bifurcations and the endings from the skeletal image through the local neighborhood examination process. Rudovitz's definition for CN has been employed as shown in Equation (1.11).

$$CN = 0.5 \sum_{i=1}^{8} |P_i - P_{i+1}|, \quad P_9 = P_1 \tag{1.11}$$

where

P_i = pixel value in the neighborhood of .

Figure 1.5 shows the 3×3 window in which the N_8P is scanned anti-clockwise; $N_8P = 8$ neighborhoods of pixel P.
The findings are as follows:

$$\text{If } P = 1 \begin{cases} \text{Has only one 1 valued neighbor, then } P = \text{Ridge ending} \\ \text{Has three 1 valued neighbors, then } P = \text{Bifurcation} \\ \text{Has two 1 valued neighbors, then } P = \text{normal pixel} \end{cases}$$

False Minutiae Removal A final step of false minutiae removal is performed. A false minutiae removal technique was developed by Tico and Kuosmanen following a minutiae validation algorithm [48]. The algorithm scans the skeletal image followed by local neighborhood examination around the minutiae point. Although the algorithm performs well, it has been observed that the fuzzy-rules-based approach presented in [49] exhibits better results and has been used for the present work. This algorithm is used to find spurious minutiae points based on well-defined fuzzy rules. Initially, the Euclidian method presented in [50] is used to generate the distance between the termination and bifurcation points. Then, the fuzzy algorithm of [49] tests the local neighborhood around the minutiae point. This is done on the thinned image for minutiae point validation. The fuzzy rules presented in [49] are as follows:

$$\text{If } (d_{\text{termination}} - d_{\text{bifurcation}} < D): \text{ Remove minutiae}$$

$$\text{If } (d_{\text{bifurcation1}} - d_{\text{bifurcation2}} < D): \text{ Remove minutiae}$$

$$\text{If } (d_{\text{termination1}} - d_{\text{termination2}} < D): \text{ Remove minutiae}$$

where

D: Threshold distance
$d_{\text{termination}}$: Termination point
$d_{\text{bifurcation}}$: Bifurcation point
$d_{\text{bifurcation1}}$: 1st Bifurcation point
$d_{\text{bifurcation2}}$: 2nd Bifurcation point
$d_{\text{termination1}}$: 1st Termination point
$d_{\text{termination2}}$: 2nd Termination point

After processing the algorithm steps mentioned earlier, the valid minutiae points are extracted. Extraction requires identification of minutiae points, which is done according to the following factors.

- x- and y-coordinate values of the minutiae points
- $O(x, y)$ discussed in Step 6 in the previous section
- type of minutiae (ridge ending or bifurcation).

1.3.2.3 Key Generation from extracted Minutiae points
This step performs the dynamic cipher key generation from the extracted minutiae (represented by (x, y)). Two vectors have been used for minutiae storage, namely M_x and M_y, representing

$$M_x = [x_1, x_2, x_3, \ldots, x_n]$$

$$M_y = [y_1, y_2, y_3, \ldots, y_n]$$

Vectors M_x and M_y are used to generate the 128-bit biometric key for the AES encryption process. The algorithm for key generation is given in Algorithm 1.2 proposed in [51]. The generated key from the key generation algorithm is used for the AES encryption process. A copy of the key K initiates the authentication process in the cloud server.

1.3.2.4 Encrypting the Biometric Fingerprint Image Using AES

AES 128-bit algorithm is used to encrypt the user fingerprint image. In this algorithm, 128 bits are involved for the input–output block and the state. The 32-bit word count in the state is represented by $N_b = 4$. The choice of cipher key for encryption can be either 128, 192, or 256. Our 128-bit encryption key was generated using a 128-bit key generation technique. As a result, $N_r = 10$ as $N = 4$ represents the number of rounds during which the AES algorithm runs. The fingerprint image is first transformed to a byte array. Therefore, the number of rounds for which the AES algorithm runs is represented by $N_r = 10$ as $N_k = 4$. The generated byte array is copied to the 2D – state array in the following form: $a_0 a_1 a_2 \ldots a_{15}$

The 128-bit sequence generates the byte and the bit ordering.

$$\text{input}_0 \ \text{input}_1 \ \text{input}_2, \ldots, \text{input}_{126} \ \text{input}_{127}$$

$$a_0 = \{\text{input}_0, \text{input}_1, \ldots, \text{input}_7\};$$

$$a_1 = \{\text{input}_8, \text{input}_9, \ldots, \text{input}_{15}\};$$

$$a_{15} = \{\text{input}_{120}, \text{input}_{121}, \ldots, \text{input}_{127}\}$$

The input byte array of the fingerprint image is copied to the 2D-state array by the following equation:

$$\text{for } (r = 0 \ to \ r < 4)$$

$$\text{for } (c = 0 \ to \ c < N_b),$$

$$s[r, c] = \text{input}[r + 4c],$$

where $N_b = 4$

The input byte array, which is used to build the 2D–state array, is given in Figure 1.6.

Figure 1.7 reflects that the state can be presented as a combination of four-word arrays as shown.

$$w_0 = S_{0,0} S_{0,1} S_{0,2} S_{0,3}$$

$$w_1 = S_{1,0} S_{1,1} S_{1,2} S_{1,3}$$

$$w_2 = S_{2,0} S_{2,1} S_{2,2} S_{2,3}$$

$$w_3 = S_{3,0} \ S_{3,1} S_{3,2} S_{3,3}$$

input$_0$	input$_4$	input$_8$	input$_{12}$
input$_1$	input$_5$	input$_9$	input$_{13}$
input$_2$	input$_6$	input$_{10}$	input$_{14}$
input$_3$	input$_7$	input$_{11}$	input$_{15}$

Figure 1.6 Input byte array.

$S_{0,0}$	$S_{0,1}$	$S_{0,2}$	$S_{0,3}$
$S_{1,0}$	$S_{1,1}$	$S_{1,2}$	$S_{1,3}$
$S_{2,0}$	$S_{2,1}$	$S_{2,2}$	$S_{2,3}$
$S_{3,0}$	$S_{3,1}$	$S_{3,2}$	$S_{3,3}$

Figure 1.7 2D-State array.

Four different byte-oriented transformations generate the round function for the pseudo-code. They are presented as follows.

- A substitution table ($S - box$) facilitates the byte substitution.
- The rows of the state array are shifted to generate different offsets.
- Each column of the State Array is shuffled for data mixing.
- Adding a Round Key to the State

Algorithm 1.2 presents the pseudo-code for the AES algorithm used in the current study.

Algorithm 1.2   Advanced Encryption Standard (AES) Algorithm

---

procedure Cipher (*byte in* $[4 * N_b]$, *byte out* $[4 * N_b]$, *word* $w[N_b * (N_r + 1)]$)
begin
    byte state$[4, N_b]$
    state = *in*
AddRoundKey (*state*, $w[0, N_b - 1]$)
    for round = 1 **step** 1 to $N_r - 1$
        SubBytes(*state*)
        ShiftRows(*state*)
        MixColumns(*state*)
        AddRoundKey(*state*, $w[round * N_b, (round + 1) * N_b - 1]$)
    end for
    SubBytes(*state*)
    ShiftRows(*state*)
    AddRoundKey(*state*, $w[N_r * N_b, (N_r + 1) * N_b - 1]$)
    out = *state*
    return *out*
end

The Key Expansion Algorithm fetches the cipher key K from the fingerprint image generating a Key Schedule. The Key Expansion process generates

N_b ($N_r + 1$) words. N_r is the iteration count and N_b is the word count in each iteration. Hence, the final Key Schedule reflects a linear array consisting of 4-byte words $w_i \, \forall \, 0 \leq i < N_b(N_r + 1)$. The pseudo-code for the key expansion algorithm is given in Algorithm 1.3.

Algorithm 1.3   Key Expansion algorithm

---

procedure Key_Expansion (*byte key*$[4 * N_k]$, *word* $w[N_b * (N_r + 1)]$, N_k)
begin

      word temp
      $i = 0$
      while ($i < N_k$)
          $w[i] = word(key[4 * i], key[4 * i + 1], key[4 * i + 2], key[4 * i + 3]$
          $i = i + 1$
      end while
      $i = N_k$
      while ($i < N_b * (N_r + 1)$]
          temp $= w[i - 1]$
          if (i mod $N_k = 0$)
              $temp = SubWord(RotWord(\text{temp}))\ xor\ Rcon[i/N_k]$
          else if ($N_k > 6$ *and i mod* $N_k = 4$)
              $temp = SubWord(\text{temp})$
          end if
          $w[i] = w[i - Nk]\ xor$ temp
          $i = i + 1$
      end while
end

Here, $N_b = 4$, $N_k = 4$, and $N_r = 10$ as we have chosen a 128-bit key.

The Cipher key K obtained using the key generation procedure is used to fill the first N_k words of the expanded key. The subsequent word $w[i]$, is generated by XORing $w[i - 1]$, and the word N_k positions earlier, $w[i - N_k]$. SubWord() is a function that takes a four-byte input word and applies the $S - box$ to each of the four bytes to produce an output word. The function RotWord() takes a word $[a_0, a_1, a_2, a_3]$ as input, performs a cyclic permutation, and returns the word $[a_1, a_2, a_3, a_4]$. The round constant word array, $R_{con}[i]$, contains the values given by $[x^{i-1}, \{00\}, \{00\}, \{00\}]$, with x^{i-1} being powers of x (x is denoted as $\{02\}$) in the field GF(2^8) [52]. The output ciphertext from the AES algorithm is used to match the user's input fingerprint image and stored fingerprint image during the authentication process. The algorithm for the authentication phase is given below. The following abbreviations given in Table 1.3 are used in Algorithm 1.4.

Table 1.3 Notation and description.

Description	Notation
Biometric sensor	BS
Biometric fingerprint image	B_{input}
Biometric fingerprint image stored on the database	B_{cloud}
Cloud server	CS
Matching unit	MU
Generated key from minutiae points	K
Encryption unit on gateway layer	EU_{edge}
Encryption unit on cloud server	EU_{cloud}
Cipher text obtained from encrypted biometric image of user	C_{user}
Cipher text obtained from stored encrypted biometric image in database	C_{cloud}
Database	DB

Algorithm 1.4   Authentication Phase of the User

procedure *Auth*orize(B_{input}, B_{stored}, K)
begin
    User requests access to the IoT network
        Establish a network connection with the CS and request to initiate the authentication process.
    B_{input} is captured using *BS*
    B_{input} is sent to the Pre-processing unit for image enhancement
    The enhanced image is used for minutiae extraction.
        Extracted minutiae points are used for generation of K using Key generation algorithm (*Algorithm* 3)
    (K,B_{input}) is sent to EU_{edge}
    A copy of K is sent to EU_{cloud} on the cloud server for encryption of B_{cloud}
    AES algorithm runs in both the encryption units.
    EU_{edge} encrypts B_{input}
    EU_{cloud} encrypts B_{cloud}
        $C_{user} = EU_{edge}(B_{input}, K)$ and $C_{cloud} = EU_{cloud}(B_{stored}, K)$ are generated by the encryption units.
    C_{user} and C_{cloud} are sent to the matching unit.
    return result of matching
end

1.4 Comparative Analysis

A comparative analysis has been performed with existing authentication frameworks in the literature. Various authentication schemes such as lightweight, decentralized blockchain-based, cloud-based IoT, key-based, and biometrics-based remote user authentication were analyzed. Existing key agreement schemes are vulnerable to denial-of-service, the man in the middle, password guessing, and parallel session attacks. Hence, we have used an improved biometric-based key generation scheme. The key generation scheme works on minutiae extraction technique to extract the minutiae points from the fingerprint image. Secure transmission of user data across the IoT network is a major concern; hence, we

Table 1.4 Comparative Analysis of the Proposed Framework with Existing Authentication Schemes.

Studies	Biometric authentication	Feature/minutiae extraction	Key generation		AES encryption
			Biometric	Non-biometric	
[53]	✓	✓	×	×	✓
[54]	✓	✓	×	×	×
[55]	✓	×	×	×	×
[36, 37]	✓	✓	×	×	×
[56]	×	×	×	✓	×
[57]	✓	✓	×	×	×
[28]	✓	×	×	✓	×
[58]	✓	✓	×	×	×
[36, 37]	✓	✓	×	×	×
[5]	×	×	×	✓	×
[23]	×	×	×	✓	×
[33]	✓	✓	×	✓	×
[26]	×	×	×	✓	×
[40]	×	✓	×	×	×
[59]	×	✓	✓	×	✓
[60]	×	×	×	✓	✓
[51]	×	✓	✓	×	✓
Proposed framework	✓	✓	✓	×	✓

have used the AES 128-bit algorithm to encrypt the fingerprint image of the user and then send it to the server-side for matching with the existing fingerprint image. Using a 128-bit key, about 2^{128} attempts are needed to break through the encrypted data by brute force attacker, which makes AES a very safe protocol for data transmission across wireless networks. We have proposed a framework that incorporates biometric authentication, fingerprint image pre-processing for effective minutiae extraction, key generation from minutiae points, and AES encryption for secure data transmission during the authentication phase. Comparative analysis of existing techniques with respect to our proposed framework is discussed in Table 1.4.

1.5 Conclusion

The last five years have witnessed a huge surge in connected IoT devices and is poised for an even further growth over the next five years. The technological advancements call for strict policies regarding the Confidentiality, Integrity, and Authentication (CIA triad) of any security system. To date, the major focus of IoT has been in the domain of performance evaluation and resource management, with recent approaches in the security dimension of IoT devices and networks. This work in this paper focuses on the authentication and the cloud-based security for an IoT framework based on dynamic cipher-text generation from human biometric input. A modified version of AES algorithm has been deployed for key generation, which is based on minutiae extraction. The fundamental approach depends on template matching for the authentication process.

References

1 Gervais, J. (2019). Retrieved from The future of IoT: 10 predictions about the Internet of Things. https://us.norton.com/internetsecurity-iot-5-predictions-for-the-future-of-iot.html.

2 Lueth, K. L. (2020, November). Retrieved from State of the IoT 2020: 12 billion IoT connections, surpassing non-IoT for the first time. https://iot-analytics .com/state-of-the-iot-2020-12-billion-iot-connections-surpassing-non-iot-for-the-first-time/

3 Kumar, J.S. and Zaveri, M.A. (2018). Clustering approaches for pragmatic two-layer IoT architecture. *Wireless Communications and Mobile Computing* 2018: 1–17.

4 Díaz, M., Martín, C., and Rubio, B. (2016). State-of-the-art, challenges, and open issues in the integration of Internet of things and cloud computing. *Journal of Network and Computer Applications* 67: 99–117.

5 Shivraj, V.L., Rajan, M.A., Singh, M., and Balamuralidhar, P. (2015). One time password authentication scheme based on elliptic curves for Internet of Things (IoT). In: *2015 5th National Symposium on Information Technology: Towards New Smart World (NSITNSW)*, 1–6. Riyadh, Saudi Arabia: IEEE.

6 Dinculeană, D. and Cheng, X. (2019). Devices, vulnerabilities and limitations of MQTT protocol used between IoT. *Applied Sciences* 9 (5): 848.

7 Sembroiz, D., Ricciardi, S., and Careglio, D. (2018). A novel cloud-based IoT architecture for smart building automation. In: *Security and Resilience in Intelligent Data-Centric Systems and Communication Networks*, 215–233. Elsevier.

8 Lee, C.-N., Huang, T.-H., Wu, C.-M., and Tsai, M.-C. (2017). The Internet of Things and its applications. In: *Big Data Analytics for Sensor-Network Collected Intelligence*, 256–279. Elsevier.

9 Zhao, G., Si, X., Wang, J. et al. (2011). A novel mutual authentication scheme for Internet of Things. In: *Proceedings of 2011 International Conference on Modelling, Identification and Control*, 563–566. Shanghai, China: IEEE.

10 Jiang, Q., Ma, J., Wei, F. et al. (2016). An untraceable temporal-credential-based two-factor authentication scheme using ECC for wireless sensor networks. *Journal of Network and Computer Applications* 76: 37–48.

11 Haripriya, A.P. and Kulothugan, K. (2016). ECC based self-certified key management scheme for mutual authentication in Internet of Things. In: *2016 International Conference on Emerging Technological Trends (ICETT)*, 1–6. Kollam, India: IEEE.

12 Kim, H., Kim, D.W., Yi, O., and Kim, J. (2019). Cryptanalysis of hash functions based on blockciphers suitable for IoT service platform security. *Multimedia Tools and Applications* 78 (3): 3107–3130.

13 BizIntellia. (n.d.). Retrieved from Benefits of Using Cloud Computing for Storing IoT Data: https://www.biz4intellia.com/blog/benefits-of-using-cloud-computing-for-storing-IoT-data/

14 Aujla, G.S., Chaudhary, R., Kumar, N. et al. (2018). SecSVA: secure storage, verification, and auditing of Big Data in the cloud environment. *IEEE Communications Magazine* 56 (1): 78–85.

15 Shi, W., Pallis, G., and Xu, Z. (2019). Edge computing. *Proceedings of the IEEE* 107 (8): 1474–1481.

16 Ai, Y., Peng, M., and Zhang, K. (2018). Edge computing technologies for Internet of Things: a primer. *Digital Communication and Networks* 4 (2): 77–86.

17 Xue, H., Huang, B., Qin, M. et al. (2020). Edge computing for Internet of Things: a survey. In: *2020 International Conferences on Internet of Things (iThings) and IEEE Green Computing and Communications (GreenCom) and IEEE Cyber, Physical and Social Computing (CPSCom) and IEEE Smart*

Data (SmartData) and IEEE Congress on Cybermatics (Cybermatics), 755–760. Rhodes, Greece: IEEE.

18 Choi, Y., Nam, J., Lee, D. et al. (2014). Security enhanced anonymous multi-server authenticated key agreement scheme using smart cards and biometrics. *The Scientific World Journal* 2014: 1–15.

19 Feng, W., Qin, Y., Zhao, S., and Feng, D. (2018). AAoT: Lightweight attestation and authentication of low-resource things in IoT and CPS. *Computer Networks* 134: 167–182.

20 Wallrabenstein, J.R. (2016). Practical and secure IoT device authentication using physical unclonable functions. In: *2016 IEEE 4th International Conference on Future Internet of Things and Cloud (FiCloud)*, 99–106. IEEE, Vienna, Austria.

21 Li, D., Peng, W., Deng, W., and Gai, F. (2018). A blockchain-based authentication and security mechanism for IoT. In: *2018 27th International Conference on Computer Communication and Networks (ICCCN)*, 1–6. Hangzhou, China: IEEE.

22 Hammi, M.T., Hammi, B., Bellot, P., and Serrhrouchni, A. (2018). Bubbles of trust: a decentralized blockchain-based authentication system for IoT. *Computers & Security* 78: 126–142.

23 Witkovski, A., Santin, A., Abreu, V., and Marynowski, J. (2015). An IdM and key-based authentication method for providing single sign-on in IoT. In: *2015 IEEE Global Communications Conference (GLOBECOM)*, 1–6. San Diego, CA, USA: IEEE.

24 Park, N., Kim, M., and Bang, H.C. (2015). Symmetric key-based authentication and the session key agreement scheme in IoT environment. In: *Computer Science and its Applications*, 379–384. Springer.

25 Ghani, A., Mansoor, K., Mehmood, S. et al. (2019). Security and key management in IoT-based wireless sensor networks: an authentication protocol using symmetric key. *International Journal of Communication Systems* 32 (16): e4139.

26 Shah, T. and Venkatesan, S. (2018). Authentication of IoT device and IoT server using secure vault. In: *2018 17th IEEE International Conference On Trust, Security And Privacy In Computing And Communications/12th IEEE International Conference On Big Data Science And Engineering (TrustCom/BigDataSE)*, 819–824. New York, USA: IEEE.

27 Wazid, M., Das, A.K., Odelu, V. et al. (2017). Secure remote user authenticated key establishment protocol for smart home environment. *IEEE Transactions on Dependable and Secure Computing* 17 (2): 391–406.

28 Dhillon, P.K. and Kalra, S. (2017). A lightweight biometrics based remote user authentication scheme for IoT services. *Journal of Information Security and Applications* 34 (2): 255–270.

29 Li, C.T., Weng, C.Y., and Lee, C.C. (2013). An advanced temporal credential-based security scheme with mutual authentication and key agreement for wireless sensor networks. *Sensors* 13 (8): 9589–9603.

30 Liu, J., Xiao, Y., and Chen, C.P. (2012). Authentication and access control in the internet of things. In: *2012 32nd International Conference on Distributed Computing Systems Workshops*, 588–592. Macau, China: IEEE.

31 Huang, X., Wang, S., Zheng, G. et al. (2010). A generic framework for three-factor authentication: preserving security and privacy in distributed systems. *IEEE Transactions on Parallel and Distributed Systems* 22 (8): 1390–1397.

32 Xue, K., Ma, C., Hong, P., and Ding, R. (2013). A temporal-credential-based mutual authentication and key agreement scheme for wireless sensor networks. *Journal of Network and Computer Applications* 36 (1): 316–323.

33 Haghighat, M., Zonouz, S., and Abdel-Mottaleb, M. (2015). CloudID: trustworthy cloud-based and cross-enterprise biometric identification. *Expert Systems with Applications* 42 (21): 7905–7916.

34 Pengfei, H., Ning, H., Qiu, T. et al. (2018). A unified face identification and resolution scheme using cloud computing in Internet of Things. *Future Generation Computer Systems* 18: 582–592.

35 Joshy, A. and Jalaja, M.J. (2017). Design and implementation of an IoT based secure biometric authentication system. In: *2017 IEEE International Conference on Signal Processing, Informatics, Communication and Energy Systems (SPICES)*, 1–13. Kollam, India: IEEE.

36 Punithavathi, P. and Geetha, S. (2019). Partial DCT-based cancelable biometric authentication with security and privacy preservation for IoT applications. *Multimedia Tools and Applications* 78 (18): 25487–25514.

37 Punithavathi, P., Geetha, S., Karuppiah, M. et al. (2019). A lightweight machine learning-based authentication framework for smart IoT devices. *Information Sciences* 484: 255–268.

38 Hong, L., Wan, Y., and Jain, A. (1998). Fingerprint image enhancement: algorithm and performance evaluation. *IEEE Transactions on Pattern Analysis and Machine Intelligence* 20 (8): 777–789.

39 Khan, M.A. (2011). *Fingerprint Image Enhancement and Minutiae Extraction*. Pomona: California State Polytechnic University.

40 Chaudhari, A.S., Patnaik, G.K., and Patil, S.S. (2014). Implementation of minutiae based fingerprint identification system using crossing number concept. *Informatica Economica* 18 (1): 17–26.

41 Tatar, F.D. and Machhout, M. (2018). Preprocessing algorithm for digital fingerprint image recognition. *International Journal of Computational Science and Information Technology* 6: 1–12.

42 Sharma, R.P. and Day, S. (2019). Two-stage quality adaptive fingerprint image enhancement using Fuzzy C-means clustering based fingerprint quality analysis. *Image and Vision Computing* 83–84: 1–16.

43 Shrein, J.M. (2017). Fingerprint classification using convolutional neural networks and ridge orientation images. In: *2017 IEEE Symposium Series on Computational Intelligence (SSCI)*, 1–8. Honolulu, HI, USA: IEEE.

44 Chen, W. and Gao, Y. (2007). A minutiae-based fingerprint matching algorithm using phase correlation. In: *9th Biennial Conference of the Australian Pattern Recognition Society on Digital Image Computing Techniques and Applications (DICTA 2007)*, 233–238. Glenelg, SA, Australia: IEEE.

45 Kaur, K. and Sharma, M. (2013). A method for binary image thinning using gradient and watershed algorithm. *International Journal of Advanced Research in Computer Science and Software Engineering* 3 (1): 287.

46 Ahmed, M. and Ward, R. (2002). A rotation invariant rule-based thinning. *IEEE Transactions on Pattern Analysis and Machine Intelligence* 24 (12): 1672–1678.

47 Ahmed, N. and Varol, A. (2018). Minutiae based partial fingerprint registration and matching method. In: *2018 6th International Symposium on Digital Forensic and Security (ISDFS)*, 1–5. Antalya, Turkey: IEEE.

48 Tico, M. and Kuosmanen, P. (2000). An algorithm for fingerprint image post-processing. In: *Conference Record of the Thirty-Fourth Asilomar Conference on Signals, Systems and Computers (Cat. No.00CH37154)*, 1735–1739. Pacific Grove, CA, USA: IEEE.

49 Stephen, M.J., Reddy, P.P., Kartheek, V. et al. (2012). Removal of false minutiae with fuzzy rules from the extracted minutiae of fingerprint image. In: *Proceedings of the International Conference on Information Systems Design and Intelligent Applications 2012*, 853–860. Berlin, Heidelberg: Springer.

50 Stephen, M.J. and Reddy, P.P. (2011). Implementation of easy fingerprint image authentication with traditional Euclidean and singular value decomposition algorithms. *International Journal of Advances in Soft Computing and its Applications* 3 (2): 1–19.

51 Srividya, R. and Ramesh, B. (2019). Implementation of AES using biometric. *International Journal of Electrical and Computer Engineering* 9 (5): 4266–4276.

52 Standard, N.-F. (2001). Announcing the advanced encryption standard (AES). *Federal Information Processing Standards Publication* 197 (1-51): 3–3.

53 Golec, M., Gill, S.S., Bahsoon, R., and Rana, O. (2020). BioSec: a biometric authentication framework for secure and private communication among edge devices in IoT and Industry 4.0. *IEEE Consumer Electronics Magazine* 1–1.

54 Wencheng Yang, S.W. (2019). A privacy-preserving lightweight biometric system for Internet of Things security. *IEEE Communications Magazine* 57 (3): 84–89.

55 Gu, T. and Mohapatra, P. (2018). BF-IoT: securing the IoT networks via fingerprinting-based device authentication. In: *2018 IEEE 15th International Conference on Mobile Ad Hoc and Sensor Systems (MASS)*, 254–262. Chengdu, China: IEEE.

56 Li, N., Liu, D., and Nepal, S. (2017). Lightweight Mutual Authentication for IoT and Its Applications. *IEEE Transactions on Sustainable Computing* 2 (4): 359–370.

57 Vijaysanthi, R., Radha, N., Shree, M.J., and Sindhujaa, V. (2017). Fingerprint authentication using Raspberry Pi based on IoT. In: *2017 International Conference on Algorithms, Methodology, Models and Applications in Emerging Technologies (ICAMMAET)*, 1–3. Chennai, India: IEEE.

58 Parkavi, R., Babu, K.C., and Kumar, J.A. (2017). Multimodal biometrics for user authentication. In: *2017 11th International Conference on Intelligent Systems and Control (ISCO)*, 501–505. Coimbatore, India: IEEE.

59 Priya, S.S.S., Karthigaikumar, P., and Mangai, N.S. (2014). Mixed random 128 bit key using finger print features and binding key for AES algorithm. In: *2014 International Conference on Contemporary Computing and Informatics (IC3I)*, 1226–1230. Mysore, India: IEEE.

60 Naif, J.R., Abdul-Majeed, G.H., and Farhan, A.K. (2019). Secure IOT system based on Chaos-modified lightweight AES. In: *2019 International Conference on Advanced Science and Engineering (ICOASE)*, 1–6. Zakho, Duhok, Iraq: IEEE.

2

Decision Support Methodology for Scheduling Orders in Additive Manufacturing

Juan Jesús Tello Rodríguez[1] and Lopez-I Fernando[2]

[1] *Universidad Autonoma de Nuevo León, Facultad de Ingeniería Mecánica y Eléctrica (FIME), Posgrado en Ingeniería en Sistemas (PISIS), Centro de Investigación y Desarrollo Tecnológico (CIDET), Av. Pedro de Alba S/N, San Nicolás de los Garza, Nuevo León 66451, México*
[2] *Autonomous University of Nuevo León, Department of Mechanical and Electronic Engineering, San Nicolás de los Garza, Nuevo León, México*

2.1 Introduction

Additive manufacturing is a family of technologies related to the manufacturing of a three-dimensional solid from a computerized model by depositing thin layers of material until the part is shaped [1]. It is considered one of the most important emerging technologies at present because it makes possible to attain benefits that are harmoniously coupled with Industry 4.0 and the digitization of manufacturing [2].

The ease of designing parts with Computer-aided design/Computer-aided manufacturing (CAD/CAM) software, cheaper development, and variety of materials (metals, paper, ceramics, polymers, etc.), improvements in scanners to "copy" pieces, sculptures, or even living organisms and body parts, among other factors [3] have given a great boost to additive manufacturing technology. In 2019, The additive manufacturing industry had a global increase of 21%, equivalent to almost 12 billion dollars [4]. Despite this increase, it is crucial for additive manufacturing companies to operate efficiently in order to remain competitive. For this purpose, the planning, scheduling, and nesting problems applied to this process have become an important topic in optimization [5]. Diverse methodologies have been developed for scheduling in additive manufacturing problems. For example, Kapadia et al. [6] present an approach using simple scheduling policies, but it does not allow complex sequencing.

The Job-shop rescheduling (JSR) problem plays a very important role in the scheduling of operations in smart industries with high demand for tailor-made

Artificial Intelligence in Industry 4.0 and 5G Technology, First Edition.
Edited by Pandian Vasant, Elias Munapo, J. Joshua Thomas, and Gerhard-Wilhelm Weber.

products [7], allowing to minimize not only the operating time but also waste and energy consumption, among other impact measures. This problem is also of great importance for the resilience of a company, especially in the present time when the immediate future is very unpredictable and the competition is getting stronger. This issue is addressed by Alami et al. [8] who developed a Mixed-Integer Linear Program (MILP) model for the JSR in the context of Industry 4.0.

In this work, a variant of the JSR problem is introduced in additive manufacturing, which has not yet been addressed in previous works reviewed, and a solution methodology is proposed based on mathematical modeling.

2.2 The Additive Manufacturing Process

Additive manufacturing is a key element in Industry 4.0 and the digital design of models and custom fabrications of single pieces. Also, additive manufacturing favors the adoption of a highly competitive business model where a very low waste of materials is achieved; it can also lead to high profit in design and manufacturing custom products. More importantly, it is a highly resilient business model for coping with an uncertain and dynamic economy.

Industry 4.0 not only offers great business opportunities but also imposes big challenges for the adopters of this business model. Many novel technologies are now cheaper and relatively easy to implement in business. Therefore, many companies, from large manufacturers to small startups, have adopted additive manufacturing technologies in their processes, creating a highly competitive environment. In this context, a quick response time to client orders and an effective use of printers are crucial parameters for a business to succeed. This is the main reason to develop optimal schedules where business and client needs are taken into account.

Typically, the additive manufacturing process contains four stages (see Figure 2.1):

1. **Sale**: The customer describes the characteristics of required parts, and an agreement of sale price is achieved. Also, the level of importance and an estimated delivery deadline are agreed upon between the customer and the business.
2. **Design**: If required, a computerized design of the parts to be manufactured is carried out. The time required for this operation depends on whether it is a new design or is a change of measurements in an existing design. Due to the complexity of the pieces, this stage of the process is carried out in a workstation with several parallel machines. Each design is assigned to one of these machines. Each machine works on a single design at a time, and its process can be interrupted at any time to be resumed later at the same point where it was stopped.
3. **Printing:** The manufacturing station has one or several parallel printing machines with differences in working speed, startup time, and work area.

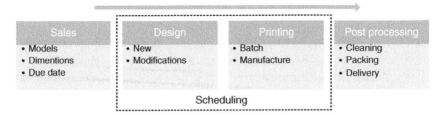

Figure 2.1 Additive manufacturing stages.

This last parameter defines the maximum number of pieces that can be manufactured at the same time in a printing machine. The designed parts of orders are scheduled in batches assigned to the printing machines, when a machine finishes its current job, a new job is loaded (following the scheduled sequence) to be manufactured. Once the printing of the pieces is started, it cannot be stopped until it is completed. If a correction is required, or the print ends with a defect, that part must be manufactured again from scratch.

4. **Post-processing**: The manufactured parts are polished and, if needed, are taken to another workstation or to a waiting area until they are packed with the rest of the parts corresponding to the order.

The design and manufacturing stages are the most critical in terms of order manufacturing time and profit. Therefore, it is of great importance to optimize the sequence of operations, allowing delivery times to be met for all cases, in addition to maximizing profits.

The whole process can be considered as a combination of flow-shop problem and job-shop problem where batch processing is also used. A scheduled sequence can be modified by the arrival of new orders with higher priority than some orders already scheduled. Typically for re-scheduling, a hybrid approach is applied, and as a result, a new schedule is generated (Figure 2.2).

In this work, we are dealing with stage 3 (printing scheduling).

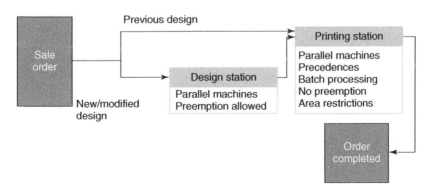

Figure 2.2 Order processing flow.

2.3   Some Background

Research regarding the optimization of planning and scheduling of additive manufacturing operations is relatively recent. Previously, optimization issues for this process focused more on printing methods or algorithms that minimize the operation time, such as developing an optimal path of the print head during the operation [9]. Zhu et al. [10] presented one of the first works that properly considers planning with this technology, in which an algorithm was developed to plan the processes corresponding to the manufacture of highly complex parts with both additive and destructive manufacturing and its inspection. Rossi et al. [11] generated precedence graphs for the scheduling of both additive and subtractive manufacturing by using an ant colony optimization metaheuristic and presented a similar more recent work. In a very similar vein, Baumung and Fomin [12] presented an optimized process for planning and controlling production in an additive manufacturing environment.

The above cited research papers have in common that they do not consider all the constraints that additive manufacturing processes entail for real-life applications [13]. Li et al [14] introduced the planning, scheduling, and nesting problems for this process for the first time, and in this work, they presented two heuristics with decision rules for the minimization of the average cost of production by volume of material in a powder bed process. Since then, various studies have been carried out on the subject, which consider similar objectives. For example, Fera et al. [15] presented a genetic multi-objective optimization algorithm in order to minimize production time and cost considering the use of a single machine and volume restrictions. The same research team presented a similar study two years later by using the same considerations, although this time they used a taboo search heuristic, which gave better results but took longer to find good solutions. Recently, Alicastro et al. [16] developed an algorithm based on reinforcement learning combined with iterated local search and achieved significant improvements compared to other published solution techniques.

Other research works considered the use of multiple parallel machines. This was investigated by Chergui et al. [17], who in their work introduced the priority rule of the nearest delivery date and made an estimation of time manufacturing. They decomposed the problem into two sub-problems: (i) in the first sub-problem, tasks were assigned to machines through nesting, and this problem was solved using quadratic integer binary programming. (ii) The second problem corresponds to the sequencing of operations, which was solved with mixed-integer programming. Dvorak et al. [18] approached the nesting sub-problem using graph theory, looking for cliques that correspond to equivalent parts for processing, and the selected parts were then added to machines with a nesting algorithm. A more complex problem is presented by Che et al. [19]; they implemented a scheduling system that simultaneously assigned order's parts to batches, determined the orientation of the parts, and allocated those batches in the machines with the objective of minimizing the make span. For this purpose, a MILP algorithm along with some heuristic methods was developed.

Other aspects studied include the acceptance or rejection of orders for additive manufacturing such as those presented by Li et al. [20]. Their algorithm supports making this decision based on time availability while maximizing the net profit per unit/time in the process. Also, Zhan and Chiong [21] addressed the effect in the resulting schedule of the orientation and location of parts within printing machines; they integrated these considerations into their evolutionary algorithm to improve their results. Kucukkoc [22] addressed the impact in the scheduling of different configurations like the use of a single machine, identical or different parallel machines; he employed mixed-integer programming seeking the optimization of various time measures in the process.

More recently, some multicriteria models have been developed such as the energy-aware schedule presented by Karimi et al. [23], which addresses a production system where process, scheduling, and electrical energy saving are integrated and controlled. In the study made by Rabsikarbum et al. [24], a multi-objective model is presented where total operational cost, machine loads, lateness, and total unprinted parts are optimized. This model considers a variety of different additive manufacturing technologies.

The JSR problem has remained relevant for more than five decades, ever since Garey et al. [25] showed that the problem is NP-complete when more than two machines are considered. Thereafter diverse solution methodologies have been developed. In deterministic models, for example, Liu et al. [26] worked in the minimization of the make-span by using buffering constraints. In a dynamic environment, which is more similar to our case study, Luo [27] developed a methodology using deep learning for the re-scheduling of new job insertions. In a similar dynamic environment, Zhang et al. [28] used both real-time and simulated data for a better prediction of available machines. Multi-objective models are also a relevant topic in this area of scheduling, as shown by Caldeira et al. [29] who considered the minimization of the make span, energy consumption, and instability of the schedule.

The development of novel algorithms is a hot area for research in the Job Shop Scheduling problem. Froehlich et al. [30] used genetic algorithms to iteratively improve some priority rules aiming to minimize the tardiness of jobs. Nouri et al. [31] worked on a Particle swarm optimization (PSO) algorithm along with single objective multi-agents in the minimization of completion time. More recently, Chen et al. [32] developed a self-learning genetic algorithm showing a great performance in the solution of the flexible JSR problem.

2.4 Proposed Approach

The decision-making process is significantly complex in the planning and sequencing of operations, also presented in portfolio selection and scheduling [33–35] due to the number of workpieces, machines, and customer needs. Among the challenges are:

- Delays in commitment dates due to an erroneous estimate of manufacturing time.
- Inactivity in machines due to lack of designs due to bad planning.
- Delays due to unexpected events such as urgent orders or part failures.
- Lack of attention to important clients.
- Slow flow of completed orders.
- Inclusion of urgent orders not planned.

The current decision-making process is summarized in Figure 2.3, which summarizes the proposed approach. The process consists of two phases: (i) *sale phase:* the characteristics of the order and a delivery commitment date are established, which also depend on the orders already in process. Once the client approves the order, the order goes on to (ii) the *planning phase*: where the priority of the order is established. This depends on the agreed order's date, the importance of the client,

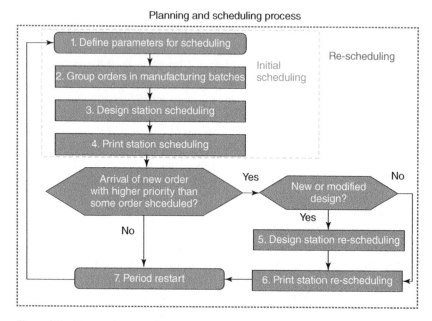

Figure 2.3 Proposed approach.

and how quickly it can be processed. Subsequently, the order is assigned to a batch that depends on the parts already programmed as well as the availability of printing machines.

The proposed process for sequencing operations (scheduling) is listed below.

1. **Define or calculate important parameters for scheduling**:
   (a) Duration of the period is defined.
   (b) Calculate available time for each station.
   (c) Assign priority to orders based on the criteria and assign a priority for each one.
2. **Group orders in manufacturing batches**:
   (d) The initial batch is created. Batches are numbered and added to the execution queue in order ascended (for two batches a and b, the batch with the smaller number is executed first).
   (e) Add orders, in descending order of priority, to the batch until there is no time available at any printing station.
   (f) If no more orders can be added to a batch, a new batch is created, and subsequent orders are added to this new batch when possible.
   (g) When a new order with higher priority arrives than at least one order already in one in the defined batches, a re-scheduling is conducted for that batch and all batches are to be executed after this one.
3. **Initial design scheduling**: Assign parts to the design machines seeking to minimize the estimated completion time of the parts (considering the time manufacturing estimate), maximize profit, and maximize importance. With this, the design release times are obtained, and from those times, the printing of designed parts can be scheduled.
4. **Initial printing scheduling**: Assign parts or sets of these to the printing machines, minimize manufacturing time (considering "real" manufacturing time, grouping of parts, release of designs, and setup time), maximize profit, and maximize importance.
5. **Design re-scheduling**: In the event of an urgent order that requires design, the sequencing is performed again, considering only the parts that are still in the design queue and urgent ones. It has the same considerations as in step 2. In this case, the process in turn could be stopped so the new urgent order can be processed; the stopped process can continue some time later exactly where it was left. The missing design time is the difference between the total design time and the progress made so far.
6. **Print re-scheduling**: Like the previous step, we work with the urgent order and the pieces that are still in the print queue. It has the same considerations as that in step 4. Due to the nature of printing, it is not possible to stop the job in progress and continue it later; it is necessary to wait for it to finish or to stop it

and start from the beginning. This step affects order completion times because the order in the print queue must wait for the release of machines.

7. **Period restart**: The process is repeated from step 1 with pending and new orders.

A mathematical model for the initial scheduling printing sequence is presented (4). In the model, it is assumed that all considered orders have been designed before and are available for printing.

2.4.1 A Mathematical Model for the Initial Printing Scheduling

The multi-objective integer programming mathematical model used for the sequencing process is presented.

For scheduling the initial printing batch, the design of all the pieces of at least one order that require design should be finished or there should exist at least one order that does not require design and can be scheduled for printing. In other words, there should exist at least one order ready for printing.

As decision variables are defined orders in batches to be scheduled, its parts and their assignments to the printing machines (these are defined in Section 2.4.3).

Are considered two objectives (Section 2.4.4): Maximize Profit and Maximize Importance (of scheduled orders). Minimize manufacturing time is not considered as an objective since it is at least weakly correlated to profit (clients should pay higher amounts for urgent orders than for non-urgent orders in general), and for that reason, it is not considered in the model presented in this work.

Also, typical scheduling constraints are considered (see Section 2.4.5).

2.4.1.1 Considerations
- It is assumed that all parts of a batch are already designed and are ready to be printed.
- An estimated printing time for each part is already known.
- Printers work at the same speed; however, they may have different setup times and work areas.

2.4.1.2 Sets
- I: set of orders i in a queue.
- S: work order for the period, it is a subset of orders I.
- J_i: set of parts j belonging to order i.
- M: set of additive manufacturing machines available for manufacturing, with a total of n_m parts.
- Q: set of objectives q.
- B: set of batches for grouping orders to be scheduled for printing.

2.4.2   Parameters

2.4.2.1   Orders
- dd_i: commitment date of the order i.
- g_i: profit of order i.
- pr_i: level of importance of order i (priority, assigned by the Decision Maker).

2.4.2.2   Parts
- a_{ij}: area of part j of order i.
- te_{ij}: estimated printing time of part j of order i.
- tm_{ij}: actual printing time of part j of order i.

2.4.2.3   Printing Machines
- A_m: total printer work area m.
- st_m: printer startup time m, it is independent of the sequence.

2.4.2.4   Process
- T: duration of the period.

2.4.3   Decision Variables

- x_{ij}^{bm} {1 if part j of order i is added to batch b of the printing machine m; 0 the otherwise}
- y_{bm} {1 if batch b on machine m is used; 0 otherwise}
- y_i {1 if order i is estimated to be completed in period; 0 otherwise}
- tp_{bm}: estimated time to start manufacturing batch b on machine m.
- •tc_{bm}: estimated manufacturing duration of batch b on machine m.
- Cp_{bm}: estimated time to complete the manufacturing of batch b in machine m.
- C_i: estimated time of completion of manufacturing for order i.

2.4.4   Optimization Criteria

In this case, two criteria are considered for making decisions in the sequencing of operations.

Maximize gain: try to include in schedule most profitable orders

$$\max \sum_{i \in S} g_i \cdot y_i \qquad (2.1)$$

Maximize importance: try to include in schedule orders with the highest priority.

$$\max \sum_{i \in S} \frac{pr_i}{dd_i} \cdot y_i \qquad (2.2)$$

2.4.5 Constrains

$$\sum_{b \in B}\sum_{m=1}^{n_m} xp_{ij}^{bm} = 1, i = 1, \dots, n_m \tag{2.3}$$

$$yp_{bm} \geq xp_{ij}^{bm}, \forall b \in B, i = 1, \dots, n_i, m = 1, \dots, n_m \tag{2.4}$$

$$tp_{bm} = st_m \cdot yp_{bm} + \sum_{i,j \in S} te_{ij} \cdot xp_{ij}^{bm}, \forall b \in B, m = 1, \dots, n_m \tag{2.5}$$

$$tc_{bm} \geq tp_{b-1m} + tc_{b-1m}, \forall b \in B, m = 1, \dots, n_m \tag{2.6}$$

$$Cb_{bm} \geq tp_{bm} + tc_{bm}, \forall b \in B, m = 1, \dots, n_m \tag{2.7}$$

$$tp_{bm} - Cd_{ij} \geq -\Psi \cdot \left(1 - xp_{ij}^{bm}\right), \forall b \in B, i = 1, \dots, n_i, m = 1, \dots, n_m \tag{2.8}$$

$$C_i - dd_{ij} \geq -\Psi \cdot (1 - y_i), \forall b \in B, i = 1, \dots, n_i, m = 1, \dots, n_m \tag{2.9}$$

$$C_i \geq Cb_{bm} - \Psi \cdot \left(1 - xp_{ij}^{bm}\right), \forall b \in B, i = 1, \dots, n_i, m = 1, \dots, n_m \tag{2.10}$$

$$\sum_{i=1}^{n_i}\sum_{j \in J_i} a_{ij} \cdot xp_{ij}^{bm} \leq A_m, \quad b \in B, m = 1, \dots, n_m \tag{2.11}$$

$$xp_{ij}^{bm}, yp_{bm} \in [0,1]; tp_{bm}, tb_{bm}, C_i, C_{bm} \geq 0 \tag{2.12}$$

Constraint (2.3) prevents the parts from being manufactured more than once, it is an assignment constraint. Constraint (2.4), where Ψ is a very large positive number, indicates the use of batch b in the machine m if at least one piece is placed in it. Constraint (2.5) calculates the manufacturing time of the batch from the parts that will be manufactured in the batch.

Constraint (2.6) prevents more than one batch from being manufactured simultaneously on the same machine; therefore, each batch must wait until the previous batch has finished. Similarly, constraint (2.7) calculates the batch completion time depending on the time it started manufacturing and its manufacturing time.

Constraint (2.8) prevents other parts from being added to some batch until its design has already been terminated. Constraint (2.9) indicates whether the order will be completed on time; Constraint (2.10) calculates the time of completion of each order. Constraint (2.11) prevents more parts from being added than the printer can handle in the printing area. Finally, constraint (2.12) indicates the nature of the variables.

2.5   Results

A real problem is being solved in this framework. The problem data are from a family business of additive manufacturing dedicated to printing custom molds for confectionery.

There are four identical printing machines with a working area A of $360\,cm^2$ and a start-up time of 10 minutes. For example, the operation corresponding to period T of three hours of work (180 minutes) per printer (12 hours in total or 720 minutes) will be planned and sequenced.

2.5.1   Orders

Initially a total of seven orders were considered, each with a different number of parts and parameters. The completion of all orders needed a period larger than T. So, a constructive heuristic was employed. This heuristic was based on a multi-objective knapsack, considering the same objectives described in the mathematical model presented above. Orders are added to subset S until no time is left to add a new one. The problem was then resolved iteratively changing the ponderation of each objective. As a result, three orders were initially selected to be scheduled for printing in T conforming set $S = \{1,4,6\}$. Tables 2.1 and 2.2 show the details of the selected orders. The order with the highest priority was 3 and the order with the least priority was 1. In a real scenario, the rest of the orders will be added to S as soon as all parts of the order in S are being printed. This process can be automated by implementing a DSS, but this is out of the scope of this work.

For solving this instance, the proposed model was implemented in CPLEX 2.13, and the bi-objective problem was solved as a mono objective one by employing weighted sum as a scalarization technique, being the new objective:

$$\lambda \cdot \sum_{i \in S} g_i \cdot y_i + (1 - \lambda) \cdot \sum_{i \in S} \frac{pr_i}{dd_i} \cdot y_i$$

Table 2.1   Parameters of orders.

Order i	Parts j	Due date dd (min)	Gain g ($)	Priority pr
1	8	420	200	1
4	2	240	60	2
6	1	380	150	3

Table 2.2 Details of orders.

Order i	Part j	Estimated printing time te (min)	Real printing time tm (min)	Area a (cm²)
1	1	60	63	18
	2	60	57	18
	3	60	78	18
	4	60	60	27
	5	60	90	24
	6	60	63	24
	7	60	52	27
	8	60	60	27
4	1	50	57	45
	2	20	34	40
6	1	120	136	300

Table 2.3 Pareto solutions.

Solution	λ_1	λ_2	f_1 ($)	f_2
1	0.1	0.9	60	0.67
2	0.9	0.1	210	0.217

For generating Pareto optimal solutions, values between 0 and 1 were assigned to λ, but only two solutions (supported) were obtained (Table 2.3), as can be seen one for $\lambda = 0.9$ and another for $\lambda = 0.1$.

Table 2.3 shows, as expected, that when λ approach 1, the solution approaches the maximum gains, while when λ approach 0, the solution approaches the maximum of orders in importance. Although "non-supported" solutions were calculated, for the purpose of this work the two supported solutions are enough to show that the model behaves as expected.

The next schedules for both solutions are shown in Tables 2.4 and 2.5, respectively.

Nevertheless, a deeper look into both solutions shows some interesting properties of both schedules (as shown in Tables 2.6 and 2.7). At first, when we look at the total manufacturing time of both solutions, it seems that there is not a significant

Table 2.4 Solution 1 schedule.

	B_1	B_2	B_3
P_1	1.7	4.2	1.3
P_2	6.1	1.5	1.1
P_3	1.8	1.2; 1.4; 1.6	
P_4	4.1		

Table 2.5 Solution 2 schedule.

	B_1	B_2	B_3
P_1	1.8	1.2; 1.5; 1.6	1.3
P_2	1.4	1.1	1.7
P_3	4.1		
P_4	4.2	6.1	

Table 2.6 Solution 1 execution time by printer.

	B_1	B_2	B_3	Total time by printer
P_1	52	34	78	194
P_2	136	90	63	319
P_3	60	63		143
P_4	57			67
			Total time + setup	723
			Total time − setup	633

difference between both solutions (only three minutes difference). But looking further, it can be seen that the effective time of use (subtracting setup times) of printers is more balanced in solution 2 than in solution 1. This is not by chance, but it is a consequence of favoring maximization of objective function "importance" of parts.

Tables 2.8 and 2.9 reveal other interesting characteristics of the model, intuitively solution 1 should have a shorter total production time than solution 2. But this "inconsistency" has also an explanation the priority assigned to orders is not supported by the information at hand, priority assignment should be consistent

Table 2.7 Solution 2 execution time by printer.

	B_1	B_2	B_3	Total time by printer
P_1	60	90	78	258
P_2	60	63	52	205
P_3	57			67
P_4	34	136		190
			Total time + setup	720
			Total time − setup	630

Table 2.8 Solution 1 orders completion times.

	Order 1	Order 4	Order 6
P_1	194	54	0
P_2	319	0	146
P_3	143	0	0
P_4	0	57	0
Completion time by order	656	111	146
	Total completion time + setup		913

Table 2.9 Solution 2 orders completion times.

	Order 1	Order 4	Order 6
P_1	258	0	0
P_2	205	0	0
P_3	0	57	0
P_4	0	34	136
Completion time by order	463	91	136
	Total completion time + setup		690

with the ratios $\frac{gain}{duedate}$ for each order being scheduled at least from a normative point of view. So, these ratios can be employed to tell the decision-makers that they could probably be inconsistent when assigning priorities to orders if the assignment does not follow the ranking derived from the ratios.

2.6 Conclusions

An approach for solving an interesting problem in additive manufacturing is proposed. The mathematical model developed solved effectively a real instance and helped to gain insight into the complex relationships of variables and objectives, thus providing an understanding of such relationships for decision-making.

The proposed approach should be fully developed, and more experimentation should be done to ensure a true impact on the related problem. In future work, an instance generator must be developed to continue studying the complexity and impact of variables in decision making. It would also be interesting to study how to integrate preferences of the DM into the model.

References

1 ISO (2015). *ASTM52900-15, Standard Terminology for Additive Manufacturing – General Principles – Terminology.* West Conshohocken, PA: ASTM International www.astm.org.

2 Savolainen, J. and Collan, M. (2020). How additive manufacturing technology changes business models? – review of literature. *Additive Manufacturing* 32: 101070. https://doi.org/10.1016/j.addma.2020.101070.

3 Bhuvanesh Kumar, M. and Sathiya, P. (2021). Methods and materials for additive manufacturing: a critical review on advancements and challenges. *Thin-Walled Structures* 159 (October): 107228. https://doi.org/10.1016/j.tws .2020.107228.

4 McCue, T.J. (2020). Additive manufacturing industry grows to almost $12 billion in 2019. *Forbes* 1: 378–392. https://www.forbes.com/sites/tjmccue/2020/ 05/08/additive-manufacturing-industry-grows-to-almost-12-billion-in-2019/ 5db205ff5678.

5 Oh, Y., Witherell, P., Lu, Y., and Sprock, T. (2020). Nesting and scheduling problems for additive manufacturing: a taxonomy and review. *Additive Manufacturing* 36 (July): 101492. https://doi.org/10.1016/j.addma.2020.101492.

6 Kapadia, M.S., Starly, B., Thomas, A. et al. (2019). Impact of scheduling policies on the performance of an additive manufacturing production system. *Procedia Manufacturing* 39: 447–456. https://doi.org/10.1016/j.promfg.2020.01 .388.

7 Liaqait, R.A., Hamid, S., Warsi, S.S., and Khalid, A. (2021). A critical analysis of job shop scheduling in context of Industry 4.0. *Sustainability (Switzerland)* 13 (14): 1–19. https://doi.org/10.3390/su13147684.

8 Alami, D. and ElMaraghy, W. (2021). A cost benefit analysis for Industry 4.0 in a job shop environment using a mixed integer linear programming model.

Journal of Manufacturing Systems 59 (February): 81–97. https://doi.org/10
.1016/j.jmsy.2021.01.014.

9 Habib, A., Ahsan, N., and Khoda, B. (2015). Optimizing material deposition
direction for functional internal architecture in additive manufacturing processes. *Procedia Manufacturing* 1: 378–392. https://doi.org/10.1016/j.promfg
.2015.09.045.

10 Zhu, Z., Dhokia, V., and Newman, S.T. (2013). The development of a novel
process planning algorithm for an unconstrained hybrid manufacturing process. *Journal of Manufacturing Processes* 15 (4): 404–413. https://doi.org/10
.1016/j.jmapro.2013.06.006.

11 Rossi, A. and Lanzetta, M. (2020). Integration of hybrid additive/subtractive
manufacturing planning and scheduling by metaheuristics. *Computers and
Industrial Engineering* 144 (April): 106428. https://doi.org/10.1016/j.cie.2020
.106428.

12 Baumung, W. and Fomin, V.V. (2018). Optimization model to extend existing
production planning and control systems for the use of additive manufacturing technologies in the industrial production. *Procedia Manufacturing* 24:
222–228. https://doi.org/10.1016/j.promfg.2018.06.035.

13 Alemão, D., Rocha, A.D., and Barata, J. (2021). Smart manufacturing scheduling approaches—systematic review and future directions. *Applied Sciences
(Switzerland)* 11 (5): 1–20. https://doi.org/10.3390/app11052186.

14 Li, Q., Zhang, D., Wang, S., and Kucukkoc, I. (2019). A dynamic order acceptance and scheduling approach for additive manufacturing on-demand
production. *International Journal of Advanced Manufacturing Technology*
105 (9): 3711–3729. https://doi.org/10.1007/s00170-019-03796-x.

15 Fera, M., Macchiaroli, R., Fruggiero, F., and Lambiase, A. (2020). A modified
tabu search algorithm for the single-machine scheduling problem using additive manufacturing technology. *International Journal of Industrial Engineering
Computations* 11 (3): 401–414. https://doi.org/10.5267/j.ijiec.2020.1.001.

16 Alicastro, M., Ferone, D., Festa, P. et al. (2021). A reinforcement learning
iterated local search for makespan minimization in additive manufacturing machine scheduling problems. *Computers and Operations Research* 131
(February): 105272. https://doi.org/10.1016/j.cor.2021.105272.

17 Chergui, A., Hadj-Hamou, K., and Vignat, F. (2018). Production scheduling
and nesting in additive manufacturing. *Computers and Industrial Engineering*
126: 292–301. https://doi.org/10.1016/j.cie.2018.09.048.

18 Dvorak, F., Micali, M., and Mathieu, M. (2018). Planning and scheduling in
additive manufacturing. *Inteligencia Artificial* 21 (62): 40–52. https://doi.org/10
.4114/intartif.vol21iss62pp40-52.

19 Che, Y., Hu, K., Zhang, Z., and Lim, A. (2021). Machine scheduling with orientation selection and two-dimensional packing for additive manufacturing.

Computers and Operations Research 130: 105245. https://doi.org/10.1016/j.cor .2021.105245.

20 Li, Q., Kucukkoc, I., and Zhang, D.Z. (2017). Production planning in additive manufacturing and 3D printing. *Computers and Operations Research* 83: 1339–1351. https://doi.org/10.1016/j.cor.2017.01.013.

21 Zhang, R. and Chiong, R. (2016). Solving the energy-efficient job shop scheduling problem: a multi-objective genetic algorithm with enhanced local search for minimizing the total weighted tardiness and total energy consumption. *Journal of Cleaner Production* 112: 3361–3375. https://doi.org/10.1016/j .jclepro.2015.09.097.

22 Kucukkoc, I. (2019). MILP models to minimise makespan in additive manufacturing machine scheduling problems. *Computers and Operations Research* 105: 58–67. https://doi.org/10.1016/j.cor.2019.01.006.

23 Karimi, S., Kwon, S., and Ning, F. (2021). Energy-aware production scheduling for additive manufacturing. *Journal of Cleaner Production* 278 (2021): 123183. https://doi.org/10.1016/j.jclepro.2020.123183.

24 Ransikarbum, K., Pitakaso, R., and Kim, N. (2020). A decision-support model for additive manufacturing scheduling using an integrative analytic hierarchy process and multi-objective optimization. *Applied Sciences (Switzerland)* 10 (15): https://doi.org/10.3390/app10155159.

25 Garey, M.R., Johnson, D.S., and Sethi, R. (1976). Complexity of flowshop and jobshop scheduling. *Mathematics of Operations Research* 1 (2): 117–129. https://doi.org/10.1287/moor.1.2.117.

26 Liu, S.Q., Kozan, E., Masoud, M. et al. (2018). Job shop scheduling with a combination of four buffering constraints. *International Journal of Production Research* 56 (9): 3274–3293. https://doi.org/10.1080/00207543.2017.1401240.

27 Luo, S. (2020). Dynamic scheduling for flexible job shop with new job insertions by deep reinforcement learning. *Applied Soft Computing Journal* 91: 106208. https://doi.org/10.1016/j.asoc.2020.106208.

28 Zhang, M., Tao, F., and Nee, A.Y.C. (2021). Digital twin enhanced dynamic job-shop scheduling. *Journal of Manufacturing Systems* 58 (April): 146–156. https://doi.org/10.1016/j.jmsy.2020.04.008.

29 Caldeira, R.H., Gnanavelbabu, A., and Vaidyanathan, T. (2020). An effective backtracking search algorithm for multi-objective flexible job shop scheduling considering new job arrivals and energy consumption. *Computers and Industrial Engineering* 149 (September): 106863. https://doi.org/10.1016/j.cie.2020 .106863.

30 Froehlich, G.E.A., Kiechle, G., and Doerner, K.F. (2019). Creating a multi-iterative-priority-rule for the job shop scheduling problem with focus on tardy jobs via genetic programming. In: *Learning and Intelligent Optimization. LION 12 2018. Lecture Notes in Computer Science*, vol. 11353 (ed. R.

Battiti, M. Brunato, I. Kotsireas and P Pardalos). Cham: Springer https://doi
.org/10.1007/978-3-030-05348-2_6.

31 Nouiri, M., Bekrar, A., Jemai, A. et al. (2018). An effective and distributed particle swarm optimization algorithm for flexible job-shop scheduling problem. *Journal of Intelligent Manufacturing* 29 (3): 603–615. https://doi.org/10.1007/s10845-015-1039-3.

32 Chen, R., Yang, B., Li, S., and Wang, S. (2020). A self-learning genetic algorithm based on reinforcement learning for flexible job-shop scheduling problem. *Computers and Industrial Engineering* 149 (June): 106778. https://doi.org/10.1016/j.cie.2020.106778.

33 Litvinchev, I. and López, F. (2008). An interactive algorithm for portfolio bi-criteria optimization of R&D projects in public organizations. *Journal of Computer and Systems Sciences International* 47 (1): 25–32. https://doi.org/10.1134/s1064230708010048.

34 Litvinchev, I., López, F., Escalante, H.J., and Mata, M. (2011). A milp bi-objective model for static portfolio selection of R&D projects with synergies. *Journal of Computer and Systems Sciences International* 50 (6): 942–952. https://doi.org/10.1134/S1064230711060165.

35 Fernandez, E., Lopez, F., Navarro, J. et al. (2009). An integrated mathematical-computer approach for R&D Project selection in large public organisations. *International Journal of Mathematics in Operational Research* 1 (3): 372–396. https://doi.org/10.1504/IJMOR.2009.024291.

3

Significance of Consuming 5G-Built Artificial Intelligence in Smart Cities

Y. Bevish Jinila[1], Cinthia Joy[2], J. Joshua Thomas[3], and S. Prayla Shyry[1]

[1]*Department of Information Technology, Sathyabama Institute of Science and Technology, India*
[2]*Sydney International School of Technology and Commerce, Sydney, Australia*
[3]*Department of Computing, UOW Malaysia, KDU Penang University College, Penang, Malaysia*

3.1 Introduction

In this era, the amount of data generated from various sectors is enormous. It is expected that the size of the data grows exponentially with time. Proper usage of the data would benefit human society, thereby enhancing the standard of human living in this world. There are tremendous technologies in the market that could make the data useful. This is carried out using technologies such as Big Data analytics, data science, and logical reasoning. Recently, advancements in technology have created a huge impact of Artificial Intelligence (AI) across various domains.

In the current decade, the advent of 5G technology and AI has been a greater breakthrough. The 5G technology is designed to enhance the network speed, thereby lowering the latency. AI is the intelligent method that helps to make wise decisions on time. AI-based algorithms are used to train computers to mimic human brain and thinking patterns, just as a human. In the previous decade, multiple works have been carried out to explore the requirement of fast data transfer, and the resultant is the 5G technology. When 5G is built up with AI, it helps to build a sustainable ecosystem and transform the lifestyle of humans.

The fuel that powers AI is data. Big Data, which consists of large, complex, and frequently changing datasets, allows machine learning (ML) applications to do what they are designed to do: learn and acquire abilities. Big Data provides the knowledge that AI algorithms need to develop and improve features and pattern-recognition capabilities. It would be impossible to design and train the sophisticated algorithms, neural networks, and predictive models that make AI such a game-changing technology without having a vast amount of high-quality data. AI and Big Data have a mutually beneficial interaction.

Artificial Intelligence in Industry 4.0 and 5G Technology, First Edition.
Edited by Pandian Vasant, Elias Munapo, J. Joshua Thomas, and Gerhard-Wilhelm Weber.

The internet of things (IoT) is driving Big Data. The IoT is simply a network of computing devices embedded in common objects that send and receive data. The number of linked devices is projected to exceed 50 billion by 2020. YouTube users upload 400 hours of fresh video every minute, while Instagram users post 2.5 million times per minute. According to a recent International Business Machines Corporation (IBM) report, 2.5 quintillion bytes of data are created every day. The IoT devices promise benefits such as increased productivity, lifespan, and enjoyment; they also bring with them a slew of security concerns for mobile service providers. The integration of IoT to 5G networks leads to mobile network attacks.

Figure 3.1 shows the impact of 5G-built AI across different domains. This has greatly stepped into multiple fields, namely health care, logistics, retail, education, agriculture, manufacturing, and smart cities. The impact of AI in healthcare has improvised the way diseases are diagnosed and predictive analytics of the occurrence of the disease is identified. In logistics, AI has a huge impact on inventory management, and in the retail sector, business needs are identified based on the current market trends, and accordingly, plans are formulated. In the education sector, AI has stepped in to improvise the requirements of the students. The present analysis shows that AI-enabled search engines and other tools assist students in education. In the field of agriculture, AI helps to make smart decisions in planting crops and in identifying suitable crops that will give more yield based on the season. This has greatly helped the farmers to decide on the plantation and the time of harvest. In smart cities, AI has enabled smart transportation and has enabled multiple applications to become smarter for comfortable and convenient travel.

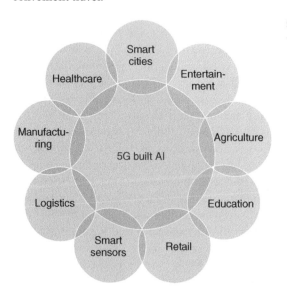

Figure 3.1 5G-built AI domains.

According to the United Nations (UN) report (https://www.un.org/ development/desa/en/news/population/2018-revision-of-world-urbanization-prospects.html), the future of the world is cities. Detailed analysis showed that in 2050, 68% of the world's population will live in cities. This will be a greater increase to approximately 2.5 billion over three decades. From the report, it is evident that the cities will face difficulties in terms of transportation, infrastructure, employment, energy, healthcare, waste management, and education. Even today, major cities are facing many issues with respect to transportation and other essentials. Improper planning of cities often has led to pollution and other discomforts for human lives.

Unmanned Aerial Vehicles (UAVs) are used in a variety of real-world scenarios, including cargo delivery, traffic monitoring, moving things in potentially risky situations, and surveillance. The deployment of UAVs in any of these applications demands the creation of viable and optimal vehicle motion trajectories. UAVs can also assist in the establishment and maintenance of communication channels between victims, ground teams, and the command center, as well as with other disaster response agencies.

Intelligent Transportation System (ITS), is an emerging area in the transportation sector that is a part of the IoT that acts as a baseline for communication between vehicles and humans. The objective of this system is to provide driver comfort, better traffic management, road safety, and better road connectivity. In the future, ITS will be an integral component for the development of smart cities. Many countries in the world are aiming for smart cities that will provide comfort and better traffic management during travel. To make the ITS possible, the vehicles are adopted to communicate themselves in the vehicular network. In addition to vehicle-to-vehicle connectivity, vehicles also communicate to the Road Side Unit (RSU) distributed in various geographical locations. Further zone-wise, RSUs are connected to Zonal Servers (ZS). The ZS is responsible to analyze the data retrieved from the RSUs and make decisions on traffic management in a particular zone [1–6]. In addition, zone-to-zone communication is required to make an entire region stay updated with traffic decisions. So, the overall management is performed by a Traffic Management Server (TMS). The TMS acts as a centralized authority and controls the entire regional network. It supports the implementation of traffic applications such as an emergency warning, congestion warning, and traffic route management. In addition, the TMS holds backend servers and decision-making agents to handle emergency shortfalls.

ITS relies on the huge exchange of data between entities, namely vehicles, RSU, ZS, and TMS. AI plays a vital role in acquiring important information from large volumes of data. The analytics tool can help decision-making faster and build a better prediction model. The decisions are made based on various factors such as the type of the application, vehicle density, weather conditions, and current traffic.

Figure 3.2 5G-built AI-enabled smart cities.

Currently, in 4G networks, in small areas where mass connectivity is required, there is often a problem in network connectivity. People often face issues when trying to post a message in a crowded area such as a concert venue or a stadium.

The advent of 5G technology has increased the connectivity and has greatly boosted the capacity of smaller networks. This has also enabled smooth connectivity between the users, better data analytics, and real-time tracking even in crowded places. The application of AI with 5G in smart cities makes urbanization smarter and thereby provides sustainable solutions to the habitats in terms of administration, traffic congestion, parking management, sanitation, and security surveillance.

With the advent of 5G, AI has become an integral part of the smart cities initiative to automate and improvise the activities and operations on travel. Figure 3.2 shows the 5G-built AI-enabled smart cities. Even though the applications differ with respect to various criteria, the main goal of the deployment of smart cities is to improvise the living conditions, make the environment sustainable for living, and create an instinct of competitiveness among the people.

To a certain extent, plans have been formalized to improve the metrics of smart cities. Recently, steps have been taken to adopt the latest technology to establish efficient, convenient, and comfortable smart cities. Right from the earlier days, urban planners have been creating plans to widen the roads to reduce congestion in cities. In the year 1960, the city management of Los Angeles used aerial photography to track demographic trends.

In continuation with the previous initiatives, steps have been taken in 2000 to establish full-fledged smart cities. The present initiative is not city planning combined with the internet or applying digital technology alone. It is the application of intelligence in cities to make them super smart. The modern idea of smart cities was incepted in the year 2005. This was initiated by the Clinton Foundation by associating with Cisco to establish sustainable cities. Cisco in association with Amsterdam, San Francisco, and Seoul launched the smart city program that focused mainly on infrastructure development, data sharing, and reduction of CO_2 levels.

The next major initiative was taken by IBM in 2008. The company launched a program called "Smarter Planet Group" to analyze the usage of sensors and networks on the planet. The main purpose of this initiative is to achieve efficient and sustainable development in society.

3.2   Background and Related Work

There is plenty of research work carried out on AI and data analytics in transportation applications over the past decade. Various mobility models, human behavioral patterns, predictive analytics, and intelligent solutions have been suggested. The recent research approaches of data analytics on various ITS applications are discussed below.

The global positioning system (GPS) trajectory data are analyzed to find the miles traveled by the vehicles on roadways (https://www.un.org/development/desa/en/news/population/2018-revision-of-world-urbanization-prospects.html). The traditional methods incur more cost on the collection of data and also require sub-collection of data from each subset and then integrating them to do a predictive analysis. This method uses a geo computing framework based on Apache Spark to estimate miles traveled by the vehicles based on the GPS trajectory data. However, a detailed statistical relationship between sample traffic and actual traffic is required, so that sampling bias can be calibrated.

The advent of autonomous vehicles, smart devices, and smart equipment has added more complexity in data storage, communication, and sharing of data. The data shared between the vehicles and between vehicles and RSU are enormous as the number of vehicles traveling daily on the road is huge [2]. A well-defined array of protocols is required to transfer the data from and to the service providers. However, a small leakage of data will greatly degrade the entire system. So, it is essential to strengthen the security of the transfer of data between entities.

As digitization has become an integral part of the transportation sector through the deployment of sensors and computational components, the involvement of AI and machine learning has given amazing solutions. Braun et al. [7] suggested that AI and other supportive technologies can greatly improve data processing. But, at the same time, human intervention is required. The intelligence is applied to the real-time data collected from the cities, which provides a deeper understanding of the deployment of smart cities and shows how various conditions can be met through the adoption of AI in smart cities [8]. The support of AI ensures sustainable solutions for comfortable living and also supports socio-economic dimensions.

Zhao et al. [9] and Zhang et al. [10] have suggested the use of AI, machine learning, and reinforcement learning to precisely monitor the traffic. Also, a detailed study on the recent development in smart cities is addressed. Veres and Moussa [11] performed a detailed study on the usage of AI and machine learning models in smart cities. Various applications such a traffic flow, traffic management, pedestrian flow estimation and accident estimation were analyzed and presented.

Ferdowsi et al. [12] performed a study on smart cities based on Deep Reinforcement Learning (DRL) techniques. In this paper, the study was based on fleet management, design of trajectory, and cyber and physical security. The impact of these applications was also analyzed and presented.

Xu et al. [13] studied the importance of AI in the telecommunication industry. In this work, AI-based solutions were applied and simulated in the telecommunication sector and the benefits were studied and presented. Morales et al. [14] applied AI-based schemes to reconfigure the traffic based on the volume and the direction suggested by neural networks. In this paper, a neural network was modeled to identify the requirement of traffic redirection by identifying the volume of traffic. Bai et al. [15] tested a decision-making technique in a heterogeneous traffic environment. In this work, to convert the data into the hyper-grid matrix, a data preprocessor is used. A Deep Neural Network (DNN) was used to extract the features, and the analysis was presented based on various traffic scenarios. Chen et al. [16] highlighted the possibility of security threats in smart cities' plan. In this paper, a reinforcement learning approach was used to understand various attackers, and a model was proposed to overcome the attacks. Comparison with existing models improved the accuracy by 6%.

Currently, the advent of 5G has created a greater impact in many applications. Recently, there are many literatures that suggests the utility of 5G in multiple domains and especially in smart cities. Literatures related to AI-based smart cities are identified and listed above. The advent of 5G and its importance are listed below. Foukas et al. [17] identified the network slicing techniques and depicted their usage from lower to higher layers. In this case, the resources are split up into virtual networks to satisfy the desired features and Service Level Agreement (SLA). But there are some challenges faced in network slicing, namely radio resource virtualization, composition of services, and management of services.

Another important survey on the utilization of 5G-enabled IoT was analyzed by Akpakwu et al. [18]. In this paper, the application of 5G in various domains, namely ITS, smart cities, industries, and health care, was analyzed. The analysis was mainly based on mobility, multicasting, and seamless services.

Klaus-Dieter Walter et al. [19] proposed the use of sensor-based AI platforms for IoT in smart cities. The authors mainly focused on decentralized, autonomous control with flexible environments and the assistance of AI.

Saurabh Singh et al. [20] introduced the convergence of AI and blockchain for sustainable smart cities. In this article, the authors listed the key factors for convergence. The issues open for the future were discussed and new security suggestions and future guidelines for a smart city ecosystem are provided.

Zaib Ullah et al. [21] presented the application of AI and Machine Learning (ML) in smart cities. This article briefly explained the prior details of ITS, cyber security, and UAV to assure the best services of 5G. Various research challenges and future directions were also presented.

Ridhima Rani et al. [22] presented the role of the IoT-cloud ecosystem in smart cities. The advent of fog computing and its utilization were discussed. The benefits of integrating fog computing and IoT in smart cities were presented.

Akshara Kaginalkar et al. [23] presented a review on air quality management in smart cities with the assistance of AI, IoT, and Cloud Computing. The primary focus of this article was on air quality management for smart cities. The authors recommended the use of the "fit for the purpose" framework to be used in emerging smart cities.

3.3 Challenges in Smart Cities

In this present era, there is an increasing demand for comfortable and fast transportation. Many systems and techniques are implemented to make travel as quick and easy as possible. ITS evolved as a solution to satisfy these requirements. However, numerous amounts of data are transferred between vehicles and RSUs to make this possible. There are several challenges faced to handle these data. The main challenge lies in the following categories.

3.3.1 Data Acquisition

Each entity in ITS collects data from various sources related to the location of the entity and traffic. The ITS uses IEEE 802.11p standard and ETSI standard and offers a variety of services, including both safety and nonsafety applications. Safety messaging in vehicular ad-hoc networks (VANET) can be periodic or event-driven. Periodic messages are informative in nature as messages are communicated among the vehicles to report the speed and position. In the case of event-driven applications, safety messaging is necessary to report the events related to traffic congestion, vehicle accidents, parking, weather conditions, etc. This is managed by the backbone spectrum allocated by dedicated short range communication (DSRC) that has two types of channels, namely the control channel and service channels.

Data are collected from a video surveillance camera [24] and vehicles are classified based on the count and the driving direction. This method uses a faster Convolution Neural Network two-stage detector to analyze the collected data. A new method for data flow collection using onboard monocular camera (OMC) [8] is a cost-effective dynamic traffic sensor. The camera used for data collection [8] uses you-only-look-once (YOLO) model and spatial transformer network (STN) for dynamic vehicle detection. The data collection done using OMC is more accurate compared to that done by other methods.

3.3.2 Data Analysis

The data acquired has to be analyzed based on various factors such as vehicle density, weather conditions, road conditions, and suitable prediction has to be made for a successful ITS. Since huge volumes of data will be transferred in a limited time, it is hectic to analyze the data based on multiple factors. However, the important factors should be considered to decide on traffic management applications such as congestion warning, route guidance, and emergency vehicle warning. A method of data analytics for cooperative-ITS [25] is introduced based on factors such as vehicle density, weather conditions, and current traffic. This scheme involves data analytics techniques and inferences are made based on the factors. For example, a high packet loss is associated with a more secure algorithm. The methodology is applied for two different applications, namely efficient route guidance application and adaptive transmission range selection.

Based on the data analytics done, decisions are made to improve traffic management.

3.3.3 Data Security and Privacy

Security and privacy are two important factors in ITS. The data being communicated in ITS should be secure and error-free. There are many possibilities where an intruder can inject false messages in this system, to collapse the entire traffic management. Further, the privacy of the user involved in this system should be preserved. Hence, it is vital to consider both these factors during the analysis of the data.

3.3.4 Data Dissemination

The data dissemination in ITS helps in improvising various applications involved in smart transportation. The RSUs disseminate a large volume of data for a specific application. Based on data analytics, consideration is given to priority-based messages. So, if there is an emergency condition such as accidents or emergency vehicle arrival, the data analytics techniques help in analyzing the priority messages.

3.4 Need for AI and Data Analytics

The data collected in ITS are complex and are in heterogeneous formats and large volumes. The acquired data should be analyzed to obtain useful inferences for efficient traffic management and to improve the reliability of communication among vehicles and RSUs in ITS. The application of efficient tools and techniques in data analytics can definitely improvise and provide a better ITS.

First, since the data acquired are enormous, simple tools for data processing and analysis are insufficient. Therefore, an effective system with efficient data collection and processing is required to analyze and draw instincts from the traffic data.

Second, as transportation has become a part of life, a short and happy journey always enriches the commuters and will help them to be energetic throughout the day. An efficient data analytics tool for transportation will surely optimize the routes so that arrival to a destination can be accomplished within a short time.

In addition, recommendations or suggestions of essential services will add benefit to the commuters. Based on the travel plan, the time of reaching the destination, analysis is done and suggestions are provided to use the essential services.

The ITS can turn the collected data into actionable knowledge, inform the commuters about their safe travel, and can further make suggestions on the essential commodities on travel.

Recently, most of the countries are striving to adopt AI-based schemes for sustainable smart cities. This could greatly benefit the commuters and also pave the way for economic development. In developing countries like India, plans are devised by the Government to adopt AI-enabled solutions to build smart cities (http://indiaai.gov.in/article/ai-in-smart-cities). A panel discussion was done on 25th March 2021 to identify the best utilization of AI globally and in India. Decisions were made to employ reasoning, understanding, and interacting based on AI algorithms. The country has initiated a smart city mission to make this possible. The transformation to classical approaches to smart cities is tremendous. Suitable measures should be taken to transform government-related services, optimization of various processes, and at the same time engage the citizens.

The adoption of AI-based schemes will incorporate a digital model that can simulate, track, and predict the future. The significant need of smart cities is sustainability and reduction in cost for maintenance and services. All the data acquired can be accumulated in the central servers or zone-wise servers and analyzed, and intelligent agents could assist in the prediction of the future.

3.5 Applications of AI in Smart Cities

The main goal of data analytics in ITS is to enhance the underlying traffic management and to provide comfort on travel. The applications of ITS are listed below and the work done earlier to solve those applications is also briefly listed. Figure 3.3 shows the AI-based smart city applications.

3.5.1 Road Condition Monitoring

Pooja et al. [25] proposed an early warning system for road safety using crowdsourcing. This work mainly focused on the identification of potholes and bumps.

Figure 3.3 AI-based smart city applications.

To make this possible, the 3-D accelerometer and GPS available in smart phones were used. The acceleration of the data is filtered through a Gaussian low-pass filter. Through vertical and longitudinal accelerations all the possible potholes and bumps are identified. The collected data are sent to the central server where the Support Vector Machine (SVM) algorithm is used to analyze and remove the false positives.

3.5.2 Driver Behavior Monitoring

The behavior of the driver should be assessed in ITS to alert the neighbors if there is any emergency situation. Saleh and Hossny [24] proposed a method for identifying the behavior of the driver. In this paper, the behavior of the driver was classified into three categories, namely normal, aggressive, and drowsiness. It is well known that a mistake caused by a driver will definitely affect and create an impact on the neighbor drivers. An aggressive driver or a drowsy driver can be a root cause of a mishap on the road. So, it is a good practice to monitor the behavior of the drivers on road. In this paper, long short-term memory (LSTM) recurrent neural networks were used to classify the behavior of the drivers. The data were collected for analysis through smart phones. The gyroscope, accelerometer, and GPS sensors in the smart phones facilitated data collection.

3.5.3 AI-Enabled Automatic Parking

Parking vehicles in crowded cities is quite challenging. People have to search for vacant slots to park the vehicle. This consumes more time and, in most cases, on crowded days it is challenging to find an empty slot and park the vehicle. Presently, in most of the parking lots, the number of vehicles inside the lot is determined by sensors deployed at the entrance of the parking lot. The sensors are used to keep track of the vehicles moving in and out of the lot. Based on this, a display board is used to denote the presence of empty parking lots available. Even though such solutions are adopted, it is often difficult to identify the location of the empty slot to park the vehicle. To commemorate this issue, AI-based schemes are adopted to exactly trace the location slot of the parking space, thereby directing the vehicle to park in the corresponding space, within a short time.

3.5.4 Waste Management

Cities are often crowded and are prone to the generation of excessive garbage. Steps should be taken to efficiently collect and segregate the waste and keep the city clean. In the present scenario, various measures have been taken to keep the city clean. Smart dustbins have been commercialized in the market for smart collection of waste. Once the dustbin is about to be filled, a notification is generated and the user is notified. This idea has been adopted in public places to keep the city clean always. Notification will be generated from multiple bins and the shortest path will be generated and given to the garbage collector. AI-based sensors will efficiently help in automatically segregating the waste during the collection process, which further improvises the current scenario and assists the users to maintain clean and smart cities.

3.5.5 Smart Governance

As smart cities have become a key point in developing a sustainable environment, e-governance services have also been streamlined to create a livelihood environment. The main goal of smart governance is to make all the services transparent to the citizens. Through the online web portals, forums, forms, and mobile applications, citizens can share their views with the government directly, rather than relying on public officials. Smart governance comes up with four different models, namely the Government to Citizen Model (G2C), Government to Business Model (G2B), Government to Government (G2G), and Government to Employee (G2E) model (https://www.smartcity.press/smart-governance-for-smart-cities/). In all these models, AI plays a vital role in the intelligent decision-making process. AI agents are deployed as a part of services to provide on-time responses to the customer. Smart Governance is a wing of smart cities deployment.

3.5.6 Smart Healthcare

The advent of digital technologies has made a greater change in the medical field. This not only helps in curing the diseases but also supports to give the right treatment to the patients on time and helps to save their lives. The idea of smart healthcare was innovated in the year 2009, and it was proposed by IBM. It is the smart deployment of sensors by which real-time data are fetched, processed, and analyzed to obtain meaningful conclusions. Liu et al. [26] proposed smart homes that come with two components, namely home automation and health monitoring. Elderly people were given special care using remote health monitoring. Qi and Lyu [27] suggested the use of a clinical decision support system, in which the doctors can give expert advice based on AI algorithms. This could greatly improvise the accuracy of diagnosis. In some critical surgeries, precise positioning is done using computer-assisted robots. They help to treat the patients promptly. AI-based schemes are employed for intelligent decision-making and thereby to treat the patients on time.

3.5.7 Smart Grid

The grid represents an electric grid, a network of transmission lines, substations, and transformers. In a smart grid, computers, automation, and novel technologies work together to control the transmission lines, substations, and transformers. The smart grid helps in maintaining efficient power transmission, quick restoration of power after minor disturbances, reduced peak demand, lower electricity charges, and improved security. The advent of AI in the smart grid helps to make intelligent decisions in efficient power transmission. Charithri Yapa et al. [28] list out the detailed evaluation on the use of blockchain technology in Smart Grid 2.0, with the goal of facilitating a smooth decentralization process. The report also delves into the blockchain-based applications of future smart grid operations, as well as the function of blockchain in each case. The article also offers a succinct examination of the blockchain integration problems, ensuring secure and scalable, decentralized operations of future, autonomous electrical networks.

3.5.8 Smart Agriculture

AI, IoT, and mobile internet are examples of advanced technology that can help solve some of the world's problems. As a result, this research focuses on innovative approaches to Smart Farming (SF) from 2019 to 2021, illustrating data collection, transmission, storage, analysis, and appropriate solutions. Elsayed Said Mohamed et al. [29] highlights the necessity of employing a 5G mobile network in the development of smart systems because it allows for high-speed data transfers of up to 20 Gbps and can connect a huge number of devices per square kilometer. Despite the fact that SF applications in underdeveloped nations face a number of problems, this study revealed various smart farming methodologies. The notion of Data Sharing Agreements (DSAs) is presented in the study by Konstantina Spanaki et al. [30]; it is a necessary step and a template for AI data management applications across multiple parties. The approach provided is based on design science concepts and develops AI-based role-based access control.

3.6 AI-based Modeling for Smart Cities

As mentioned in the previous sections, the applications in smart cities are data-driven. This requires AI-based methods that make intelligent decision-making for making wise decisions on time. The AI-based methods are categorized into various methods, namely machine learning, deep learning, expert system, Natural Language Processing (NLP), and knowledge representation. Figure 3.4 shows an AI-based expert system.

3.6.1 Smart Cities Deployment Model

Figure 3.5 shows the AI-based smart cities deployment model. The model differs from the traditional approach in its continuous monitoring and optimization. The

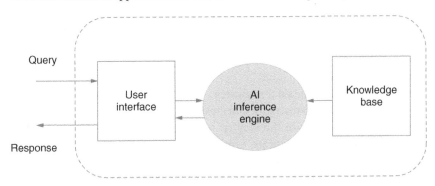

Figure 3.4 AI-based expert system.

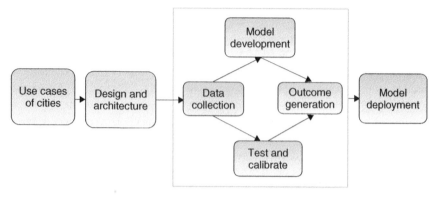

Figure 3.5 Smart cities deployment model.

key approach in this deployment model is to continuously monitor and identify the changes required and thus train the system to satisfy the requirements.

The first component represents the requirement with respect to smart cities. In this phase, the data required for formulating the machine learning model are understood and collected. As a next step, the appropriate design and architecture are chosen to implement the model. After the selection of the working architecture, and identifying the use cases of smart cities, a suitable machine learning model is developed to perform in an iterative manner. This model deployment is crucial since iterative modeling happens when the system is actually used by the user. Hence, the accuracy of the model increases progressively over time. When the model reaches the optimal level, it is deployed. Even after deployment, continuous monitoring is done, to verify the requirement for further changes in the model.

Figure 3.6 shows the 5G-built AI-enabled models for smart cities. The network is formulated with 5G technology to enable high-speed communication possible. When used in conjunction with smart cities, the performance will highly exceed the classical approach. As in the existing mobile networks, there are macro cells and micro cells. The micro cells are responsible for making the transfer of information possible in fine-grained applications and the macro cells are responsible for accumulating the fine-grained information and consolidating it into coarse-grained information.

As the network is grouped into clusters, based on the location, a Cluster Head (CH) is elected to represent a particular cluster. In the CH, the AI agent is deployed to understand and analyze the requirement and act accordingly. Multiple CHs are accumulated to communicate with the RSU. An AI agent is deployed in the RSU to understand, analyze and provide intelligent decisions based on the smart city application. Further, if there is any situation that the RSU cannot decide, the request is sent to the TMS. In addition, all the important decisions are updated to

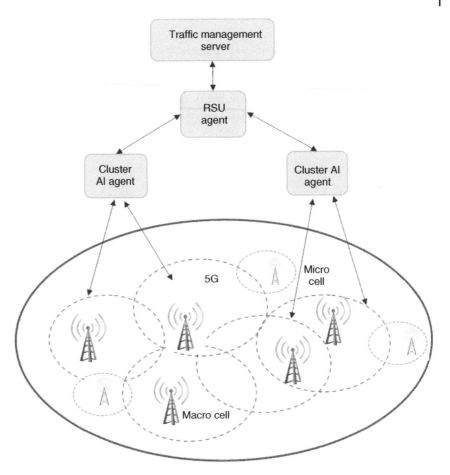

Figure 3.6 5G-built AI-enabled model for smart cities.

TMS, and based on the decisions of the TMS, the model is trained using suitable machine learning models.

3.6.2 AI-Based Predictive Analytics

The predictive analytics model helps to find or forecast the occurrence of events in the future. This model identifies the future requirements well and suggests suitable corrections to prevent the event occurrence. In ITS, the predictive models play a major role in predicting the future in several applications. In particular, the role of predictive analytics becomes significant in safety applications, as this may avoid mishaps and provide safety for commuters. There are many popular predictive models such as clustering models, classification models, and neural networks.

3.6.3 Pre-processing

Data pre-processing is the primary step in predictive analytics. The raw data that is fetched from the source cannot be processed to derive gainful insights. So, it is important to cleanse the data before processing it further. In the pre-processing phase, the null values are cleared, standardization is applied through a standard scaler, categorical values are handled, and then encoding is applied through one-hot encoding and multicollinearity.

3.6.4 Feature Selection

The primary attribute for predictive analytics is feature selection. It determines which variable has the maximum predictive value. There are various methods for feature selection, namely the filter methods and wrapper methods.

The most common method used for feature selection is the filter method. In this method, features are selected based on their computed scores in statistical tests. The correlation coefficient is the basic term. The correlation coefficient (r) quantifies the degree to which two variables are related. When $r = 0$, no variables are related. When the value of "r" is positive, there is a strong association between variables and if the value of "r" is negative, there is an inverse relationship. Based on the type of features, a different statistical method is used. In the case of wrapper methods, a subset of features will be selected and the model is trained. In this case, the best subset of features is selected.

Even though wrapper methods are computationally expensive, they select the subset of best features through cross-validation. There are various wrapper methods, namely forward selection, backward elimination, and recursive feature elimination. In the case of forward selection, no feature is selected first. A new feature is added to improve the model. This process is continued until there is no change in the performance. Figure 3.7 shows the filter methods vs. wrapper methods.

3.6.5 Artificial Intelligence Model

The AI agents are deployed in CHs and RSUs to make intelligent decisions, thereby reducing most of the workload. AI can be used to control and optimize the data collected from varied sensors, and it can also be used to monitor the physical infrastructure [31]. The machine learning models are used because of their nonlinearity and ability to learn from the historical data. According to Prieto et al. [32] and Torres et al. [2], the Artificial Neural Network (ANN) or DNN show high accuracy in solving problems related to classification, association, and prediction. The basic ANN has a single input layer, a single hidden layer, and a single output layer. For processing complex data, it is always desirable to have multiple hidden layers. In each hidden layer, the computation consists of a summation function. This

Figure 3.7 Filter methods vs wrapper methods.

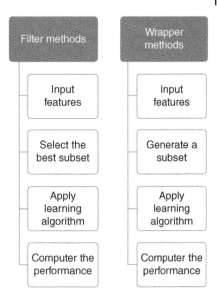

function finds the weighted average of inputs and the activation function is used to compute the output of the hidden layer. The activation function can be a linear or nonlinear function. The computation is given by,

$$y_c = f\left(\sum_{i=1}^{n} \mathbf{w_i x_i}\right)$$ (3.1)

where y_c represents the output computed, $\mathbf{x_i}$ denotes the elements of the input vector $\mathbf{X}$, $\mathbf{w_i}$ represents the elements of the weight vector $\mathbf{W}$, and "n" represents the number of hidden layers. Regularization is an important step, in which the weight is regularized using the dropout method. This enables the generalization better and the drop-out rate is 0.1. The activation function applied to the hidden layer is the sigmoidal function. The sigmoid function is a nonlinear activation function that generates the output continuously in the range 0 to 1. The sigmoid function is computed as shown in Eq. (3.2).

$$a = \frac{1}{1 + e^{-z}}$$ (3.2)

Sigmoid is a differentiable function and is more or less similar to the step function. However, the output is continuous and prevents unusual jumps that happen in the step function. The output layer adopts a linear function. In each hidden layer, optimization is applied using the Adam algorithm where the learning rate is fixed to 0.001 and the maximum training epochs is 150.

Based on the training done in TMS, the information is appropriately shared with the RSU agent, and based on the regions or zones, it is distributed further to the CH agents. The CH agents are responsible to keep track of the real-time applications, mainly with respect to traffic-related applications, namely traffic monitoring, road condition monitoring, emergency vehicle tracking, and so on.

3.7 Conclusion

In this paper, the challenges and needs of AI in smart cities are addressed. The advent of 5G and its integration with AI-enabled schemes has put up the future with innovative, convenient, and comfortable applications that could help humans. In addition, the challenges and needs of 5G-built AI are addressed. Further, a 5G-built AI-enabled model on smart cities is depicted. In the future, this model can be further improvised by adopting multiple hidden layers and it can be tested with real-time data. Further comparative analysis can be done to show the efficacy of the proposed model.

References

1 Bevish Jinila, Y. and Komathy, K. (2015). Rough set based fuzzy scheme for clustering and cluster head selection in VANET. *Elektronika Ir Elektrotechnika* 21 (1): 54–59, ISSN: 1392-1215.

2 Torres, J.F., Galicia, A., Troncoso, A., and Martínez-Álvarez, F. (2018). A scalable approach based on deep learning for big data time series forecasting. *Integrated Computer-Aided Engineering* 25 (5): 1–14.

3 Neha Roy, Y. and Jinila, B. (2015). A survey on security challenges and malicious vehicle detection in vehicular ad hoc networks. *Contemporary Engineering Sciences* 8 (5–8): 235–240, ISSN: 1313-6569.

4 Bevish Jinila, Y. and Komathy, K. (2014). Distributed and secured dynamic pseudo ID generation for privacy preservation in vehicular ad hoc networks. *Journal of Applied and Theoretical Information Technology* 66 (1): 126–134, ISSN: 1992-8645.

5 Sam Mathews, Y. and Jinila, B. (2014). An effective strategy for pseudonym generation and changing scheme with privacy preservation for VANET. In: *International Conference on Electronics and Communication Systems (ICECS) 2014*, ISBN –978-1-4799-2321-2, 1–6. IEEE.

6 Bevish Jinila, Y. and Komathy, K. (2013). A privacy preserving authentication framework for safety messages in vanet. In: *4th International Conference on Sustainable Energy and Intelligent System (SEISCON 2013), December 12-14, 2013*, 456–461. IET.

7 Braun, T., Fung, B.C.M., Iqbal, F., and Shar, B. (2018). Security and privacy challenges in smart cities. *Sustainable Cities and Society* 39: 499–507.

8 Allam, Z. and Newman, P. (2018). Redefining the smart city: culture, metabolism & governance. *Smart Cities* 1: 4–25.

9 Zhao, Z., Chen, W., Wu, X. et al. (2017). LSTM network: a deep learning approach for short-term traffic forecast. *IET Intelligent Transport Systems* 11 (2): 68–75.

10 Zhang, J., Zheng, Y., and Qi, D. (2017). Deep spatio-temporal residual networks for citywide crowd flows prediction. In: *Thirty-First AAAI Conference on Artificial Intelligence*. AAAI.

11 Veres, M., Moussa, M. et al. (2019). Deep learning for intelligent transportation systems: A survey of emerging trends. *IEEE Transactions on Intelligent Transportation Systems* 21 (8).

12 Ferdowsi, A., Challita, U., and Saad, W. (2019). Deep learning for reliable mobile edge analytics in intelligent transportation systems: an overview. *IEEE Vehicular Technology Magazine* 14 (1): 62–70.

13 Xu, G., Mu, Y., and Liu, J. (2017). Inclusion of artificial intelligence in communication networks and services. *ITU J. ICT Discoveries, Special* 1: 1–6.

14 Morales, F., Ruiz, M., Gifre, L. et al. (2017). Virtual network topology adaptability based on data analytics for traffic prediction. *IEEE/OSA Journal of Optical Communications and Networking* 9 (1): A35–A45.

15 Bai, Z., Shangguan, W., Cai, B., and Chai, L. Deep reinforcement learning based highlevel driving behavior decision-making model in heterogeneous traffic. In: *2019 Chinese Control Conference, CCC*, vol. 2019, 8600–8605. IEEE.

16 Chen, Y., Zhang, Y., Maharjan, S. et al. (2019). Deep learning for securemobile edge computing in cyber-physical transportation systems. *IEEE Network* 33 (4): 36–41.

17 Foukas, X., Patounas, G., Elmokashfi, A., and Marina, M.K. (2017). Network slicing in 5G: survey and challenges. *IEEE Communications Magazine* 55 (5): 94–100.

18 Akpakwu, G.A., Silva, B.J., Hancke, G.P., and Abu-Mahfouz, A.M. (2017). A survey on 5G networks for the internet of things: communication technologies and challenges. *IEEE Access* 6: 3619–3647.

19 Walter, K.-D. (2019). AI-based sensor platforms for the IoT in smart cities, Chap. 7. In: *Big Data Analytics for Cyber-Physical Systems* (ed. G. Dartmann, H. Song and A. Schmeink), 145–166. Elsevier, ISBN 9780128166376.

20 Singh, S., Sharma, P.K., Yoon, B. et al. (2020). Convergence of blockchain and artificial intelligence in IoT network for the sustainable smart city. *Sustainable Cities and Society* 63: 102364, ISSN 2210-6707.

21 Ullah, Z., Al-Turjman, F., Mostarda, L., and Gagliardi, R. (2020). Applications of artificial intelligence and machine learning in smart cities. *Computer Communications* 154: 313–323, ISSN 0140-3664.

22 Rani, R., Kashyap, V., and Khurana, M. (2020). Role of IoT-cloud ecosystem in smart cities : review and challenges. *Materials Today: Proceedings* 49: 2994–2998. ISSN 2214-7853.

23 Kaginalkar, A., Kumar, S., Gargava, P., and Niyogi, D. (2021). Review of urban computing in air quality management as smart city service: an integrated IoT, AI, and cloud technology perspective. *Urban Climate* 39: 100972, ISSN 2212-0955.

24 Saleh, K., Hossny, M., and Nahavandi, S. (2017). Driving behaviour classification based on sensor data fusion using lstm recurrent neural networks. In: *2017 IEEE 20th International Conference on Intelligent Transportation Systems (ITSC)*, 1–6. IEEE https://doi.org/10.1109/ITSC.2017.8317835.

25 Pooja, P.R. and Hariharan, B. (2017). An early warning system for traffic and road safety hazards using collaborative crowd sourcing. In: *2017 International Conference on Communication and Signal Processing (ICCSP)*, 1203–1206. IEEE https://doi.org/10.1109/ICCSP.2017.8286570.

26 Liu, L., Stroulia, E., Nikolaidis, I. et al. (2016). Smart homes and home health monitoring technologies for older adults: a systematic review. *International Journal of Medical Informatics* 91: 44–59.

27 Qi, R.J. and Lyu, W.T. (2018). The role and challenges of artificial intelligence-assisted diagnostic technology in the medical field. *Chinese Medical Device Information* 24 (16): 27–28.

28 Yapa, C., de Alwis, C., Liyanage, M., and Ekanayake, J. (2021). Survey on blockchain for future smart grids: technical aspects, applications, integration challenges and future research. *Energy Reports* 7: 6530–6564, ISSN 2352-4847.

29 Mohamed, E.S., Belal, A.A., Abd-Elmabod, S.K. et al. (2021). Smart farming for improving agricultural management. *The Egyptian Journal of Remote Sensing and Space Science*, ISSN 1110-9823. (Pages 971–981).

30 Spanaki, K., Karafili, E., and Despoudi, S. (2021). AI applications of data sharing in agriculture 4.0: a framework for role-based data access control. *International Journal of Information Management* 59: 102350, ISSN 0268-4012.

31 Augusto, J.C., Nakashima, H., and Aghajan, H. (2010). Ambient intelligence and smart environments: a state of the art. In: *Handbook of Ambient Intelligence and Smart Environments*, 3–31. Springer.

32 Prieto, A., Prieto, B., Martinez Ortigosa, E. et al. (2016). RojasNeural networks: an overview of early research, current frameworks and new challenges. *Neurocomputing* 204: 242–268.

4

Neural Network Approach to Segmentation of Economic Infrastructure Objects on High-Resolution Satellite Images

Vladimir A. Kozub[1,2], Alexander B. Murynin[2,3], Igor S. Litvinchev[3], Ivan A. Matveev[3], and Pandian Vasant[4]

[1]*Moscow Institute of Physics and Technology, Department of Control and Applied Mathematics, Dolgoprudny, Moscow Region, 141701, Russia*
[2]*AEROCOSMOS Research Institute for Aerospace Monitoring, Gorohovsky per., 4, Moscow, 105064, Russia*
[3]*Federal Research Center "Computer Science and Control" of Russian Academy of Sciences, Dorodnicyn Computing Center, Vavilov str., 44/2, Moscow, 119333, Russia*
[4]*Ton Duc Thang University, Modeling Evolutionary Algorithms Simulation & Artificial Intelligence (MERLIN), Faculty of Electrical & Electronic Engineering, Ho Chi Minh City 700000, Vietnam*

4.1 Introduction

In the last decade, due to the active development of digital technologies for aerospace imagery, the possibility of automated collection and processing of satellite information has emerged. This information can be used to predict the yield of agricultural land, monitor deforestation and water resources, evaluate the tree species reserves, etc. In the case of artificial objects – buildings, cars, roads, and railways – the analysis of satellite information allows assessing the population of the area, helps in logistics, gives hints on possible hazards, etc. In achieving these goals, three-dimensional (3D) models of urban economic infrastructure can be of significant help.

The literature describes many different techniques for reconstructing the 3D shape of objects [1, 2]. Reconstruction of a digital terrain model is a complex and still poorly understood task. There are no universal approaches to its solution; moreover, various methods rely on data of a different nature: lidar, unmanned aerial vehicles, satellite imagery, etc. If the task is to obtain regularly updated terrain models of a large area, then the required method should allow reconstructing the terrain model using only satellite images and metadata. The development of such an approach will improve the quality of available information in Geo-Informational Systems (GIS); therefore, it is an essential problem of modern computer vision in the field of satellite information processing.

Artificial Intelligence in Industry 4.0 and 5G Technology, First Edition.
Edited by Pandian Vasant, Elias Munapo, J. Joshua Thomas, and Gerhard-Wilhelm Weber.
© 2022 John Wiley & Sons, Inc. Published 2022 by John Wiley & Sons, Inc.

In the set of works [3–6] the method was developed for building 3D digital models of economic infrastructure focused on its housing, rail-road, and road components. A complex solution has been developed, consisting of increasing the resolution of satellite images, extracting semantic information, and building a digital terrain model. This work is aimed at creating an algorithm for the extraction of semantic information or, in other words, for *semantic segmentation* of satellite images.

4.2   Methodology for Constructing a Digital Terrain Model

This section shortly describes the method presented in [3–6]. It aims on building a 3D digital model of rigid objects of economic infrastructure using only one satellite image of the area and metadata. Economic infrastructure refers to buildings, railways, trains, and roads.

The proposed method is focused on working with data from Russian spacecraft of the *Resurs-P* series. These apparatuses can operate in panchromatic mode, receiving data with a resolution of about 0.7 m/px or in multispectral mode, receiving data in narrower spectral ranges with a resolution of $2 - 3$ m/px. A detailed description of spectral channels can be found in [7]. It is assumed that the input data of the method are four satellite images: low-resolution images corresponding to the red, green, and blue parts of the spectrum as well as a high-resolution panchromatic image.

At the first stage, the *data are aggregated*, see the principle scheme in Figure 4.1. In the beginning, resolution of the panchromatic image is doubled with the help of the Generative Adversarial Neural Network (GAN) approach. Then multispectral images are aligned with panchromatic images. The alignment is performed using an affine transformation. Coefficients of transformation are chosen to align pairs of points extracted using the SURF algorithm. Data aggregation is finalized by pansharpening of the enhanced panchromatic image with aligned multispectral images using the method [8]. As a result of these manipulations, an RGB color image with the increased spatial resolution is obtained. In the case of using the data of Resource-P, this resolution is about $35 - 40$ cm/px.

The second stage is *semantic segmentation*. It is proposed to perform segmentation in two steps, called integral and local analysis. At the step of integral analysis, the following classes are distinguished: roofs of buildings, walls of buildings, shadows of buildings, area of railway infrastructure, railways, railway carriage and locomotives, and roads. At the second step, more fine details are segmented, such as wall edges, roof edges, lamp posts along railways and roads, and the shadows of the lamp-posts. Local analysis can be used to refine the shape of buildings as

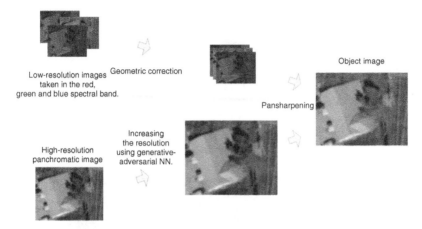

Figure 4.1 The scheme of the first stage of the method.

well as calculate scaling factors and vertical and shadow direction on the image without metadata, but this step is not obligatory. Both stages of segmentation are performed using neural network methods.

The final stage of the method is *constructing 3D models of objects* using segmentation masks and metadata. The scaling factors and direction vectors are evaluated first. The horizontal, vertical, and shadow scaling factors are estimated as well as the direction of the vertical lines in the image (if the image was taken slantly) and the direction of the shadow fall. Then, based on segmentation masks, the graph of roads and railways is restored. Further, roads, carriages, and lamp-posts are modeled using prepared parametric models. At the final stage, the building is modeled. The foundation contour is vectorized, the height is estimated and the shape of the roof is modeled. Depending on the quality of segmentation and survey, a level of detail from LoD0 to LoD2 [9] is possible.

So, a method for constructing 3D models of urban areas from satellite images is briefly outlined, see Figure 4.2. Here the semantic segmentation of infrastructure objects is considered in detail.

4.3   Image Segmentation Problem

Primary segmentation is the division of an image into several areas (segments). In digital images, this means that each pixel is assigned some label from a set of labels. Pixels with the same labels form a segmentation area. With no loss of generality, one can state that the union of the segments covers the entire image, considering unlabelled pixels as parts of "background segment." When

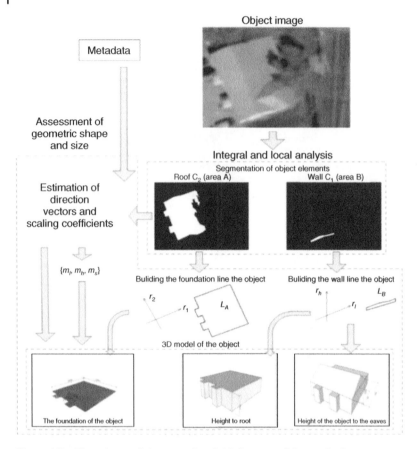

Figure 4.2 The scheme of the second and third stages of the method.

developing segmentation algorithms, it is assumed that pixels in the same segment should have similar visual characteristics or properties.

The next task is *semantic segmentation*. In this case, class labels should determine not only similar pixels but also a certain class of objects. Such classes can be plants, houses, cars, etc. An image of an object belonging to the same semantic class can consist of several parts that are different from each other. For example, a building may contain a roof, walls, and stairways that differ greatly in color and texture, but they should be recognized as parts of a single semantic class – the building.

Even more complex is *instance segmentation* where each object should receive two labels: one label for semantic class and another for the order number of the object in this class. Thus, the result of object segmentation allows not only to distinguish objects of different classes but also to distinguish two different objects within the same class.

Here we explore the second problem in this list, i.e. the semantic segmentation.

4.4 Segmentation Quality Assessment

For simplicity, we will consider the case of segmentation into two classes: object and background. The result of the segmentation of image I having size $H \times W$ can be considered as a binary matrix of the same size, called as *segmentation mask*. The components of this matrix are zero for pixels marked as background and unit for pixels marked as an object.

We denote the ground-truth segmentation mask for the image I as $M = \|m_{ij}\|_{i,j=1,1}^{H,W}$, and the one obtained using an algorithm as $\hat{M} = \|\hat{m}_{ij}\|_{i,j=1,1}^{H,W}$. There are several values defining their similarity.

Precision is the ratio of correctly defined pixels of an object to all pixels that were marked as objects:

$$Q_P = \frac{\sum\limits_{i,j=1}^{H,W} [m_{i,j} = 1 \wedge \hat{m}_{i,j} = 1]}{\sum\limits_{i,j=1}^{H,W} [\hat{m}_{i,j} = 1]}$$

where $\wedge$ is a logical conjunction, $\vee$ is a logical disjunction, $[P]$ is Iverson's notation: the expression takes the value 1 if P is true, and 0 otherwise. *Recall* is the ratio of the correctly defined pixels of the object to the total number of pixels of the target class that are actually present in the image:

$$Q_R = \frac{\sum\limits_{i,j=1}^{H,W} [m_{i,j} = 1 \wedge \hat{m}_{i,j} = 1]}{\sum\limits_{i,j=1}^{H,W} [m_{i,j} = 1]}$$

Accuracy is the percentage of correctly predicted pixel's labels:

$$Q_A = \frac{1}{WH} \sum\limits_{i,j=1}^{H,W} [m_{i,j} = \hat{m}_{i,j}]$$

Considering Precision or Recall separately does not allow us to evaluate the quality of segmentation. Indeed, if the algorithm marks almost all pixels as a background and only a small number of pixels obtain labels (correctly), then the value of Precision is close to one, but in fact there is under-segmentation. If the algorithm marks almost all pixels as an object, the opposite situation will occur when the value of Recall will be close to its possible maximum, despite significant over-segmentation, which is also useless. If there is an imbalance of classes, which is often found in practice, Accuracy also cannot be a sufficiently representative measure of quality, because the over-segmentation of a more frequently encountered class will not significantly reduce its value. Some other values may be used solely. These are F_1-score:

$$F_1 = \frac{2Q_P Q_R}{Q_P + Q_R}$$

and Jaccard measure, currently known as intersection-over-union:

$$J = \frac{\sum\limits_{i,j=1}^{H,W} [m_{i,j} = 1 \wedge \hat{m}_{i,j} = 1]}{\sum\limits_{i,j=1}^{H,W} [\hat{m}_{i,j} = 1 \vee m_{i,j} = 1]}$$

In both adverse cases (under- and over-segmentation), the values of these metrics will be low, so these measures are the most preferable in evaluating and comparing algorithms.

If one needs to evaluate an algorithm for segmentation for multiple (more than two) classes, these measures are calculated for each class separately. If it is necessary to evaluate the quality of the algorithm with a single number, then measures for all classes can be averaged with some weights.

4.5 Existing Segmentation Methods and Algorithms

There are numerous methods for image segmentation as well as their classifications [10–13]. This section contains short descriptions of some segmentation methods that are of the greatest interest for solving a considered problem. We will use the following notation:

- H, W are image height and width in pixels.
- $\vec{p} = (p_1, p_2)$ is pixel of the image, given by its coordinates, $p_1 \in \overline{1, W}$, $p_2 \in \overline{1, H}$. The coordinates are the ordinal numbers in the row and column in the image matrix, counted from the upper-left corner.
- Image can be represented as a set of its pixels: $F = \{\vec{p}_n\}$, $n \in \overline{1, WH}$.
- Image can be represented as a function $I : \vec{p} \rightarrow L^d$, which assigns d values from set L to each pixel. The set L is governed by color depth. Common cases are $L_8 = \{0, 1, ..., 2^8 - 1\}$ and $L_{16} = \{0, 1, ..., 2^{16} - 1\}$. It is also possible to represent brightness as a floating-point number. d is the dimension of the color space; $d = 1$ for grayscale images, $d = 3$ for RGB images.

- Spatial distance between pixels is $\rho(\vec{p}, \vec{p}') : (F, F) \to \mathbb{R}_+$, $\rho(\vec{p}, \vec{p}') = \sqrt{(p_1 - p_1')^2 + (p_2 - p_2')^2}$.
- Spatial the neighborhood of the pixel $\vec{p}$ with radius r is $B(\vec{p}, r) : (F, \mathbb{R}) \to 2^F$, $B(\vec{p}, r) = \{\vec{p}' \mid \rho(\vec{p}, \vec{p}') \leq r\}$.

We will also call the pixel of $\vec{p}$ local r-minimum of some function $g : F \to 2^{\mathbb{R}}$ if:

$$\vec{p} = \underset{\vec{p}' \in B(\vec{p}, r)}{\operatorname{argmin}} g(\vec{p}')$$

The local $\sqrt{2}$-minimum is simply called the local minimum.

4.6 Classical Methods

One of the most popular segmentation methods is the *watershed transform* [14]. There are several ways to implement it. Conceptually, it can be done by the following steps:

1. From image I one can construct a function I' and treat it as a height map $I(p_1, p_2), p_1 \in \overline{1, W}, p_2 \in \overline{1, H}$. In the case of a grayscale image, the height map can be the image itself, where the lighter areas of the image have higher brightness values and, therefore, are higher than the darker areas. In the case of an RGB image, the original image I reduced to grayscale can be used as a height map. An edge map may also be used. In this case, pixels in monotonous areas are located lower, and the area boundaries are close to the top. In some cases, the distance to some selected points or areas (for example, to a pre-selected image background) can be used as a height map.
2. Define a set of sources $S \subset F$, through which "water" will "fill" the image. Usually, the local minima of the function I' is used as S. However, there are often situations when there are too many such minima, which, in turn, can lead to a large number of small segments. In such cases, one can use local r-minima with $r > \sqrt{2}$, or do pre-processing of the image (for example, using Gaussian or Laplace blur, or making a projection of the image brightness L on a set of smaller dimensions).
3. "Filling the area with water" is performed. This process simulates iteratively, starting from the global minimum. Water enters the area through sources, and with each iteration, its level increases. A certain number of pools are formed. At some iteration, when the water level is already high enough to combine some two pools, we put a barrier between them. This barrier is the boundary of the segmented sections.

Another common segmentation method is *mean shift*. It is an equivalent of K-means clustering. The key idea of the method is to find N centroids in a certain feature space $\mathbb{Z}$, which should separate the pixels of the image in the best way. The quality of the separation is determined by the functional:

$$\sum_{\vec{p} \in F} \min_{\vec{c} \in C} K(\vec{c}, g_I(\vec{p}))$$

where $C = \{\vec{c} \in \mathbb{Z}\}$ $|C| = N$ is the set of centroids, $g_I : \vec{c} \to \mathbb{Z}$ is the function that maps a pixel to its set of features (usually these are spectral brightness and the ordinal number of pixel column and row), and K is a function of the distance, usually the weighted Euclidean distance.

After minimizing the functional, the nearest centroid is assigned to each pixel. Thus, the segmentation of the image into N segments is fixed. There are many varieties of the mean shift method. This approach is often used to divide the image into so-called "super pixels." It is assumed that each super pixel is approximately homogeneous in its characteristics, and represents only one object or part of it, so when solving the semantic segmentation problem, you can first divide the image into super pixels and then classify super pixels. The number of superpixels is less than the number of pixels, thus the calculation amount is reduced and more elaborate methods may be applied.

A variation of the mean shift method is the SLIC [15] algorithm. It assumes regular initialization of cluster centers (at the nodes of a certain grid) and subsequent refinement of the centers, which takes into account the location of the clusters, which allows reducing the search when determining the nearest cluster. Due to this, the complexity of the algorithm is estimated as $O(|F|)$ i.e. it is linear in the number of pixels and does not depend on the number of clusters.

The methods discussed above solve the problem of dividing an image into segments of the same type, but they do not allow you to assign an object class to each segment. This should be solved by some auxiliary methods. Segmentation by the method of *conditional random fields* can be directly applicable to the problem of semantic segmentation. In this method two sets are considered: the set of observed variables X, which is simply the set of image brightness values, and the set of hidden variables $Y = \{y_p \in C \in | p \in F\}$, where C is the set of labels. The segmentation problem is formulated as:

$$Y = \arg_Y \max P(Y \mid X) \tag{4.1}$$

where $P(Y \mid X)$ is the probability of Y labels under the condition of the observed values of X.

To take into account the dependencies between variables, a graph-based model is used. That means that some graph $G = (V, E)$ is mapped to the image I. Each pixel $\vec{p} \in I$ is associated with a vertex $v_p \in V$. If there is an edge $(v_{p1}, v_{p2}) \in V$, this means that the variables x_{p1}, x_{p2} and y_{p1}, y_{p2} are dependent. If there is no edge, then x_{p1}, x_{p2} and y_{p1}, y_{p2} are considered conditionally independent. In this case, conditional independence means the following: if we define any set $D \subset V$ that divides v_{p1} from v_{p2} (i.e. such a set that any path from v_{p1} to v_{p2} will pass through at least one of the vertices of the set D) and fixes the values y_p, x_p for all $p \in D$, then the variables in pairs x_{p1}, x_{p2} and y_{p1}, y_{p2} will be independent. The triple (X, Y, G) is called a conditional random field. Various structures of the graph G are possible. Most often either a dense structure is used, or a structure where only neighbouring pixels are connected, see Figure 4.3.

For a conditional random field the Hammersley-Clifford theorem is valid:

Theorem 1: The joint distribution of a conditional random field can be factorized into the product of non-negative factors Φ_i defined on the maximum cliques D_i of the graph G, i.e.:

$$P(Y \mid X) = \frac{1}{Z} \prod_{i=1}^{n_D} \Phi_i(D_i \mid X) , \quad \forall i : \Phi(D_i \mid X) > 0$$

where Z is the normalizing factor:

$$Z = \sum_{Y \in 2^{C|F|}} \prod_{i=1}^{n_D} \Phi_i(D_i \mid X)$$

Each factor may be represented as $\Phi(D \mid X) = \exp(-\varepsilon(D \mid X))$, where the function $\varepsilon(D)$ is called energy, potential, or penalty. Then the maximization problem $P(Y \mid X)$ turns into an energy minimization problem:

$$E = \sum_{i=1}^{n_D} \varepsilon(D_i \mid X)$$

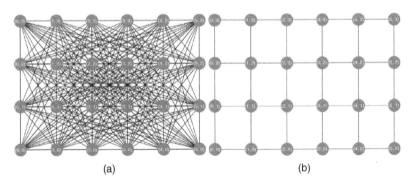

(a) (b)

Figure 4.3 Possible graph structures for the method of conditional random fields. (a) is a dense graph; each pixel depends on all the others. (b) is a non-dense graph; each pixel depends only on its neighbors.

Usually D_i are sets of one and two elements (the so-called unary and paired factors). Factors of higher orders turn out to be computationally much more expensive.

As unary factors, an estimate of the similarity of a pixel by a color model is used. For binary factors, it is common to use estimation proportional to the difference in spectral brightness and inversely proportional to the distance of pixels.

The method of conditional random fields can be applied to the segmentation of objects, including semantic segmentation, provided that the segmented classes have more or less stable color models. Before the advent of convolutional neural networks (CNN), this approach allowed us to obtain the highest quality of segmentation in complex tasks. There are efficient iterative algorithms for finding the solution of (1). Moreover, it was shown that such an iterative algorithm can be represented in the form of a recurrent neural network [16].

In practice, it turns out that the aforementioned methods cannot give satisfactory results in the problem of segmentation of artificial objects on satellite images due to the large color and texture diversity of these objects (even within a single object). However, they can be used for post-processing to refine segmentation masks.

4.7 Neural Network Methods

Recently, deep learning and CNN have become *de facto* standards in solving semantic segmentation problems. They allow achieving significantly better accuracy and speed of processing, which makes it possible to apply segmentation algorithms in problems. There is a large number of neural network segmentation architectures. This section will briefly describe only some of them that are of the greatest interest.

One of the successful early architectures proposed for the segmentation is *DeconvNet* [17], introduced in 2015. It consists of an encoder and decoder. The encoder repeated the architecture of VGG-16 [18], and the decoder was a mirror image of the encoder, in which the max pooling layers were replaced with layers of the max unpooling type. In the same year *U-net* [19] architecture was proposed, quite similar to DeconvNet, but with several important differences. First, it is fully convoluted, that is, there are no dense layers. Second, it contains additional connections between the corresponding layers of the decoder and the encoder, see Figure 4.4. These additional connections allow using the information of different levels of aggregation in the encoder, since the last layers of the network usually contain more complex features, and the initial ones contain less complex ones. This architecture is very popular up to this day, successfully solving a wide range

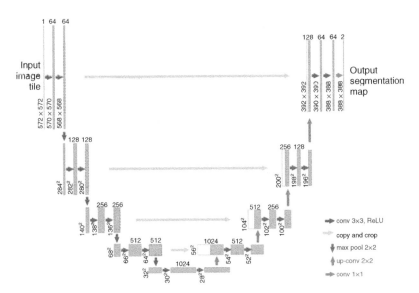

Figure 4.4 U-net [19] architecture. After each step of increasing spatial resolution in decoder, output is concatenated with the feature map from the encoder (gray arrows).

of tasks. In the subsequent works, this architecture was expanded and refined in various ways.

The use of *Residual blocks* and batch normalization allowed the building of similar U-net architectures with a significantly larger number of layers. In addition, U-net variants with multiscale convolutional blocks [20, 21] were proposed. In such blocks, there are several parallel paths through which data are processed. Each path contains convolutions covering a different level of context, which allows taking into account information of different scales. Usually, different paths contain either convolutions of different sizes or so-called atrous convolutions, which allows covering a larger context without increasing the number of convolution parameters.

Another modification of the Unet is obtained by replacing the original convolutional blocks with *Dense blocks*, see Figure 4.5. Dense blocks are in a sense evolution of the idea of Residual blocks. Within one block, the concatenation of the input and all intermediate feature maps is fed to the next convolutional layer. This approach also partially solves the problem of aggregation of features of different scales. Such modification of U-net is called *FC-DenseNet* [22].

An alternative approach to segmentation involves using *region proposal network* (RPN)-based architectures. The RPN is trained to evaluate the probability of the presence of an object in a certain region and to clarify the location of the region in case it does not completely cover the object. The number of analyzed regions, their positions and dimensions are fixed and the same for each image. As a rule, there are about several thousand of possible region (called anchors). This architecture is usually used in conjunction with a Region of interest (ROI) pooling. ROI pooling is used to convert parts of the feature map that are different in shape and size to a single size. Thus, the entire segmentation process is divided into three stages. At the first stage, a map of image features is extracted, then RPN is used to determine the probability of meeting objects of specified classes in the specified regions. The final stage is the use of a segmentation subnet to analyze the N areas with the highest probability of encountering an object. One of the most popular architectures using RPN is the Mask R-CNN [23]which was introduced in 2017. This network allows simultaneous detection and segmentation of objects (including instance segmentation), but it also has a number of disadvantages. First, such a network is more difficult to train, because it has many outputs and, accordingly, a more complex structure of the error function, which must take into account errors in all subtasks. Second, the use of anchors makes it difficult to use such an architecture for the segmentation of elongated objects and objects with holes.

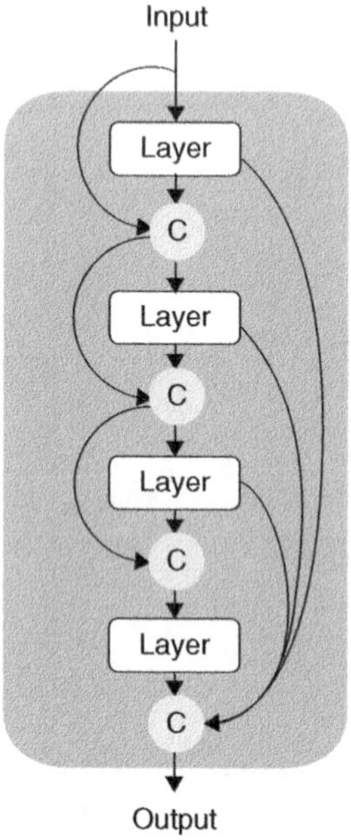

From the above analysis, an architecture based on FCDenseNet was chosen to solve the problem, in which the order of the activation layer and the batch normalization layer was changed. This architecture allows creating deep networks with a relatively fewer number of parameters and has proven itself well in many similar tasks. It is described in more detail in Section 6.4.

Figure 4.5 Dense-block with four convolutional layers [22]. c is a concatenation operation. The output of each layer is added to the inputs of all subsequent layers.

4.7.1 Semantic Segmentation of Objects in Satellite Images

The most significant features of the problem are:

1. *Arbitrary orientation of objects*: Houses, roads, and cars can have an arbitrary orientation in the image.
2. The inability to perform high-quality segmentation without taking into account the context of the image. Artificial objects can usually have an arbitrary color and texture (for example, buildings can be made of different materials).

3. *A variety of forms*. Roads and some buildings have a great variety of shapes; they can be very elongated or represent some nonconnected area in shape. Moreover, some objects may obscure others or be obscured by vegetation.

4. *The presence of shadows*. Some objects may be partially or completely shaded, which leads to strong differences in spectral characteristics between shaded and unshaded areas.

All these factors make the task difficult even to solve using neural network methods. Many works in the literature are devoted to the segmentation of objects on satellite images. This section highlights the most relevant ones.

In the beginning, we will consider examples of classical methods that do not use neural network techniques. An example of solving the segmentation problem without using neural networks was proposed by Grinas, Panagiotakis, and Tziritas [24]. They developed a set of structural features and used Markov random fields to highlight roads and buildings on satellite images with four spectral channels (R, G, B, and NIR were used). A detailed analysis of images of Attica (a city in Greece) was carried out and fairly high quality of segmentation was shown. In the work [25], the authors also determine the presence of houses in images. The proposed algorithm uses spectral characteristics, shape of objects, and an estimate of the heights of objects above the surface obtained using a stereo pair. The segmentation is performed using the IKONOS [26] and Quickbird [27] algorithms.

Let's now turn to the analysis of neural network approaches.

Wen and his team proposed to use the modified Mask R-CNN architecture to contouring houses on satellite images [28]. The main contribution of this work relates to improving the receptive block. This block determines the areas of the image for which the presence of buildings in them is likely. Unlike the standard receptive block, where the rectangles had sides directed in parallel to the image sides, the proposed receptive block allows you to determine rectangular areas that may have an angle of inclination to the image sides.

In [29], Kaiser et al. use a modification of the fully convolutional network VGG [18] to segment roads and buildings on open satellite data from Google Maps. Information from the Open Street Maps GIS was used as markup. Obtaining in such a way image markup has less precision than manual one: there are discrepancies in the markup up to 11 pixels. However, this approach allows to get a large amount of training data. Images of Paris, Zurich, Berlin, and Chicago were used for training, and images of Tokyo were used for validation. The quality score as measured by F1 has values ranging from 0.77 to 0.85.

Yanin et al. in their article, [30] described the training of the DeepResUnet neural network for segmenting houses on ultra-high-resolution images. The dataset created by the authors and consists of open images of the city of Christchurch in

New Zealand. According to the authors, the value of $F1\_score$ measure exceeds 0.93, which is very high quality.

There are a number of works in which the segmentation problem is only part of the problem of restoring three-dimensional models of objects. In such cases, the classes of segmented objects are determined primarily by the 3-D reconstruction algorithm, and the segmentation stage itself serves to highlight the features of objects in the image. For example, in [31], Alidoust, Hossein and Federico proposed an algorithm for constructing 3-D models of buildings. Two neural networks are used: one to determine the roof lines (it should distinguish three different types: eaves (edge lines of the roof), skates (ridges) and lines that connect the eaves and skates (hip lines)), and the second one to restore the height map (trained according to lidar data). In both cases, the same architecture a mechanically switched damped capacitor network (MSCDN) is used, which was specially developed by the authors for this task. The algorithm can automatically determine the roof lines, restore the roof structure (one of 11 predefined types) from them, and then use the height data to create a 3D model.

Another example of segmentation as an intermediate step is the work of [32]. The authors propose a method for recognizing houses based on remote sensing data, followed by determining one of the three types of roofs. First, with the help of selective search, areas of interest are selected, and only then two neural networks are applied to them sequentially. The first network gives the probability of finding a building in a given area of interest, and the second tries to predict the type of roof. Pre-trained networks based on the VGG architecture are used. Specially created datasets were used for training.

From the above review, we can conclude that there is a great variety in the data used and types of classes. Unfortunately, there are no examples of sufficiently similar tasks in the literature, where, first, images of the Russian territory would be used, and second, classes related to the railway infrastructure would be segmented.

4.8 Segmentation with Neural Networks

This section describes the basic principles of building and training neural networks in relation to classification and segmentation tasks. The main attention is paid to full-convolutional neural networks. The object will be considered as a tensor-multidimensional matrix.

The lower indexes of the tensor will denote the numbers of the corresponding components, and the number of indexes will correspond to the dimension (rank) of the tensor. For example, $y_{0,1}$ means the component of the tensor y with the value of the first index equal to 0 and the second index equal to 1.

The size of the tensor y will is the vector $s \in \mathbf{N}^{dim(y)}$, where s_i is the number of different values that the i-th component of the tensor y can take. Unless otherwise specified, it is assumed that the indexes take values in the form of integers from 0 to $s_i - 1$.

The size and dimension of the tensor y will be denoted by $size(y)$ and $dim(y)$, respectively.

In the case of the object being an image, the object tensor is 3D, the first component is responsible for the corresponding spectral channel, and the second and third components are responsible for the spatial location of the image pixel. Therefore, when describing three-dimensional tensors (not only images), the second and third components will be called spatial components. The concepts of a layer or channel of an image will denote the components of a three-dimensional tensor at some fixed value of the first component.

Any neural network consists, generally, of a set of parametric and nonparametric functions applied in a certain sequence to the input data to obtain a result. The most common block of a neural network is a linear parametric function, followed by a nonparametric and nonlinear function (called the activation function):

$$y = g(W \cdot x + b) = g(L(x, W, b)) = H(x, W, b)$$

where y is the output data, x is the input data, W is the weight coefficients, b is a constant additional vector, g is a nonlinear activation function, and the operation "·" is a tensor convolution. x, y, W are tensors of some sizes, W and b are learnable parameters. In this case, $L(x, W, b)$ maps the tensor y to the tensor x, where each component of y is a linear combination of the components of the tensor x with some constant addition, which is a component of the tensor b. H is usually called a neural network layer, i.e. a set of neurons, where each component of the tensor y is responsible for the output of one artificial neuron.

In the majority of architectures function g is the rectified linear unit (ReLU) and its various variations. Much less often, an exponential function or a hyperbolic tangent function is used. These and some other activation functions are shown in Figure 4.6.

It is important that the function H is differentiable with respect to all three arguments, which allows adjusting the parameters using the gradient descent method.

The simplest neural network is the combination of several layers of the type H. The output of one layer of the neural network is fed to the input of another layer. The outputs of different layers can be combined in different ways. The input data are also called the input layer of the neural network (input layer). The intermediate layers in which calculations are performed are called hidden layers, and the result of the network operation is called the output layer. When solving the problem of object classification, it should be possible to understand from the output layer of the neural network to which class the object most likely belongs. Neural networks

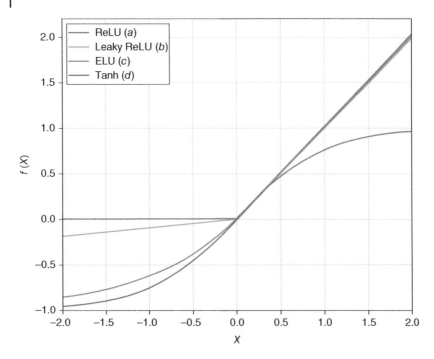

Figure 4.6 Examples of activation functions: (a) is a rectified linear function, (b) is a rectified linear function with a "leak," (c) is an exponentially linear function, (d) is a hyperbolic tangent. The angles of inclination of the positive parts of the functions (a), (b), and (c) were chosen differently for clarity, but they can (and usually are) the same.

are designed in such a way that each class corresponds to one of the components of the output tensor of the neural network. The process of training the network, in turn, is organized in such a way that the values of the output component are proportional to the probability of assigning the object to the corresponding class. To convert the output values of the neural network to probabilities, the *SoftMax* function is used:

$$SofMax(y)_i = \frac{\exp(y_i)}{\sum\limits_{j=1}^{N} \exp(y_j)}$$

where y is a one-dimensional tensor, $dim(y) = 1$, $size(y) = N$, N, number of classes.

SoftMax is usually the activation function of the last hidden layer. A schematic view of the neural network, discussed above, for the case when the input and intermediate layers are one-dimensional tensors, is illustrated in Figure 4.7.

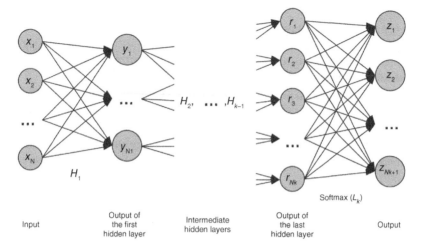

Figure 4.7 Schematic representation of a simple fully connected neural network with k hidden layers.

4.9 Convolutional Neural Networks

If the input object is an image, different components of its tensor correspond to different pixels. The existence of a pixel position in the image means that there is an order relation between the individual components of the input tensor. One pixel can be to the left (to the right) or above (below) than the other. This allows us to define the following concept of convolution:

$$Conv(w, y)_{ij} = \sum_{k=-k_l}^{k_h} \sum_{r=-r_l}^{r_h} \sum_{c=1}^{C} w_{c,k,r} \cdot y_{c,k+i,r+j} \qquad (4.2)$$

where y is an input tensor (for example, an image tensor) having dimension 3, the first component of which is responsible for the channel (layer) of the image, and the second and third components are responsible for the spatial position of the pixel on the image. The size of the tensor y: $[C, H, W]$. w are convolution weights, also represented as a tensor. For convenience, the second and third indices of the weight tensor begin to be numbered from some negative number. The size of the tensor w: $[C, k_l + k_h + 1, r_l + r_h + 1]$.

This definition of convolution is very limited compared to the one used in mathematical analysis, but it fully describes the convolution operation used in neural networks.

When summing, a situation may occur when $k + i \notin \{0, \ldots, H-1\}$ or $r + j \notin \{0, \ldots, W1\}$. In this case, the corresponding non-existent component of the tensor can be taken by some constant (for example, zero), or it can be extrapolated in some way. Such an extension of the original tensor with new components is called padding.

The convolution, as given in Eq. (4.2), is also called a filter. The numbers $k_l + k_h + 1$ and $r_l + r_h + 1$ determine the size of the filter. It is worth noting that the filter sizes are usually odd numbers.

The filter can be applied not to every combination of values of the second and third index, but with a certain step, called stride. Stride can be different for each spatial coordinate. The larger the stride size, the smaller the spatial size of the output tensor is obtained. (4.2) is written in the case when the stride is an equal unit for each coordinate. In a more general case, the formula can be rewritten as follows:

$$Conv(w, y)_{ij} = \sum_{k=-k_l}^{k_h} \sum_{r=-r_l}^{r_h} \sum_{c=1}^{C} w_{c,k,r} \cdot y_{c,k+s_H \cdot i, r+s_W \cdot j} \tag{4.3}$$

where s_H and s_W denote the stride size by the corresponding spatial indexes.

The standard layer of a CNN consists of applying several filters with the same size, stride and padding to the input tensor. In this case, the resulting tensors also have the same spatial dimensions. They can be combined into one 3D tensor and passed through the activation function, see Figure 4.8. The output tensor of the convolutional layer is also called a feature map.

Convolutional layers are a very important part of neural network architectures in the field of image processing. The following advantages distinguish them favorably from fully connected layers [33]:

- Reduction of the number of parameters in one layer of the neural network. Unlike fully connected layers, where each component of the input tensor affects each component of the output tensor, in convolutional layers, only a small number of components of the input tensor affects the composition of the resulting feature map. In practice, relatively small filters are usually used (compared to

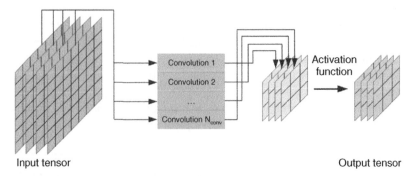

Input tensor   Output tensor

Figure 4.8 Schematic image of the convolutional layer.

the size of the input image and intermediate layers). This allows you to increase the number of layers of the neural network while maintaining the total number of parameters the same.

- Sharing of parameters. In the linear part of the fully connected layer, different weights were used for each output component. In the case of a convolutional layer, the same convolution weights are used for each output component. This leads to more stable learning and a decrease in the amount of memory required to store the model weights.
- When using small filters, the convolutional layer has a shift-invariance to the input data. This property leads from the previous one.
- A single convolutional layer can be used to process data of different sizes. This follows from the definition of the convolution operation. It does not impose any restrictions on the size of the input tensor.

Another important part of convolutional neural networks is pooling. Pooling is similar to convolution, however, instead of weighing the components of the input tensor in some square-sized spatial region, pooling takes predefined statistics from them. Usually, this is the minimum, maximum, or average value. Unlike convolution, which affected all the layers of the input tensor at once, pooling works with each layer separately. In relation to pooling, the concept of stride is also applicable, so its use can reduce the spatial size of the tensor:

$$Pool(y)_{c,i,j} = Stat(\{y_{c,k+s_H \cdot i, r+s_W \cdot j} \mid k \in \{-k_l,..,k_h\} \mid r \in \{-r_l,..,r_h\}\}) \quad (4.4)$$

where y is the input tensor, $dim(y) = 3$, the sets $\{-k_l,..,k_h\}$ and $\{-r_l,..,r_h\}$ determine the size of the area for which statistics are taken, s_H and s_W are the size of stride, *Stat* is the statistics function.

When solving image classification problems using convolutional networks, the sequential use of convolutional layers and pooling is used, so that the sizes of intermediate feature maps have smaller and smaller spatial dimensions and at the same time increase the number of channels. When the spatial dimensions of the feature map become quite small (or completely collapse), its components are flattened into a tensor of dimension 1 and passed through several fully connected layers (see Figure 4.9).

When there is a segmentation task, we must determine the class of each pixel of the image from the output of the neural network. A generally accepted approach is that the output of a segmentation neural network is a three-dimensional tensor Y, having sizes $[C, H, W]$, where C is the number of segmentation classes. If H, W do not match the dimensions of the original image, then an additional interpolation is performed.

In this case, the *SoftMax* function is not applied to the entire tensor at once, but separately for each pixel-with a fixed second and third component, along with the first component.

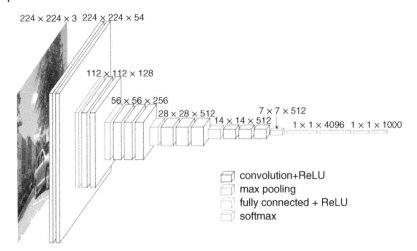

Figure 4.9 VGG 16 architecture [34].

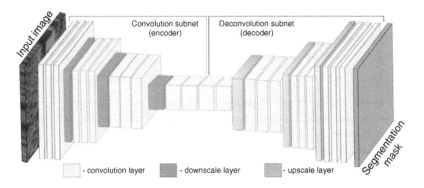

Figure 4.10 Schematic representation of the segmentation neural network architecture. Source: Alaudah et al. [35].

Most of neural networks architectures for segmentation consist of two parts, see Figure 4.10. The first part is the encoder (or convolutional subnet) extracts the characteristic features of the image. This part repeats the architecture of the classification neural network. However, it may not contain fully connected layers in its final part. The second part, the decoder, uses the features extracted by the encoder to build a segmentation map.

In the decoder, transformations occur in which the number of layers in the intermediate feature maps gradually decreases and the spatial dimensions increase. To increase the spatial size of the feature map, there are several approaches. The most popular approaches are unpooling and transposed convolutions.

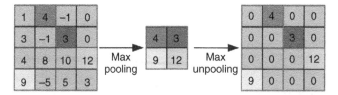

Figure 4.11 An illustration of how the max pooling and max unpooling functions works.

Unpooling is an operation that is inverse of the pooling operation. In this case, each individual component of the input tensor is associated with a certain square region-a tensor, the components of which are filled according to a certain rule. Let's consider an example of using the max unpooling operation, which maps a tensor of size [2,2] to each component of the source tensor. Max unpooling works in conjunction with the max pooling operation, see Figure 4.11. The result of using unpooling is a tensor with the same number of layers, but with twice the spatial dimensions. At the pooling stage, the tensor components in which the maximum in their area was located are remembered. During the unpooling operation, the values of the components of the input tensor are placed in the places of the maxima, and the remaining components of the output tensor are set to zero.

Unpooling is a non-parametric way to increase spatial resolution. A more "smart" approach is to use transposed convolution, also called inverse convolution, non-integer step convolution (non-unit stride), or deconvolution. The easiest way to understand how a transposed convolution works to consider this operation as a matrix transformation [36]. It turns out that it is possible to write a similar matrix transformation, only with a transposed matrix, working in opposite direction.

The main points of building convolutional neural networks are described above. However, there are a number of well-known and frequently used technical solutions that can improve the quality and speed of training. The following parts of the section will be devoted to them.

4.10 Batch Normalization

When training deep networks, problems of explosions and decay of gradients occurs. We say that the network is deep when the number of its layers is 50 or more. The problem arises due to the gradient of the parameters on each layer having multiplicative dependence on the parameters of all subsequent layers. Therefore, if the absolute values of the parameters are small enough, the gradient of the parameters of the initial (closest to the input data) layers will be smaller the deeper the network is used. Therefore, in deep networks, the gradients of the parameters of the first layers will be extremely small, which will eventually lead to long training (or under-training). If the absolute values of the parameters are large enough, then the gradient will grow on the contrary. This can lead to the situation that after the next step of gradient descent, the absolute values of the parameters of the model will become larger, which makes the gradient even larger at the next step. This quickly leads to the model weights moving beyond the possible maximum values and further training becomes impossible. Conducted reasoning shows that the training of deep neural networks is very sensitive to the choice of the gradient descent step size. However, even if the optimal step is chosen, networks with a large number of layers may show a worse final quality than less deep networks.

The use of [37] batch normalization helps to avoid such problems. Batch normalization can be considered as another layer of the network that normalizes the average value and variance of the components of the input tensor. In the case of using convolutional neural networks, where the input tensor and feature maps have dimension 4 (the first component of these tensors is responsible for the number of the object inside the batch), normalization works as follows:

1. $\mu_c = \frac{1}{BHW} \sum_{b=1}^{B} \sum_{i=1}^{H} \sum_{i=1}^{W} y_{bcij}$, $c = 1, .., C$ i.e. the average values of the tensor elements for each channel are calculated.

2. $\sigma_c^2 = \frac{1}{BHW} \sum_{b=1}^{B} \sum_{i=1}^{H} \sum_{i=1}^{W} (y_{bcij} - \mu_c)^2$, $c = 1, .., C$ i.e. the variances for each channel are calculated.

3. $\hat{y}_{bcij} = \frac{y_{bcij} - \mu_c}{\sigma_c + \varepsilon}$ i/e/ the elements of the input tensor are normalized.

4. $\hat{z}_{bcij} = \gamma \cdot \hat{y}_{bcij} + \beta$ i.e. additional linear scaling is performed.

In practice, the averages and variances are determined not for a single batch, but for the entire training sample during training. The parameters γ and β are also trainable. Sergey Laffe and Christian Sideli, who proposed the idea of batch normalization, claim that it allows reducing the internal covariance shift - the difference in statistical characteristics between different objects of the training sample. There are works showing that this may not be the case [38]. However, batch normalization greatly simplifies the training of neural networks, allowing to worry less about the selection of the training step and initialization of the weights.

4.11 Residual Blocks

Another effective technical solution for training deep neural networks is the use of residual blocks [39] (Figure 4.12). These blocks consist of several ordinary convolutional layers, and it ends with a feature map of the last layer is summed with

Figure 4.12 Residual block [39].

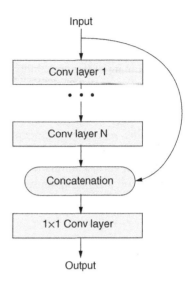

the input tensor. When summing, the input tensor may not undergo any additional processing (in this case, the output tensor of the last convolutional layer must have the same size as the input layer), or it may be subject to additional linear transformation (for example, convolution).

4.12 Training of Neural Networks

In this section, we will consider the neural network *Net* as a function of two arguments: $y = Net(x, \Theta)$, where x is the input of the neural network, Θ is the model weights and y is the output of the neural network. It is assumed that for training a neural network, there is a training dataset $D = \{d_i = (x_i, a_i), | i = 1, \dots, N\}$, consisting of various input objects and true answers, which is divided into two disjoint parts: D_{train} and D_{test}. The first one is used to adjust the weights of Θ; this is the task of training neural network. The second part of the sample is D_{test}, and it is used to estimate how well the outputs of the neural network $Net(x_i) = y_i$ correspond to the true answers of a_i

4.13 Loss Functions

The loss function is called the function loss, which is defined on the set of pairs of true answers and possible outputs of the neural network and returns a scalar. The larger the scalar should be, the more the output of the neural network differs from the true response. Thus, by selecting the parameters Θ that minimize the values of loss(a_i, Net(x_i, Θ)), we can train the network for the existing training sample. It is worth noting that in segmentation case, the output of the neural network is an estimate of the probability that each pixel belongs to each class, while the true answer a is a binary tensor, in which the component a_{cij} is equal to one if the (i, j)-th pixel of the image belongs to the class with the number c and zero otherwise.

Consider several loss functions used in the segmentation problem. We will assume that the training batch consists of a single object. In this case, the component corresponding to the object number in the batch can be omitted, and so, the input and output tensors of the neural network have three dimensions.

Dice Loss:

$$\text{DiceLoss}(A, Y, w, \gamma, sm) = \sum_{c=1}^{C} w_c \cdot \left(1 - \frac{\sum\limits_{i,j=1,1}^{H,W} A_{cij} Y_{cij}}{\left(\sum\limits_{i,j=1,1}^{H,W} A_{cij} \right)^{\gamma} + \left(\sum\limits_{i,j=1,1}^{H,W} Y_{cij} \right)^{\gamma} + \varepsilon} \right)$$

(4.5)

where Y is the output of the neural network, A is the true answer, w is the vector of weights, *gamma* and ε are some positive scalars.

The minimum of DiceLos is reached, obviously, when $A = Y$. w and γ can be used in the case of an unbalanced sample to increase the error in poorly represented classes. ε is used to smooth the function. If not a single pixel from the target data class is represented in the training sample, the coefficient helps to avoid uncertainty during dividing.

Focal Loss:

$$\text{DiceLoss}(A, Y, w, \gamma) = -\frac{1}{HW} \sum_{c=1}^{C} w_c \cdot \sum_{i,j=1,1}^{H,W} [A_{cij} = 1](1 - Y_{cij})^{\gamma} \log Y_{cij}$$

(4.6)

where Y is the output of the neural network, A is the true answer, w is the vector of class weights, and γ is some positive scalar.

When $\gamma = 0$, this error function is also called Cross-Entropy Loss. Choosing $\gamma > 0$ allows you to reduce losses for well-recognized classes, which improve learning.

4.14 Optimization

The purpose of training a neural network is to minimize the average losses on the training set of objects:

$$\text{Loss} = \frac{1}{N_{train}} \sum_{o=1}^{N_{train}} \text{loss}(a_o, \text{Net}(x_o, \Theta)) \to \min_{\Theta}$$

(4.7)

It is not possible to find the minimum analytically. Fortunately, it is not necessary to look for a global minimum, and just finding a good local one is usually enough. Gradient optimization methods cope well with this task, and they can be applied due to the fact that all layers of the neural network are differentiable functions. Gradient methods are iterative methods. For them to work, you must first initialize in some way, for example, randomly, the initial parameters of the network Θ_0 and the size of the optimization step λ. Next, T optimization steps are performed according to the following rule:

$$\Theta_{t+1} = \Theta_t - \lambda \frac{\partial Loss(\Theta)}{\partial \Theta}\bigg|_{\Theta=\Theta_1} \tag{4.8}$$

The gradient shows the direction of the fastest growth of the function. In case, the loss function must be minimized, the displacement of the parameter vector Θ is performed in the opposite direction to the gradient.

In practice, it is not possible to calculate the gradient of the loss function immediately over the entire dataset. Therefore, at each step, a small subset (batch) of objects is selected, and the optimization step is carried out only with these objects in mind. Such optimization method is called stochastic gradient descent:

$$Loss_B = \frac{1}{|B|} \sum_{o \in B} loss(a_o, Net(x_o, \Theta)), \quad B \subset D_{train} \tag{4.9}$$

$$\Theta_{t+1} = \Theta_t - \lambda \frac{\partial Loss_{B_t}(\Theta)}{\partial \Theta}\bigg|_{\Theta=\Theta_1} \tag{4.10}$$

Some improvement of stochastic gradient descent is the algorithm Adam [40]. For its operation, in addition to initializing the weights Θ_0 and λ, the following values are additionally initialized: $m_0 = 0$ is the first moment, $v_0 = 0$ is the second moment, and $\beta_1, \beta_2 \in [0, 1)$ are the decay rates of the first and second moments, respectively.

The optimization step of the algorithm consists of the following steps:

$$g_t = \frac{\partial Loss_{B_t}(\Theta)}{\partial \Theta}\bigg|_{\Theta=\Theta_1}$$

$$m_{t+1} = \beta_1 \cdot m_t + (1 - \beta_1) \cdot g_t$$

$$v_{t+1} = \beta_2 \cdot v_t + (1 - \beta_2) \cdot g_t^2$$

$$\hat{m}_{t+1} = m_{t+1} / \left(1 - \beta_1^{t+1}\right)$$

$$\hat{v}_{t+1} = v_{t+1} / \left(1 - \beta_2^{t+1}\right)$$

$$\Theta_{t+1} = \Theta_t - \lambda \frac{\hat{m}_{t+1}}{\sqrt{\hat{v}_{t+1} + \epsilon}} \tag{4.11}$$

where ε is a small smoothing constant. This algorithm has proven itself well for solving many practical problems.

4.15   Numerical Experiments

This section describes the process of learning a segmentation algorithm that is focused on using the method of digital terrain models construction, which has been described in Section 1. The training sample and segmentation classes are defined in accordance with the requirements of this method.

4.16   Description of the Training Set

Due to the parallel development of different stages of the methodology, it was not possible to use immediately prepared data from a Resurs-P satellite for testing the second stage. Therefore, images from the Google Earth GIS were used as the closest in quality and available for use alternative.

There are 502 non-intersecting sections were selected in the Moscow and Tula regions and on the Black Sea coast along the railway track between Tuapse and the Nizhnemeretinskaya Bay, see Figure 4.13. To reduce the imbalance of classes, the regions were selected in the way that all of them necessarily contain railway infrastructure areas. Each region is represented by an RGB image with a color channel depth of 8 bits and a size of about 4800×2736 pixels, which covers an area of approximately $1000m \times 600m$. These images have a higher resolution than can be obtained after the Resource-P satellite data aggregation proposed in section. Therefore, they were previously roughened to a resolution of 35 cm/px.

All regions were manually marked up. Objects of the following classes were allocated:

- *Building*: Area of the image containing roofs and walls belonging to the same building;
- *Roof*: Area of image bounded by the horizontal eaves of the building roof;
- *Building shadow*: Area of the image of the earth's surface adjacent to the building in the places where the shadow falls from it;
- *Railway infrastructure*: Includes images of the railway subsystems (railway tracks, rolling stock on the tracks, conducting power lines, support poles, etc.), as well as objects of the station subsystem (railway platforms, observation buildings, etc.), limited by protective screens or lying in the exclusion zone of the railway subsystem;
- *Rails*: The area of the image of one railway track, taking into account the rolling stock and cars placed on these rails;

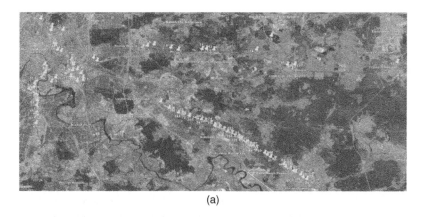

(a)

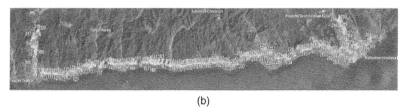

(b)

Figure 4.13 Regions of the dataset on the map. The pushpin icon marks the centers of the regions. The numbers near to the icons indicate the serial number of the region. (a) Regions in the Moscow region and (b) regions on the Black Sea coast.

- *Carriage*: The area of the image of a freight or passenger railway carriage of a particular type;
- *Road*: The area of the image of the road (paved, concrete or unpaved);
- *Excluded objects*: Objects that may belong to previous classes, but are excluded from the training sample for one some reason (for example, poor visibility)

Background pixels are all pixels that are not included in any of the above classes. Walls are the area of the image, the visible part of the building wall adjacent to the roof of the building, is defined as the difference between buildings and their roofs. It is worth noting that the building class is not the target class for segmentation. This class is a union of wall and roof classes, thus it can be obtained by combining the segmentation results for these two classes.

The markup was performed using the ENVI 5.2 software [41] and was presented in the form of XML files. This representation allows to store the semantic description of the image in a compact form. However, it is difficult to use it in training neural networks. Before training, the markup was previously rastrarized. The translation of vector masks into a raster was carried out using the Shapely python library.

4.17 Class Analysis

The set of classes described above has two features, causing problems in processing. One problem is that the classes are not mutually exclusive. First, the shadows of buildings can be overlapped on any object. Second, carriage and railways by definition must be located on the territory of the railway infrastructure. This makes it impossible to apply a single SoftMax function to the output of a neural network to estimate the probabilities of pixels belonging to different classes.

Another problem is the imbalance of classes. The sizes of classes in percentage are shown in Table 4.1. The Table shows a significant predominance of the background over the other classes. In addition, the classes "shadows of buildings" and "carriages" make up less than one percent of the total training sample. Such an imbalance of classes can lead to poor network training especially in relation to poorly represented classes.

4.18 Augmentation

The set of classes described above provides various ways for augmentation. The initial regions can be rotated at an arbitrary angle, stretch along the axes, and be subjected to different kinds of noise.

A special method of data augmentation was developed, which allows simultaneously increasing the amount of training data and partially balancing the training data.

This method works as follows. First, the number of augmentation epochs N is determined. Then, within the i-th augmentation epoch for each image from the training set (see Figure 4.14):

1. The image is rotated by a random angle in the range from $360 \cdot i/N$ to $360 \cdot (i+1)/N$ degrees;
2. S random square areas are sampled, with a width in the range $[w_l, w_h]$ and a height in the range $[h_l, h_h]$. Sampling performed one area at a time. For each sample, it is checked that the proportion of informative pixels is at least P_1. If this is not the case, then the sample is discarded. In order to samples cover the

Table 4.1 The initial ratio of the classes in the dataset.

Class	Roofs of buildings	Shadows of buildings	Walls of houses	Roads	Railway tracks	Carriages	Railway infrastructure	Background
Share of pixels (%)	5.86	2.07	0.63	5.53	1.99	0.47	11.47	72.28

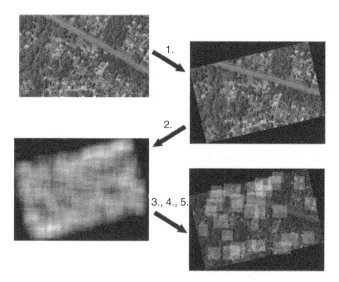

Figure 4.14 Scheme of the main augmentation steps.

image most uniformly, the probability of choosing the point x_s, y_s as the center of the sample was set inversely proportional to the number of samples that cover this point.

3. For each sampled area s, its significance is calculated: $I_s = \sum_{k=1}^{C} \frac{p_k}{Pr_k^\gamma}$, where p_k is the percentage of pixels from the class k in the sample, Pr_k is the percentage of pixels of the class k in the entire dataset and γ- some positive constant.

4. Significance of samples $I = [I_1, ..., I_S]$ are normalized: $\hat{I} = I/\mathrm{sum}(I)$.

5. M pieces are randomly selected from all samples with probabilities $\hat{I}$. As in step 1., one sample is selected at a time, after which it is checked that its intersection with the rest of the selected samples is no more than P_2. If the condition is not satisfied, the sample is discarded.

6. The selected samples are scaled to the size $[h_f, w_f]$ and added to the training set.

In relation to the existing dataset, the method was applied with the following parameters: $N = 10$, $S = 1000$, $M = 30$, $[h_l, h_h] = [380,550]$, $[w_l, w_h] = [380,550]$, $[h_f, w_f] = [256,256]$, $P_1 = 0.7$, $P_2 = 0.5$, $\gamma = 1$.

An important advantage of this method is the possibility of introducing additional conditions at different stages of sampling. As it was mentioned earlier, excluded objects were also marked when marking the sections. In addition, when rotating region in the step 1. empty areas appeared at the corners. Both of these types of uninformative areas can get into samples, but they are of little use from for training. To minimize the occurrence of such uninformative areas in the

Table 4.2 The ratio of classes in the dataset after augmentation.

Class	Roofs of buildings	Shadows of buildings	Walls of houses	Roads	Railway tracks	Carriages	Railway infrastructure	Background
Amount of pixels (%)	11.27	4.07	1.72	6.42	3.95	1.30	22.13	51.17

training samples, in step 2. an additional check was introduced. Also, to increase the variety of data, the degree of intersection with the other samples was checked for each sample at step 5.

As a result of applying this method, 300 samples of the size of 256×256 pixels were obtained from each region, which increased the total number of pixels to train by about one and a half times. The ratio of class sizes in the augmented dataset is shown in Table 4.2.

As shown in the table, there was a partial balancing of classes due to the fact that samples with rarer classes were selected more often. It is also worth noting that it is hardly possible to achieve a better balance in this case by using only linear transformations. Indeed, segmented classes have some context. The walls are located near to the roofs of buildings, roads and railways are usually surrounded by a landscape related to the background, etc. Thus, if the samples are large enough, then in addition to the object, they will capture its environment, which in the vast majority of cases belongs to other classes. Using samples of a smaller size will not allow you to take into account the context of objects, which can significantly affect the quality of segmentation. The use of nonlinear transformations, in this case, is unjustified because these objects have a strict geometry, which can be broken with such an augmentation.

4.19 NN Architecture

To implement the segmentation algorithm, a convolutional neural network based on the FCDenseNet architecture was used. The general scheme of this family of neural networks is shown in Figure 4.15a. The architecture of FCDenseNet-103 was taken as a basis, see Figure 4.15b. The foundation of this architecture is a Dense block, which consists of several Dense layers (see Figures 4.5 and 4.16a). Each Dense layer is a sequence of the ReLU activation function, batch normalization, 3×3 convolution, and DropOut of the layer with a probability of zeroing 0.2. ReLU and batch normalization are in reverse order as compared to the original construction of the Dense layer from [22].

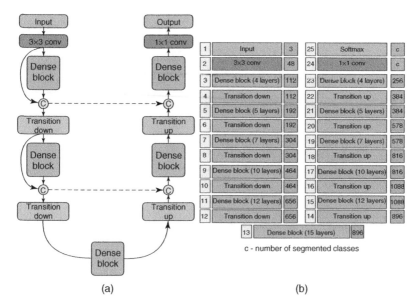

Figure 4.15   (a) is the scheme of the FCDenseNet architecture family; (b) is the FCDenseNet-103 architecture.

When building a network, the growth factor k is determined. The feature map at the output of all Dense layers of the network has k channels. The number of input channels of the Dense layer grows linearly within one block. As follows from the diagram (Figure 4.5), the first layer accepts a feature map with an arbitrary number m of channels, the second-with $m + k$ channels, etc. In the encoder, the output of the Dense block is the concatenation of all the layers of the block, in the decoder only the feature map of the last Dense layer is returned (to reduce the total number of parameters and computational costs). To change the spatial dimensions of the feature map, the transition up and transition down blocks are used, see Figures 4.5 and 4.16.

FCDenseNet-103 and MultiResUnet [42] were also trained for comparative analysis.

4.20   Training and Results

As a test part of the dataset, 35 regions were selected that quite completely represented the diversity of training data. The training part was divided into a training set and validation in a ratio of $9 : 1$.

The gradient descent method *Adam* with an initial step of $4 \cdot 10^{-4}$ and a decay coefficient of 0.9995, applied after every thousand training steps was used.

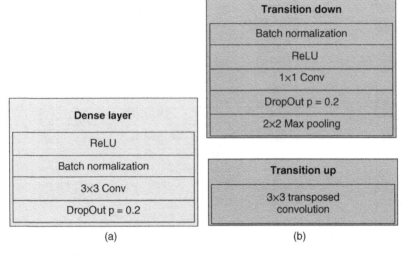

Figure 4.16 The structure of dense layer (a), and transition layers (b).

The training lasted 30 epochs. After that, ten more similar training epochs were conducted with an initial step $1 \cdot 10^{-5}$.

Since the segmented classes are not mutually exclusive, the SoftMax function was applied separately to the following class groups: {roofs, walls, roads, railway tracks, carriages, background}, {shadows, background}, {railway infrastructure, background}. Threshold filtering was used to convert probability maps into binary masks.

The software part was implemented in the Python 3.6 programming language using the PyTorch 1.8.1 software platform. For training, a computing node based on the IBM Power System AC922 server platform was used with the following characteristics: 2*CPU Power 9 (2.1 GHz, 20 Core), 1024 Gb RAM, 2*10G Ethernet, 2*100G InfiniBand, 4*GPU Nvidia Tesla V100. The research was carried out using the Infrastructure of the Shared Research Facilities ≪ High-Performance Computing and Big Data≫ (CKP ≪Informatics≫) of FRC CSC RAS (Moscow).

The quality measures of the proposed algorithm in comparison with some other architectures are presented in Tables 4.3, 4.4, and 4.5; ROC curves are presented in Figures 4.17 and 4.18.

It can be noted that the proposed architecture is slightly behind in the quality of segmentation of walls, carriages and railway tracks, but has a higher quality in the case of segmentation of roofs, roads and the area of railway infrastructure.

Some examples of the proposed architecture's work are shown in Figure 4.19. When analyzing the results of the algorithms, the following can be noticed:

Table 4.3 Comparison of ROC AUC values of different segmentation architectures.

NetworkClass (innerwidth = 3.1 cm)	Roofs of buildings	Shadows of buildings	Walls of buildings	Roads	Railway tracks	Carriages	Railway infra-structure
MultiResUnet	0.959	0.948	0.966	0.934	0.962	0.962	0.935
FCDenseNet-103	0.946	0.962	0.957	0.944	0.975	**0.998**	0.940
FCDenseNetR	**0.968**	**0.979**	**0.977**	**0.964**	**0.982**	0.998	**0.986**

Table 4.4 Comparison of F1-score values of different segmentation architectures.

NetworkClass [innerwidth = 3.6 cm]	Roofs of buildings	Shadows of buildings	Walls of buildings	Roads	Railway tracks	Carriages	Railway infra-structure
MultiResUnet [4]	0.750	0.703	0.575	0.623	0.582	0.554	0.750
FCDenseNet-103 [5]	0.778	0.704	**0.578**	0.680	0.606	**0.588**	0.794
Proposed	**0.793**	**0.713**	0.570	**0.702**	**0.607**	0.584	**0.823**

Table 4.5 Comparison of IoU values of different segmentation architectures.

NetworkClass [innerwidth = 3.6 cm]	Roofs of buildings	Shadows of buildings	Walls of buildings	Roads	Railway tracks	Carriages	Railway infra-structure
MultiResUnet [4]	0.636	0.559	0.414	0.514	0.442	0.420	0.712
FCDenseNet-103 [5]	0.663	0.558	**0.418**	0.527	**0.461**	**0.433**	0.741
Proposed	**0.674**	**0.564**	0.411	**0.555**	0.455	0.430	**0.762**

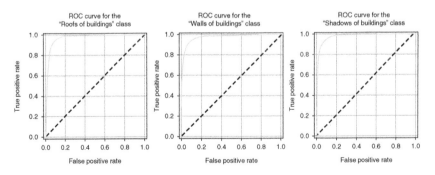

Figure 4.17 ROC curves for classes "Roofs," "Walls of buildings," "Shadows of buildings."

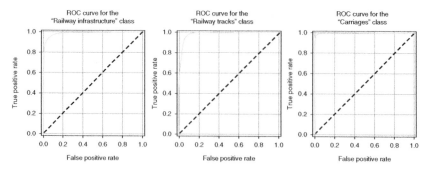

Figure 4.18   ROC curves for classes "Railroad infrastructure," "Railways," "Carriages."

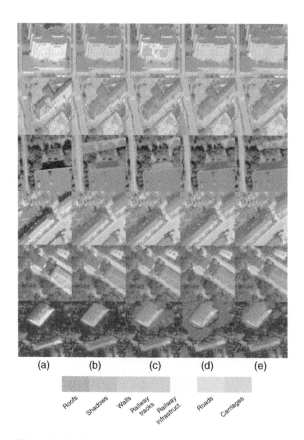

Figure 4.19   Some examples of how algorithms work. (a) The original image; (b) Manual markup; (c) MultiResUnet; (d) FCDenseNet-103; (e) The proposed architecture.

- It is not possible to train the network in such a way that it predicts the part of the object that is obscured (for example, by vegetation). Perhaps, for this purpose, it is worth marking the obscured parts of the object to a separate class.
- The neural network does not distinguish well garages and some of the artificial terrain irregularities from houses.
- The most difficult class for segmentation was the wall class. Most likely, this is due to the fact that many buildings in the presented sample have a small number of floors or were captured almost at the nadir. In this way, the shape of the walls turns out to be long and narrow. In addition, the approach in which the contours of houses and the roofs of houses are marked separately leads to inaccuracies in the class of walls, which also reduces the quality of segmentation of this class.

4.21 Conclusion

The analysis of classical and neural network segmentation methods was performed. The works considering the semantic segmentation of artificial objects on satellite images were also studied. A training sample set was created based on open data from the Google Earth geoinformation system. Data augmentation was carried out using a specially developed method of probabilistic augmentation, which made it possible to partially balance the data. A neural network segmentation model was trained on the prepared sample. The quality of the results was evaluated, and the errors of the trained model were also analyzed.

Acknowledgments

The work is supported by the Russian Foundation of Basic Research, grant no. 20-01-00609 and by the Ministry of Science and Higher Education of the Russian Federation, agreement no. 075-15-2020-776.

References

1 Mou L. and Zhu X.X. (2018). *IM2HEIGHT: Height Estimation from Single Monocular Imagery via Fully Residual Convolutional-Deconvolutional Network.* https://arxiv.org/abs/1802.10249.

2 Srivastava, S., Volpi, M., and Tuia, D. (2017). Joint height estimation and semantic labeling of monocular aerial images with CNNs. In: *Proceedings of 2017 IEEE International Geoscience and Remote Sensing Symposium (IGARSS)*, 23–28. Fort Worth, TX: IEEE.

3 Gvozdev, O., Kosheleva, N., Murynin, A., and Richter, A. (2020). 3D-modeling infrastructure facilities using deep learning based on high resolution satellite images. In: *20th Internship Multidisciplinary Scientific Geoconference (SGEM)*.

4 Gvozdev, O.G., Kozub, V.A., Kosheleva, N.V. et al. (2020). Construction of three-dimensional models of rigid objects from satellite imagesof high spatial resolution using convolutional neural networks. *Issledovanie Zemli iz kosmosa* 5: 78–96.

5 Gvozdev, O.G., Kozub, V.A., Kosheleva, N.V. et al. (2021). Neural network method for constructing three-dimensional models of rigid objects from satellite images. *Mekhatronika, Avtomatizatsiya, Upravlenie* 22 (1): 48–55. (in Russian).

6 Gvozdev, O.G., Kozub, V.A., Kosheleva, N.V. et al. (2020). Constructing 3D models of rigid objects from satellite images with high spatial resolution using convolutional neural networks. *Izvestiya, Atmospheric and Oceanic Physics* 56: 1664–1677.

7 http://russianspacesystems.ru/bussines/dzz/resurs-p.

8 Gorokhovskiy, K.Y., Ignatiev, V.Y., Murynin, A.B., and Rakova, K.O. (2017). Parameters optimization of the novel probabilistic algorithm for improving spatial resolution of multispectral satellite images. *Journal of Computer and Systems Sciences International* 56 (6): 1008–1020.

9 Filip, B., Ledoux, H., Stoter, J., and Zhao, J.-Q. (2014). Formalisation of the level of detail in 3D city modelling. *Computers and Environment and Urban Systems* 48: 1–15.

10 Khanykov, I.G. (2018). Classification of image segmentation algorithms. *Journal of Instrument Engineering* 61: 978–987. (in Russian).

11 Novik, V., Matveev, I., and Litvinchev, I. (2020). Enhancing iris template matching with the optimal path method. *Wireless Networks* 26 (7): 4861–4868.

12 Matveev, I., Novik, V., and Litvinchev, I. (2018). Influence of degrading factors on the optimal spatial and spectral features of biometric templates. *Journal of Computational Science* 25: 419–424.

13 Matveev, I. (2010). Detection of iris in image by interrelated maxima of brightness gradient projections. *Applied and Computational Mathematics* 9 (2): 252–257.

14 Roerdink, J. and Meijster, A. (2001). The watershed transform: definitions, algorithms and parallelization strategies. *Fundamenta Informaticae* 41: 187–228.

15 Achanta R., Shaji A., Smith K., Lucchi A., Fua P., Susstrunk S. (2010). SLIC superpixels EPFL. *Technical Rep. 149300*.

16 Zheng, S., Jayasumana, S., Romera-Paredes, B. et al. (2015). Conditional random fields as recurrent neural networks. *Proceedings of IEEE Internship Conference Computer Vision (ICCV)* 1529–1537.

17 Noh, H., Hong, S., and Han, B. (2015). Learning deconvolution network for semantic segmentation. *Proceedings of 2015 IEEE International Conference on Computer Vision (ICCV)* 1520–1528.

18 Simonyan, K. and Zisserman, A. (2015). Very deep convolutional networks for large-scale image recognition. *arXiv:1409.1556*.

19 Ronneberger, O., Fischer, P., and Brox, T. (2015). U-Net: convolutional networks for biomedical image segmentation. *arXiv:1505.04597*.

20 Diakogiannis, F., Waldner, F., Caccetta, P., and Wu, C. (2020). ResUNet-a: a deep learning framework for semantic segmentation of remotely sensed data. *arXiv:1904.00592*.

21 Chattopadhyay, S. and Basak, H. (2020). Multi-scale Attention U-Net (MsAUNet): A Modified U-Net architecture for scene segmentation. *arXiv:2009.06911*.

22 Jegou, S., Drozdzal, M., Vazquez, D. et al. (2017). The one hundred layers tiramisu: fully convolutional densenets for semantic segmentation. *Proceedings of IEEE Conference Computer Vision and Pattern Recognition Workshops (CVPRW)* 1175–1183.

23 He, K., Gkioxari, G., Dollar, P., and Girshick, R. (2017). Mask R-CNN. *International Conference on Computer Vision (ICCV)* 2980–2988.

24 Grinias, I., Panagiotakis, C., and Tziritas, G. (2016). MRF-based segmentation and unsupervised classification for building and road detection in peri-urban areas of high-resolution satellite images. *ISPRS Journal of Photogrammetry and Remote Sensing* 122: 145–166.

25 Sefercik, U., Karakis, S., Bayik, C. et al. (2014). Contribution of normalized DSM to automatic building extraction from HR mono optical satellite imagery. *European Journal of Remote Sensing* 47 (1): 575–591.

26 Xiao, P., Zhao, S., and She, J. (2007). Multispectral IKONOS image segmentation based on texture marker-controlled watershed algorithm. *Proc. SPIE* 6790.

27 Alkan, M., Marangoz, A.M., Karakis, S., and Buyuksalih, G. (2006). Verification of automatic and manual road extraction methods using Quickbird imagery. *ISPRS Archives* 36.

28 Wen, Q., Jiang, K., Wang, W. et al. (2019). Automatic building extraction from Google Earth images under complex backgrounds based on deep instance segmentation network. *Sensors* 19 (2): 333.

29 Kaiser, P., Wegner, J.D., Lucchi, A. et al. (2017). Learning aerial image segmentation from online maps. *IEEE Transactions of Geoscience and Remote Sensing* 55 (11): 6054–6068.

30 Yi, Y., Zhang, Z., Zhang, W. et al. (2019). Semantic segmentation of urban buildings from VHR remote sensing imagery using a deep convolutional neural network. *Remote Sensing* 11: 15.

31 Alidoost, F., Arefi, H., and Tombari, F. (2019). 2D Image-To-3D model: knowledge-based 3D building reconstruction (3DBR) using single aerial images and convolutional neural networks (CNNs). *Remote Sensing.* 11 (19): 2219.

32 Alidoost, F. and Arefi, H. (2018). A CNN-based approach for automatic building detection and recognition of roof types using a single aerial image. *Journal of Photogrammetry, Remote Sensing and Geoinformation Science* 86: 235–248.

33 Goodfellow, I., Courville, A., and Bengio, Y. (2016). *Deep Learning.* Cambridge, MA: MIT Press.

34 Garcia-Garcia, A., Orts-Escolano, S., Oprea, S.O. et al. (2017). A review on deep learning techniques applied to semantic segmentation. *arxiv:1704.06857.*

35 Alaudah, Y., Michalowicz, P., Alfarraj, M., and AlRegib, G. (2019). A machine learning benchmark for facies classification. *SEG Interpretation Journal* 7 (3): 175–187.

36 Dumoulin, V. and Visin, F. (2016). A guide to convolution arithmetic for deep learning. *arxiv:1603.07285.*

37 Ioffe, S. and Szegedy, C. (2015). Batch normalization: accelerating deep network training by reducing internal covariate shift. *Proceedings of 32nd Internship Conference Machine Learning* 37: 448–456.

38 Santurkar, S., Tsipras, D., Ilyas, A., and Madry, A. (2018). How does batch normalization help optimization? *NIPS'18: Proceedings of 32nd Internship Conference Neural Information Processing Systems* 2488–2498.

39 He, K., Zhang, X., Ren, S., and Sun, J. (2016). Deep residual learning for image recognition. *Proceedings of IEEE Conference Computer Vision and Pattern Recognition (CVPR)* 770–778.

40 Kingma, D.P. and Ba, J. (2014). Adam: a method for stochastic optimization. *arXiv:1412.6980.*

41 ENVI. https://www.l3harrisgeospatial.com/Software-Technology/ENVI.

42 Ibtehaz, N. and Rahman, S. (2020). MultiRes UNet: rethinking the U-Net architecture for multimodal biomedical image segmentation. *Neural Networks* 121: 74–87.

5

The Impact of Data Security on the Internet of Things

Joshua E. Chukwuere and Boitumelo Molefe

North-West University, Department of Information Systems, South Africa

5.1 Introduction

The Internet of Things (IoT) is a network of different physical and technological devices that communicate through the Internet [24]. IoT also involves interconnecting computing devices that use the Internet to send or receive data [49]. The IoT mainly refers to the network of physical devices such as a vehicle, machine, and home appliances, which use sensors and Application Programming Interfaces (APIs) through the Internet to connect and share information [47]. The Internet is exposed to cyber threats and, because IoT mainly uses the same platform, it is also subjected to the same threats [47]. The security challenge affects users directly; privacy of personal information on the Internet is not guaranteed. The personal information of the users, including their behavior on the Internet, is collected and stored to edify their personal service preferences [57]. The access to and ability to store users' personal information can pose challenges, such as the information being exploited by hackers, thus putting individual/s at risk.

Data security in IoT is important to protect the personal information, confidentiality, and the safety of the user [49]. The IoT device users need to trust the development of the product before using it; making the users aware of the security threats and how they can avoid being victims is very important [49]. The assurance of security in IoT will bring in more users of such technology and will decrease the number of critics of the IoT development [54]. The evolution of IoT is in line with social and economic digitalization being powered by the fourth industrial revolution (4IR), faster connectivity, automation with no human interaction, and smart world (smart cities, smart home, and smartphones; these are all part of IoT, which is the center of 4IR [19]). The 4IR provides great and innovative opportunities and advantages for the society and the economy, but

Artificial Intelligence in Industry 4.0 and 5G Technology, First Edition.
Edited by Pandian Vasant, Elias Munapo, J. Joshua Thomas, and Gerhard-Wilhelm Weber.
© 2022 John Wiley & Sons, Inc. Published 2022 by John Wiley & Sons, Inc.

security in technology, in general, is a major concern [49]. The security aspect of modern technological devices remains a challenge. Therefore, this chapter aimed to investigate the impact of data security on the IoT. To achieve this, the objectives are as follows: (i) investigate whether or not users are aware of data security on the IoT, (ii) determine the security threats that are involved in the IoT, (iii) determine how the identified security threats affect users (people) of IoT devices, and (iv) determine different ways to assist with keeping the IoT device safer from security threats.

5.2 Background of the Study

The Internet has widely grown to be the most important part of our everyday lives, and this had led to important information about businesses or our personal lives being vulnerable on the Internet [48]. The interconnection, communication, and sharing of information or data through the Internet have grown widely [15]. According to Gillwald et al. [22], South Africa is also experiencing this extensive growth of technology, which affects the economic growth of the country positively and has also contributed toward individuals being more productive through the help of the IoT and artificial intelligence (AI). However, the issue of a reduced labor force remains a challenge as it will result in a higher rate of unemployment [22].

Information security is focused on protecting user information from unintentional or malicious harm [35]. Security is an important aspect of information technology, especially if working with revealed or authorized information [35]. The obligation to shield information from unauthorized access, confidentiality vulnerability and integrity, tampering of data, and guaranteed secure transfer is an important piece of an information management strategy [35]. Research done in South Africa by the Council for Scientific and Industrial Research together with the University of Venda has indicated that the local communities' population has not been fully exposed to technology, internet access, or cyber threats and security; therefore, these local communities do not have the ability to deal with such threats [25].

According to Abomhara and Køien [5], data security issues are not new to IoT; yet, as IoT will be totally interrelated in our lives and social orders, it is getting to be important to move up and pay attention to data security. Thus, there is a genuine need to protect IoT, bringing about a need to totally understand the dangers and issues on IoT substructure and security [5]. The global growth in technology makes it difficult for a country like South Africa to control the use of IoT devices, and data security remains a reality; therefore, individuals who are not aware of cybercrimes remain victims, as they fail to protect their information or devices [25].

5.3 Problem Statement

The growth and advancement of technology have escalated, and South Africa was not excluded from this paradigm [40]. IoT has been used effectively on devices like water meter and electricity meter, and Global Positioning System (GPS) trackers; however, Paul Williams [56], country manager at Fortinet, Johannesburg, South Africa, reported that due to the security threats on these devices, the costs of cybercrimes are estimated at over R3 trillion annually. It was further explained that maintaining the security of its devices is a recent challenge that is faced by the technology environment [56]. IoT relies on the Internet; therefore, it is experiencing the security threats that affect the Internet as a whole; these issues are escalating as technology is also growing [57]. IoT users are unaware of their vulnerability to cybercrime caused by Information Communication Technology (ICT) security breaches [18]. The users of IoT devices are not aware of security threats that are involved in using these devices and what can be done to avoid them [7]. Therefore, this study aims to investigate the type of impact in data security of IoT, which affects the users, and then discusses different measures to protect users from vulnerabilities to security threats.

5.4 Research Questions

The following questions were derived through the research objectives: (i) Are IoT users aware of the data security on the IoT?, (ii) What are the security threats involved in IoT?, (iii) How do the identified security threats affect users' (people) IoT devices?, and (iv) What are the different ways to assist with keeping IoT devices safe from security threats?

5.5 Literature Review

IoT is making things easier and convenient; however, its security remains a challenge, just like in any technological computing field [6]. IoT is referred to as a connection of different devices to the Internet and some devices connecting one another [7]. While the shift and growth are happening, data security in IoT is an important requirement; therefore, the manufacturing companies of IoT devices and companies that use those devices must ensure data security of IoT [54].

5.5.1 The Data Security on IoT

Data security is a process of protecting digital information from unauthorized access, theft, or corruption of data [54]. IoT contains huge amounts of data from different streams (Big Data), including sensitive data of different organizations and personal data meant to be private [46]. Ensuring that data used by IoT devices is protected can be an advantage that increases consumer trust and results in the system being accepted and used by consumers because of assured integrity [50].

The main reason why data protection is important is that any organization's data security, whether big, established organizations or small, medium enterprises, can make or break that organization [38, 43]. Protecting data is important, so organizations use authentication methods like providing a username and password to ensure identity before accessing the system to protect the business data [54]. Once the data security has been breached and cybercrimes have victimized users, the IoT systems lose the reputation of secured devices, which hinders the usage of IoT systems [11]. The users need to be assured that the IoT and its devices are safe, which means that the information contained therein is safe and will not be vulnerable to any theft or cybercrimes [21]. It benefits the organization as it keeps information secure because it is regarded as a valuable asset [16].

The manufacturing companies of the IoT devices must be in full control of ensuring that those devices are secured as they are produced [49]. Organizations need to develop and implement better data security measures beyond firewall protection, backup, and recovery or the different antiviruses [43]. The safer the IoT is declared, the more the users will be comfortable using its technology, but if IoT exposes sensitive data to any vulnerability, it results in users not trusting IoT devices [21]. Most of the IoT devices are controlled by different users and that opens a gap for attacks and these devices are also not physically protected, which leaves room for vulnerability [13]. Figure 5.1 shows the different stages of how data travel through the use of IoT devices and in all these steps, data need to be protected from any cybercrime.

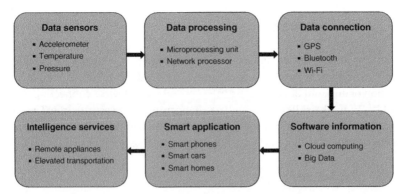

Figure 5.1 Transmission of data on an IoT device. Source: Based on Kamburugamuve et al. [34].

5.5.2 The Security Threats and Awareness of Data Security on IoT

IoT devices in smart cities or smart homes use mobile phones for remote controlling like switching on and off lights at home; therefore, if the mobile phone is not secured with a strong password, it can easily be attacked [12]. There are many other different devices, applications, and services that fall under IoT and are almost used in our daily lives. These include AI machines, computers, air conditioning systems, smartphones, lighting systems, and different household gadgets like cooking ovens and refrigerators. These devices are mainly exposed to security threats because they are connected to cloud computing for data transition they use for interactions, which increases their vulnerabilities [53]. Maintaining the security of IoT platforms is a challenge globally that needs technical innovators to devise relevant assessments to resolve it [36]. The other factor that can increase the security threat on IoT devices is if the network services for the IoT device are not secured enough with challenges like Denial-of-Service (DoS)or buffer overflow [44]. Security threats to devices may also include the physical attack of the device, which refers to instances such as natural disasters or if the device was not confirmed or authorized to be legitimate as expected to be carefully configured by the manufacturing companies of these IoT devices [36].

IoT users must be aware of the security threats that affect IoT [12]. The privacy and security of IoT users are constantly at risk and are violated for several reasons because of the lack of implementation of reliable security techniques [13]. The users of IoT should be informed about how to use the technology to know what to do to protect their information exposed online [39]. A user of IoT should be aware of how the device they use can be attacked. Outdated device software can allow attacks on the device; therefore, it is important to regularly update your device [12]. In South Africa, not all communities or users can fully operate their smartphones, making them vulnerable and exposed to cybercrimes [21]. When the users of IoT are aware of how important data security is then, it will be easier to take precautions for the safety of their confidentiality and privacy of information [15].

5.5.3 The Different Ways to Assist with Keeping Your IoT Device Safer from Security Threats

Different IoT smart devices come with a default password set when produced. Most of those passwords are either the same default passwords or are easy to crack; therefore, changing the default password into a strong password is important to deny unauthorized access [17]. Connecting to the Internet is vulnerable; therefore, it is much safer to connect IoT devices on a separate network, in a network where only IoT devices get connected [32]. Physical access to the device can also expose it to security threats such as theft; therefore, ensuring that the device is physically secured where physical access is given only to designated people in order to protect it from security threats and, thus, accountability lies on few people [1]. Securing an IoT device can also be done through a Virtual Private Network (VPN). Though mostly used for convenience, VPN can also be used to make the device more secure from attacks like botnets or eavesdropping; it is assumed that the data, which are used within an IoT device through VPN, are secured [51]. It is important or safe to switch off an IoT device when it is not being used; this can help minimize security threats like erasing some recent malware attack because the malware can be stored in the memory and easily erased by the action of power cycle, which is initiated by switching the device on and off [29].

5.6 Research Methodology

Research methodology is described as a path of solving a research problem that follows certain procedures [37]. It clearly explains how the study collected data and how the data were analyzed, determining the reliability. The quantitative approach refers to the usage of a large quantity of collected data from a population's ideas, perceptions, behaviors, and opinions numerically, which can be converted into statistical information to solve a research problem [3]. The chapter problem and the population of the study need to be specifically stated so that a reader can understand them[2] clearly. The quantitative approach was used for this study [4].

The philosophical framework, which was followed for this study, is positivism to help evaluate the research objectives. The approach used in this chapter is inductive, which allows starting with the data collection using questionnaires to develop a theory. The chapter involves students from the North-West University, Mafikeng campus. The study population had a sample of students; out of the entire population of the university, only a few students of the Information System department (from the first-year undergraduates to postgraduates) were asked to participate because of their understanding of the topic. The chosen data collection tool is the questionnaire, a set of close-ended questions relating to the chapter topic. The students answered the questions regarding the study and their IoT experiences.

5.6.1 Population and Sampling

When conducting a study, the researcher gathers data from participants who are considered as population, and these are people with the feature of interest; hence, the results from these participants are generalized [9]. All the people who meet the standard specified in the study can participate, and the people can either have something in common or not [41]. This study was conducted using the student population of North-West University, Mafikeng campus.

When conducting a study, it is virtually not possible to reach every member of the specified population; therefore, the small number of people who participate in the study are referred to as the sample of the population [41]. There are two methods of sampling. First is the probability method or random sampling, which is not ideal for use in a qualitative study. Still, it is the most accurate sampling method. Second is the non-probability method, and it is mostly used to choose the population for a qualitative study; it is not specific to the population, which makes it difficult to generalize the findings [45]. For the current study, the first method was used because a survey was deployed, and the sampling was done on the population using statistical analysis. In the population of North-West University students, a sample of students (first-year undergraduates to postgraduates) in the Department of Information Systems was used. The students possessed good knowledge about technologies and the topic under investigation.

5.6.2 Data Collection

Data collection is defined as the method used to collect data from different sources to answer a particular research question and analyze the findings. There are two methods of data collection: primary and secondary data collection methods [20]. In primary data collection, the researcher collects data for the first time, the data are objective and authentic, and are collected for a particular study. The first-time experience of the data ensures the quality of the information gathered [33]. According to Dudovskiy [20], primary data collection in quantitative research is collecting data using tools like questionnaires, interviews, and experiments, which cost less to execute compared to qualitative research, and at the end of collecting data, it becomes easy to compare the findings. For this study, the data were collected for the first time using questionnaires from 206 students in the Department of Information Systems; therefore, the primary data collection method was used. A questionnaire to achieve the objectives of the study. The questionnaire was designed to determine the impact of data security on the IoT. The questionnaire was distributed to North-West University, Mafikeng campus, specifically to Information Systems students, from first-year undergraduate to postgraduate students. The questionnaire was given to participants to fill at their own time and return the same.

The questionnaire used for this study was structured and closed-ended, and divided into five sections. The first Section A is the demographic information, using the nominal scale, then Sections B to D were using the ordinal scale. Microsoft Excel was used to capture the data collection before analysis. The study used the Statistical Package for the Social Sciences (SPSS) software to analyze the data collected from participants using questionnaires to reach the findings.

5.6.3 Reliability and Validity

The extended measure of accuracy and quality of the study, viz. the assumption that if the same study instrument, the questionnaire in this instance, was taken again by the same population sample, the same results will be achieved [28]. The questions during data collection were ensured to be relevant to the study, and this resulted in good quality of the study. The reliability of the questionnaire was tested using Cronbach's alpha. A sample of 20 participants attempted the questionnaire, which had 35 items. The results were > 60%, as shown in Table 5.1, proving the reliability of the questionnaire. The reliability coefficient of Cronbach's alpha is between 0 and 1; the closer the coefficient is to 1 (≥ 0.6), the higher internal consistency or reliability it shows [23].

5.7 Chapter Results and Discussions

This chapter is aimed at the data analysis and discussions of the findings. It used primary data collected from students of the Department of Information System, North-West University, Mafikeng campus, using a structured questionnaire. Data were analyzed using SPSS to evaluate the impact of data security on IoT devices these students use. This chapter is organized in different sections: demographic information of respondents, awareness of users about data security of IoT, the security threats that are affecting IoT devices, the effects of security threats on users' IoT devices , and different ways to assist with keeping IoT smart devices safer from security threats.

The study's sample size was 254 for data collection, and questionnaires were distributed mainly at the Information System computer lab, which was used by first-, second-, third-, and fourth-year undergraduates, and postgraduates. Out of the 254 questionnaires that were distributed, only 206 (81% of the total number of distributed questionnaires) were returned.

Table 5.1 Reliability Table.

Reliability statistics		
Cronbach's alpha	Cronbach's alpha based on standardized items	N of items
0.687	0.811	35

5.7.1 The Demographic Information

5.7.1.1 Age, Ethnic Group, and Ownership of a Smart Device

The results indicate that all the 206 participants were aged 18-35 years; there were no students <18 years or > 35 years. The results of the ethnic group indicate that all (100%) of the students who responded were black, and the questionnaires were distributed fairly without any specific preference of the ethnic group. This was done to determine whether the respondent owns any technological smart device like a smartphone or laptop. According to the results, all respondents owned a smartphone or laptop.

This section aimed to determine the gender of the respondents. The findings demonstrate that 109 (52.9%) respondents were male and 97(47.1%) were female. These results indicate that the majority of respondents were male students. This section was aimed to determine the academic level of the respondents, the questionnaires were distributed, and no preference was given to a specific level of study. The results in Figure 4.1 indicate that 56 (27.2%) respondents were first-year students, 71 (34.5%) were second-year students, 48 (23.3%) were third-year students, 23 (11.2%) were fourth years, and 8 (3.9%) were postgraduates students (which include honors, masters', and Ph.D. students). The results show that the dominant respondents were second-year students.

5.7.2 Awareness of Users About Data Security of the Internet of Things

The first question aimed to evaluate whether the respondents were aware that it becomes vulnerable to security threats when their devices are connected to the Internet. The results showed that 3 (1.5%) strongly disagreed that they were not aware, 7 (3.4%) disagreed that they were also not aware, 124 (60.2%) agreed that they were aware, and 72 (35%) strongly agreed that they were aware. Most respondents were aware that connecting to the Internet with their IoT device makes the device vulnerable to security threats. The second question focused on determining whether or not the respondents were aware that if their IoT device is not regularly updated, it becomes vulnerable to security threats. The results show that 7 (3.4%) strongly disagreed that they were not aware, 73 (35.4%) disagreed that they were not aware, 106 (51.5%) agreed that they were aware, and 20 (9.7%) strongly agreed that they were aware. These results indicate that most respondents were aware that they become vulnerable to security threats if their IoT devices are not regularly updated.

The third question aimed to determine whether the respondents were aware that the privacy policy of applications on their devices could expose their IoT devices to security threats. The findings show that 4 (1.9%) strongly disagreed, meaning they were not aware of this, 73 (35.4%) disagreed that they were not aware, 108 (52.4%) agreed that they were aware, and 15 (7.3%) strongly agreed that they were aware. Therefore, the results of this question show that most people were aware that the privacy policies of applications on IoT devices expose devices to security threats.

This question intended to understand whether the respondents knew that connecting to an open or free Wi-Fi that is not secured could expose their IoT device to security threats. The results showed that 4 (1.9%) strongly disagreed that they didn't know, 32 (15.5%) disagreed that they didn't know, 101 (49%) agreed that they knew, and 69 (33.5%) strongly agreed that they knew. The results, therefore, show that most respondents were aware that when they are connecting to an open or free Wi-Fi, it makes their device vulnerable to security threats. Another question was on awareness of easy or common passwords. The purpose of this question was to determine whether the respondents knew that using a common or easy password on their IoT device can allow security threats on that device. The results showed that 8 (3.9%) strongly disagreed that they didn't know, 22 (10.7%) disagreed that they did not know, 58 (28.2%) agreed that they knew, and 118 (57.3%) strongly agreed that they knew. These results indicate that most of the respondents agreed that using an easy or common password on their IoT devices can allow security threats on their devices.

The respondents were asked to determine awareness of location exposure. This question was focused on evaluating whether the respondents were aware that keeping the location on and allowing applications to access the location of an IoT device can expose it to security threats. The results show that 1 (0.5%) respondent strongly agreed that he/she was not aware, 28 (13.6%) disagreed that they were also not aware, but 102 (49.5%) agreed that they were aware, and 75 (36.4%) strongly agreed that they were aware of the threats. An illustration of the results shows that most respondents generally agreed that they were aware that keeping the location of your IoT device on and allowing applications to access their location can expose them to security threats. Another question aimed to determine whether the respondents knew that hackers could gain access to information through an audio mic, a camera of an IoT device, or a connected printer. Results indicate that 3 (1.5%) strongly disagreed that they did not know, 30 (14.6%) disagreed that they never knew, while 149 (72.3%) agreed that they knew, and 24 (11.7%) strongly agreed that they knew. The findings show that most respondents agreed that they knew that hackers access information through an audio mic, a camera, or a connected printer to the IoT device.

The awareness of proximity questions was asked. The question focused on evaluating if the respondents were aware that leaving the Bluetooth or Wi-Fi of an IoT device in a public place can automatically share information due to the proximity, which is a security threat. Table 4.9 shows that 12 (5.8%) strongly disagreed that they were not aware, 48 (23.3%) disagreed that they did not know too, but 128

(62.1%) agreed that they were aware, and 18 (8.7%) strongly agreed that they were aware. The majority of the respondents showed that they were aware that leaving the Bluetooth or Wi-Fi on of an IoT device in a public place can leave it to share information due to the proximity automatically, which is regarded as a security threat.

5.7.3 The Security Threats that are Affecting the Internet of Things Devices

In this chapter, there are different questions. The first question aimed to determine whether DoS is a security threat that affects the IoT devices by making services unavailable. The results show that 1 (0.5%) strongly disagreed, 19 (9.2%) disagreed, 174 (84.5%) agreed, and 12 (5.8%) strongly agreed. According to these results, most respondents have agreed that DoS is indeed a security threat toward IoT devices. Another question focused on hackers – to determine if data theft from IoT devices by hackers is considered a security threat. Results showed that 19 (9.2%) respondents disagreed, 114 (55.3%) agreed, and 73 (35.4%) strongly agreed. The results generally indicate that most respondents agree that the theft of data by hackers on IoT devices is regarded as a security threat.

The purpose of this question was to find out the respondents' view on IoT devices that are attacked through remote controlling, like using a camera of an IoT device without physical access to gain information that is regarded as a security threat. The results show that out of the total number of respondents, 2 (1%) strongly disagreed, 38 (18.4%) disagreed, 145 (70.4%) agreed, and 21 (10.2%) strongly agreed. The results in Table 5.2 show that most respondents agreed that when an IoT device is attacked through remote control, using a camera of the device without physical access can be regarded as a security threat. Also, a question was used to find out from respondents if ransomware can be used in emails to attack IoT devices, which stops the device's functionalities. The results from respondents show that 11 (5.3%) strongly disagreed, 22 (10.7) disagreed, 108

Table 5.2 Remote recording.

	Frequency	Percentage	Valid percentage	Cumulative percentage
Strongly disagree	2	1.0	1.0	1.0
Disagree	38	18.4	18.4	19.4
Agree	145	70.4	70.4	89.8
Strongly agree	21	10.2	10.2	100.0
Total	206	100.0	100.0	

(52.4%) agreed, and 65 (31.6%) strongly agreed. According to these results, the majority of respondents agreed that ransomware could use emails to attack IoT devices, which can make the functionalities of the device stop.

This question was intended to determine whether the passwords on IoT devices or online transactions could be hacked for authentication into the network or for physical access to the smart IoT device. The results showed that 9 (4.4%) respondents strongly disagreed, 14 (6.8%) disagreed, 148 (71.8%) agreed, 35 (17%) strongly agreed. The results showed that most respondents agreed that passwords on IoT devices or online transactions could be hacked for authentication into the network or for physical access to the smart IoT device.

5.7.3.1 The Architecture of IoT Devices

Another question was focused on determining whether the architecture or infrastructure of IoT smart devices, such as centralized databases, exposes devices' vulnerability. The results show that 11 (5.3%) strongly disagreed, 27 (13.1%) disagreed, 114 (55.3%) agreed, and 54 (26.2%) strongly agreed. Therefore, these results show that most of the respondents agreed that the architecture or infrastructure of IoT smart devices, such as centralized databases, exposes devices' vulnerability.

5.7.3.2 The botnets Attack

The question was meant to find out whether the respondents think that botnets can attack an IoT device to gain information or exploit online banking transactions. The results showed that 11 (5.3%) strongly disagreed, 20 (9.7%) disagreed, 157 (76.2%) agreed, and 18 (8.7%) strongly agreed. Thus, most respondents agreed that botnets can attack an IoT device to gain information or exploit online banking transactions.

5.7.4 The Effects of Security Threats on IoT Devices that are Affecting Users

5.7.4.1 The Slowness or Malfunctioning of the IoT Device

This question aimed to find out whether the respondents knew that the IoT devices become slow or malfunction once they are exposed to security threats. The results showed that only 3 (1.5%) respondents disagreed and 109 (52.9%) agreed, while 94 (45.6%) strongly agreed. Therefore, these results show that most respondents agreed that IoT devices become slower or they malfunction once exposed to security threats. The security threats negatively affect IoT. This question was focused on finding out from respondents if they think that the security threats negatively affect the IoT smart devices. The results show that out of the total number of respondents, 1 (0.5%) strongly disagreed, 17 (8.3%) disagreed, 69 (33.5) agreed, and

119 (57.8%) strongly agreed. The results conclude that most respondents agreed that the security threats affect IoT smart devices negatively.

5.7.4.2 The Trust of Users on IoT

This question aimed to determine whether the respondents think that security threats that attack the IoT devices also affect users' trust in these devices as products. According to results, 2 (1%) strongly disagreed, 10 (4.9%) disagreed, 111 (53.9%) agreed, and 83 (40.3%) strongly agreed. Further, most respondents agreed that the security threats affecting IoT devices also affect users' trust in these devices as products. IoT reduced life span. The question was meant to find out from respondents whether they think that the security threats that affect IoT devices can reduce the device's life span. The results showed that 7 (3.4%) strongly disagreed, 14 (6.8%) disagreed, 108 (52.4%) agreed, 77 (37.4%) strongly agreed. These results showed that most of the respondents agreed that the security threats could reduce the life span of the IoT device.

5.7.4.3 The Safety of Users

The question was aimed to determine whether security threats on IoT devices can compromise the safety of users. The results showed that out of the total number of respondents, only 7 (3.4%) strongly disagreed, while 102 (49.5%) agreed, and 96 (46.6%) strongly agreed. These results show that majority of the respondents agreed that the security threats on IoT devices could compromise the safety of users of these IoT devices.

The question focused on determining whether security threats negatively affect the service delivery and customer satisfaction of smart devices. According to the results shown in Table 5.3, out of the total respondents, only 10 (4.9%) disagreed, while 92 (44.7%) agreed and 104 (50.5%) strongly agreed. These results show that most respondents agreed that security threats negatively affect service delivery and customer satisfaction of IoT smart devices.

Table 5.3 Service delivery and customer satisfaction.

	Frequency	Percentage	Valid percentage	Cumulative percentage
Disagree	10	4.9	4.9	4.9
Agree	92	44.7	44.7	49.5
Strongly agree	104	50.5	50.5	100.0
Total	206	100.0	100.0	

5.7.4.4 The Guaranteed Duration of IoT Devices

This question aimed to determine whether IoT smart devices damage quicker than the guaranteed functional time because of the attacks from security threats. The results indicate that only 10 (4.9%) disagreed, while 92 (44.7%) agreed and 104 (50.5%) strongly agreed. According to these results, the majority of the respondents agreed that IoT smart devices do damage quicker than the guaranteed functional time because of the attacks from security threats.

5.7.5 Different Ways to Assist with Keeping IoT Smart Devices Safer from Security Threats

This first question on this research seeks to understand the regular update of IoT devices. The question aimed to determine whether it is safer to regularly update the software of the IoT device to keep it safer from security threats. The results show that 3 (1.5%) disagreed, while 119 (57.8%) agreed and 84 (40.8%) strongly agreed. The results showed that most respondents agreed that it is safer to regularly update the IoT device to keep it safer from security threats.

5.7.5.1 The Change Default Passwords

This question aimed to find out if it is safer to change the default passwords that the IoT smart devices come with from the manufacturer. According to the results from respondents, 3 (1.5%) strongly disagreed, 8 (3.9%) disagreed, 74 (35.9%) agreed, and 121 (58.7%) strongly agreed. The results indicate that most respondents agreed that it is safer to change the default passwords that the IoT smart devices come with from the manufacturer.

5.7.5.2 The Easy or Common Passwords

The question was focused on finding out whether it is safer to use easy, common, or default passwords like "1234." Among the respondents, 125 (60.7%) strongly disagreed, 62 (30.1%) disagreed, 14 (6.8%) agreed, and 5 (2.4%) strongly agreed. The results showed that majority of the respondents disagreed that it is not safer to use an easy, common, or default password on the IoT device.

5.7.5.3 On the Importance of Reading Privacy Policies

The question aimed to determine whether it is important to read the privacy policies for applications and deny permission if necessary. According to the results, 5 (2.4%) respondents out of 206 strongly disagreed, 17 (2.3%) disagreed, 136 (66%) agreed, and 48 (23.3%) strongly agreed. The results then show that most of the respondents agreed that it is important to read the privacy policies for applications and, if necessary, to deny permission.

Table 5.4 Internet security software.

	Frequency	Percentage	Valid percentage	Cumulative percentage
Strongly disagree	1	0.5	0.5	0.5
Disagree	7	3.4	3.4	3.9
Agree	134	65.0	65.0	68.9
Strongly agree	64	31.1	31.1	100.0
Total	206	100.0	100.0	

Questions given in Table 5.4 aimed to determine whether it is best to install internet security software on IoT devices to protect them from security threats from the Internet. The results show that 1 (0.5%) strongly disagreed, 7 (3.4%) disagreed, while 134 (65%) agreed and 64 (31.1%) strongly agreed. Table 4.28 shows that the majority of the respondents think it is best to install the security software on IoT devices to protect them from security threats from the Internet.

5.7.5.4 The Bluetooth and Wi-Fi of IoT Devices
The focus of the question was to find out whether it is much safer to switch off the location, Bluetooth, and Wi-Fi of the IoT devices if one is not using them. The results from respondents revealed that 6 (2.9%) strongly disagreed, 17 (8.3%) disagreed, 104 (50.5%) agreed, and 79 (38.3%) strongly agreed. Therefore, the results show that the dominant respondents agreed that it is much safer to switch off the IoT devices' location, Bluetooth, and Wi-Fi if one is not using them.

5.7.5.5 The VPN on IoT
The question aimed to find out whether it is best to use Virtual Private Networking to help protect the transmitted data through private or public Wi-Fi. Table 4.30 shows that 1 (0.5%) strongly disagreed, 20 (9.7%) disagreed, 142 (68.9%) agreed, and 43 (2.9%) strongly agreed. The results indicate that most respondents agreed that it is best to use VPN to help protect the transmitted data through a private or a public network.

5.7.5.6 The Physical Restriction
The question was aimed to find out whether it is much safer for IoT smart devices to be physically secured and physical accessibility should be restricted. The results then show that 9 (4.4%) disagreed, while 79 (38.3%) agreed and 118 (57.3%) strongly agreed. The results show that most respondents agreed that it is much safer for IoT smart devices to be physically secured and physical accessibility should be restricted.

5.7.5.7 Two-Factor Authentication

This question aimed to determine whether using two-factor authentication helps with strengthening security on an IoT device. The results show that 4 (1.9%) strongly disagreed, 16 (7.8%) disagreed, 100 (48.5%) agreed, and 86 (41.7%) strongly agreed. Also, most respondents agreed that using a two-factor authentication helps strengthen the security of the IoT device.

5.7.5.8 The Biometric Authentication

This question was meant to determine whether it is safer to use a strong password plus biometric authentication to make the device more secure. Results show that 7 (3.4%) disagreed, 98 (47.6%) agreed, and 101 (49%) strongly agreed. Therefore, the results show that most respondents agree that it is much safer to use a strong password plus biometric authentication to make the device more secure.

5.8 Answers to the Chapter Questions

The main objective of this study was to determine the impact of data security on the IoT, and this main objective was broken down into four objectives; therefore, the results are discussed according to the objectives below.

5.8.1 Objective 1: Awareness on Users About Data Security of Internet of Things (IoT)

This objective aimed to determine the awareness that IoT users have on the IoT devices' data security. This objective was broken into eight different questions. This study indicates that most respondents are aware of the security of data on the IoT. It is always important for manufacturing companies to ensure that before consumers or customers purchase their product or service, they are well informed about its security and privacy threats, either on the user manual or the packaging label of the product [31], because results show that several respondents are not aware that the privacy policies of applications, which can be downloaded, can expose them to security threats or vulnerability of data. The results indicated that most respondents were aware that connecting to the Internet is a security threat. The literature stated that connecting to the Internet is the foundation of IoT as well as security threats [5]. The results also showed that respondents are aware that connecting the IoT device to an open or free Wi-Fi is not safe, and Baig [10] also stated that public Wi-Fi is not safe. The results also showed that respondents are aware that using easy or common passwords is not safe. It was also stated as a precaution of safety, using a strong password is important [52]. The results indicated that respondents are aware that an IoT device has to be regularly updated, as it was mentioned in the analysis that it helps keep the IoT device safe [26]. This study has also shown that overall only a small number of respondents are not aware of data security on the IoT, but the majority know or are aware of this data security on the IoT, and this will lead to the IoT devices being handled appropriately in order to secure data.

5.8.2 Objective 2: Determine the Security Threats that are Involved in the Internet of Things (IoT)

This objective intended mainly to understand different types of security threats involved in the IoT; during data collection, this objective was turned into Section B, which was broken down into seven questions. The results from respondents and the literature review, according to Lerner (2019), indicated that most of the respondents agreed that botnets, ransomware, and Denial of Service are some of the security threats, which attack the IoT. Results also show that 81.5% of the respondents agreed that the architecture or infrastructure of these devices, like a centralized database, can expose IoT devices to security threats.

5.8.3 Objective 3: The Effects of Security Threats on IoT Devices that are Affecting Users

This objective aimed to determine the effects of security threats on IoT devices that affect the users; when data were collected, this objective was turned into Section C, divided into seven questions. The results showed that the participants agreed that once a device is affected by security threats, the device becomes slow or malfunctions; the security threats can even decrease the life span of the device [8]. The results also demonstrate that respondents agree that users' trust in IoT is affected as the devices continue to be attacked by security threats [30]. The results also show that security attacks on the IoT device can compromise the safety of users by exposing their location [42]. Lastly, most respondents have agreed that security threats on IoT devices generally affect these devices negatively [8].

5.8.4 Objective 4: Different Ways to Assist with Keeping IoT Devices Safer from Security Threats

This objective ensured that users knew what to do to keep their devices safer from security threats. This objective was set as Section D when data were collected and were also broken down into 10 questions, and respondents gave feedback accordingly. According to results from the respondents, they show that it is much safer not to use easy or common passwords like "1234," and it is much safer to change the default passwords of new IoT devices to keep them safe from security threats [17]. The results agree with the literature review that an IoT device must be updated regularly to keep it safe from security threats [55]. These results and Hassija et al. [27] also indicate that it is important and safer to use biometric authentication to protect the device from security threats. The findings show that it is safer to use VPN to protect data stored and transmitted through the IoT device network. This agrees with the literature review that it's safer to use VPN on IoT devices [51]. The results also show that it is safer to switch off the location of your IoT device, Bluetooth, and Wi-Fi when not in use; this is supported in the literature review [29]. According to results and Corser [1], it is much safer to keep IoT devices physically safe, and access to IoT devices should be restricted.

5.8.5 Other Descriptive Analysis (Mean)

The descriptive analysis is given below, where the questionnaires had four options for respondents to choose from with 1 = strongly disagree, 2 = disagree, 3 = agree, and 4 = strongly agree. The mean for the below finding is 2, meaning that results between 1 and 2 have disagreed and results between 2.1 and 4 have agreed. The results below Mean 1–4.

5.8.5.1 Mean 1 – Awareness on Users About Data Security on IoT

Tables 5.5, 5.6, 5.7, uses the [14] as a online survey form., shows data on how to determine whether the respondents were aware of the security threats affecting the data security of IoT and the mean is between 2.67 and 3.39. The respondents agreed that they are aware of security threats affecting them; the respondents also agree that they are aware that using easy or default passwords like "1234" can expose one to security threats.

This analysis aimed at determining the security threats, which affect the IoT devices, and according to results in Table 5.6, the mean for this section is between 2.88 and 3.26, which means that respondents agreed that Denial of Service, hackers, ransomware, botnets are some of the security threats, which are affecting IoT devices.

5.8.5.2 The Effects of Security Threats on IoT Devices that are Affecting Users

Table 5.7 aims at finding out the effects of security threats on IoT devices that are affecting users and shows the mean average for this section was between 3.24 and 3.49, which indicates that most respondents have agreed that security threats

Table 5.5 Descriptive analysis (Mean 1).

		Are you aware that when your IoT device is connected to the Internet it becomes vulnerable to security threats?	Are you aware that the privacy policy of applications that you download on your IoT device can expose it to security threats?	Are you aware that if your IoT device is not updated regularly it becomes vulnerable to security threats	Do you know that connecting to an open or free Wi-Fi that is not secured can expose your device to security threats?	Do you know that using an easy or common password on your device like "1234" can allow security threats on your IoT device?	Are you aware that keeping your location on and allowing your applications to access your location can expose you to security threats?	Do you know that hackers can gain access to information through an audio mic, a camera of IoT device or a connected printer?	Are you aware that when you leave your Bluetooth or Wi-Fi of your IoT device on in public places it can automatically share information due to the close proximity and that is a security threat?
N	Valid	206	206	206	206	206	206	206	206
	Missing	0	0	0	0	0	0	0	0
Mean		3.29	2.62	2.67	3.14	3.39	3.22	2.94	2.74
Std. Deviation		0.601	0.693	0.696	0.742	0.829	0.688	0.564	0.698

Table 5.6 Security threats that are affecting IoT devices (Mean 2).

		Denial of Service is one of the security threats that affects the IoT devices by making services unavailable.	Hackers can break the data breaches and steal data and this is a security threat of theft on IoT devices.	IoT devices are attacked through remote recording like using a camera of smart IoT devices to gain information	Ransomware can use emails to attack IoT devices which stops the operation or functionalities of the device	The password on IoT devices or transactions can be hacked for authentication into the network or for physical access to the smart IoT device	Architecture or infrastructure of IoT smart devices such as centralized database exposes many devices' vulnerability.	Botnets can attack an IoT device in order to gain information or to exploit online banking transactions
N	Valid	206	206	206	206	206	206	206
	Missing	0	0	0	0	0	0	0
Mean		2.96	3.26	2.90	3.10	3.01	3.02	2.88
Std. Deviation		0.411	0.616	0.562	0.793	0.644	0.780	0.622

Table 5.7 Security threats effects on IoT (Mean 3).

		The IoT devices become slow or malfunction once it is exposed to security threats	The security threats negatively affect IoT smart devices	The security threats that attack the IoT devices also attacks the trust of users on these devices	The security threats can reduce the life span of IoT devices	Security threats on IoT smart devices can compromise the safety of users	Security threats affect the service delivery and customer satisfaction of IoT smart devices negatively	The IoT smart devices damage quicker than the guaranteed functional time due to attacks from security threats
N	Valid	206	206	206	206	206	206	206
	Missing	0	0	0	0	0	0	0
Mean		3.44	3.49	3.33	3.24	3.42	3.46	3.46
Std. Deviation		0.526	0.668	0.616	0.724	0.585	0.589	0.589

negatively affect the IoT smart devices, and that these security threats also affect the life span of these IoT devices.

5.8.5.3 Different Ways to Assist with Keeping an IoT Device Safer

Table 5.8 focuses on determining different ways to assist with keeping IoT smart devices safer from security threats, and the mean for the results in Table 4.37 ranges between 3.10 and 3.52, which are the question in this section where respondents have agreed that these are the ways by which one can keep the IoT devices safe, like regularly updating the software of the devices and physically securing an IoT device and restricting the accessibility of the device. However, there's one question with a mean of 1.51, which means respondents have disagreed with – that it is safer to use easy or common passwords like "1234." According to these results, respondents are saying that it is rather not safe.

5.9 Chapter Recommendations

Future studies should consider why manufacturing companies of IoT devices are not completely securing these devices in totality and ensuring that attackers cannot crack in. Future studies should consider determining the packing or manual of IoT devices to indicate the possibility of security threats that might attack the IoT device and suggestions on what to do to prevent it, like ensuring that users understand the importance of regularly keeping up to date with firmware updates. Future studies should encourage IoT device users to change the default passwords or easy and common passwords like "1234." Users also need to read to understand the privacy policies of the applications they download on their IoT devices.

5.10 Conclusion

This study aimed to determine the impact of data security on the IoT. There were four objectives to reach the goal of this study. The introduction, the problem statement, and the review literature of other authors, and results. Results from participants response show that people are aware of security threats that are affecting the IoT devices but do not know what to do in order to keep these devices safe. A quantitative research method was used to conduct this research, and data collection, analysis, and discussions of findings and finally the conclusion and future recommendations were made. Overall, it is concluded that different security threats threaten data security on IoT devices. These threats have a negative impact on the functionality of these IoT devices and IoT users.

Table 5.8 Descriptive analysis of section (Mean 4).

	It is safer to regularly update the software your IoT smart devices safer from security threats	It is safer change the default passwords that the IoT smart devices come with from the manufacturer	It is safer to use easy, common, or default passwords like "1234"	It is important to read the privacy policies for applications and if necessary to deny permission	It is best to install internet security software on devices in order to protect them from security threats from the Internet	It is much safer to switch off the location, Bluetooth, and Wi-Fi of your smart devices if you are not using them	It is best to use Virtual Private Network to help to protect the transmitted data through a private or public Wi-Fi	It is much safer for IoT smart devices to be physically secured and that physical accessibility should be restricted	Using two-factor authentication helps with strengthening security on your IoT smart device	It is much safer to use strong password plus biometric authentication to make your device more secured
N Valid	206	206	206	206	206	206	206	206	206	206
Missing	0	0	0	0	0	0	0	0	0	0
Mean	3.39	3.52	1.51	3.10	3.27	3.24	3.10	3.53	3.30	3.46
Std. Deviation	0.519	0.646	0.731	0.636	0.542	0.725	0.562	0.582	0.696	0.564

References

1 Corser, G. et al. (2017). Internet of things (IoT) security best practices. *IEEE Internet Technology Policy Community White Paper* 1 (1): 1–12.

2 Kumar, R. (2018). *Research Methodology: A Step-by-Step Guide for Beginners.* Sage.

3 Bernard, H.R. (2017). *Research Methods in Anthropology: Qualitative and Quantitative Approaches.* Rowman & Littlefield.

4 Williamson, K. (2004). *Research Methods for Students, Academics and Professionals: Information Management and Systems.* Elsevier.

5 Abomhara, M. and Køien, G. (2015). Cyber security and the internet of things: vulnerabilities, threats, intruders and attacks. *Journal of Cyber Security* 4: 65–88.

6 Aldowah, H., Rehman, S., and Umar, I. (2018). Security in internet of things: issues, challenges and solutions. *Research Gate* 1 (1): 1–38.

7 Aman, F., Maryam, M., Bashir, S., and Batool, S. (2016). Public awareness and attitude about IoT and its impact. *Research Journal of Recent Sciences* 5 (8): 31–38.

8 Arora, G. (2019). Security Boulevard – The growing presence (and seurity risks) of IoT. https://securityboulevard.com/2019/11/the-growing-presence-and-security-risks-of-iot/ (accessed 13 December 2019).

9 Asiamah, N., Mensah, H.K., and Oteng-Abanyie, E.F. (2017). General, target, and accesible population: demystifying the concepts for effective sampling. *The Qualitative Rrport* 1: 1607–1621.

10 Baig, A. (2018). Global Sign Blog- Tips for staying safe on public WIFI Nteworks. https://www.globalsign.com/en/blog/staying-safe-using-public-wifi/ (accessed 18 November 2019).

11 Barot, S. (2018). DZone. https://dzone.com/articles/why-is-data-security-important-for-everyone (accessed 26 June 2019).

12 Barry, C. (2019). BARRACUDA – Cybersecurity awareness and IoT security. https://blog.barracuda.com/2019/10/02/cybersecurity-awareness-and-iot-security/ (accessed 23 November 2019).

13 Bertino, E. (2016). *Data Security and Privacy in the IoT. s.l.* Open Proceedings.

14 Bhat, A. (2018). QuestionPro -Research Design; Definition, Characteristics ans types. https://www.wufoo.com/integrations/questionpro/ (accessed 31 October 2019).

15 Bourgeois, D.T. and Bourgeois, D. (2014). Press Books. https://bus206.pressbooks.com/chapter/chapter-5-networking-and-communication/ (accessed 02 May 2019).

16 Brunt, A. (2019). Sales-I sell smart. https://www.sales-i.com/the-importance-of-data-security (accessed 30 July 2019).

17 Chang, Z. (2019). Trend Micro. https://www.trendmicro.com/vinfo/us/security/ news/internet-of-things/inside-the-smart-home-iot-device-threats-and-attack-scenarios (accessed 22 Novemeber 2019).

18 CS Interative Training (2017). ITWEB. http://www.itweb.co.za/content/ XGxwQDM1dyL7IPVo (accessed 16 May 2019).

19 Das, A. (2019). ItProPortal - 5G and 4IR has unleashed the Internet of Things. https://www.itproportal.com/features/5g-and-4ir-has-unleashed-the-internet-of-things/ (accessed 16 January 2020).

20 Dudovskiy, J. (2018). The Ultimate Guide to writing a Dessertation in Business studies.

21 Gercke, M. (2019). *Understanding Cybercrime: Phenomena, Challenges and Legal Responses*. CYBER CRIME -ITU Publications.

22 Gillwald, A., Mothobi, O. and Rademan, B. (2018). The State of ICT in South Africa. http://researchICTafrica.net.

23 Gliem, J.A. and Gliem, R.R. (2003). Calculating, interpreting, and reporting Cronbach's alpha reliability coefficient for likert-type scales. *Midwest Research to Practice Conference in Adult, Continuing and Community Education* 1 (1): 82–87.

24 Gokhale, P., Bhat, O., and Bhat, S. (2018). Introduction to IOT. *International Advanced Research Journal in Science Engineering and Technology* 4 (1): 41–44.

25 Grobler, M., van Vuuren, J.J., and Zaaiman, J. (2013). *Evaluating Cyber Security Awareness in South Africa. Research Gate*.

26 Hasan, M. (2019). Ubuntu pit. https://www.ubuntupit.com/25-most-common-iot-security-threats-in-an-increasingly-connected-world/ (accessed 01 July 2019).

27 Hassija, V. et al. (2019). A survey on IoT security: application areas, security threats, and solution architectures. *IEEE Access* X (2): 64–72.

28 Heale, R. and Twycross, A. (2015). Validity and reliability in quantitative studies. *Evid Based Nurs* 18.

29 Hossain, M., Hasan, R., and Skjellum, A. (2017). Securing the internet of things: a meta-study of challenges, approaches, and open problems. In: *IEEE 37th International Conference on Distributed Computing Systems Workshops (ICDCSW)*, 220–225. https://doi.org/10.1109/ICDCSW.2017.78.

30 Husamuddin, M. and Qayyum, M. (2017). Internet of things :a study on security and privacy threats. *Research Gate* 1 (1): 90–96.

31 Jentzen, A. (2019). Proofpoint: Are Your users' IoT purchases a security threat?. https://www.proofpoint.com/us/security-awareness/post/are-your-users-iot-purchases-security-risk (accessed 28 November 2019).

32 Joshi, N. (2019). BBN Times. https://www.bbntimes.com/technology/8-types-of-security-threats-to-iot (accessed 22 November 2019).

33 Kabir, S. M. (2016). Methods oF Data Collection: An Introductory Approach for all Disciplines.

34 Kamburugamuve, S., Christiansen, L., and Fox, G. (2015). *A Framework for Real-Time Processing of Sensor Data in the Cloud.* School of Informatics and Computing and Community Grids Laboratory.

35 Kizza, J.M. (2013). *Guide to Computer Network Security,* 2nde. Springer.

36 Koduah, S.T., Skouby, K.E., and Tadayoni, R. (2017). Cyber security threats to iot applications and service domains. *Wireless Personal Communications* 95 (1): 169–185.

37 Kothari, C. (2004). *Research Methodology- Methods and Techniques.* s.l.: New Age International Publishers.

38 Lerner, S. (2019). Enterprice Digitilization- Digital transformation insight and Analysis. https://www.enterprisedigi.com/iot/articles/iot-security-challenges (accessed 14 December 2019).

39 Maple, C. (2017). Security and privacy in the Internet of things. *Journal of Cyber Policy* 2 (2): 155–184.

40 McKane, J. (2019). My Broadband trusted in Tech. https://mybroadband.co .za/news/internet-of-things/294358-how-iot-is-quietly-taking-over-south-africa .html (accessed 15 June 2019).

41 Mohsin, A., 2016. A manual for selecting sampling techniques in research.

42 Mookyu, P., Haengrok, O., and Kyungho, L. (2019). Security risk measurement for information leakage in IoT-based smart homes from a situational awareness perspective. *Sensors (Basel)* 19 (2): 60–95.

43 Neaves, H. (2018). Why data secueity is important for every Business. *Insights for Professionals.*

44 Pal, A. (2019). CISCO-The Internet of Things (IoT) – Threats and Counter-measures. https://www.cso.com.au/article/575407/internet-things-iot-threats-countermeasures/ (accessed 22 October 2019).

45 Phrasisombath, K. (2009). Sample size and sampling methods: Training Course in Reproductive Health Research. University of Health Sciences, (September), 1–34. Retrieved from https://www.gfmer.ch/Activites_internationales_Fr/Laos/ PDF/Sample_size_methods_Phrasisombath_Laos_2009.pdf.

46 Pollmann, M., 2017. IoT Agend. https://internetofthingsagenda.techtarget.com/ blog/IoT-Agenda/IoT-data-is-growing-fast-and-security-remains-the-biggest-hurdle (accessed 25 June 2019).

47 Porkodi, D. and Velumani, B. (2014). The Internet of Things (IoT) applications and communication enabling technology standards: an overview. In: *2014 Iinternation Conference on Intelligent Computing Applications,* 324–329. IEEE.

48 Ranger, S. (2018). ZD Net. https://www.zdnet.com/article/what-is-the-internet-of-things-everything-you-need-to-know-about-the-iot-right-now/ (accessed 02 May 2019).

49 Rizvi, S., Kurtz, A., Pfeffer, J., and Rizvi, M. (2018). Securing the Internet of Things (IoT): a security taxonomy for IoT. In: *2018 17th IEEE International Conference On Trust, Security And Privacy In Computing And Communications/12th IEEE International Conference On Big Data Science And Engineering (TrustCom/BigDataSE)*, 163–168. IEEE.

50 Rottigni, R. (2018). ReadWrite. https://readwrite.com/2018/11/27/why-iot-data-protection-has-become-more-important-than-ever/ (accessed 25 June 2019).

51 Rottigni, R. (2019). How a VPN Can Enhance Your IOT Devices Security - ReadWrite. https://readwrite.com/2019/02/06/how-a-vpn-can-enhance-your-iot-devices-security/ (accessed 23 November 2019).

52 SANS (2016). Security Awareness. https://www.sans.org/security-awareness-training/ouch-newsletter/2016/internet-things-iot (accessed 15 October 2019).

53 Skouby, K., Tadayoni, R., and Tweneboah-Koduah, S. (2017). *Cyber Security Threats to IoT Applications and Service Domains. Wireless Personal Communications*. Springer Science+Business Media.

54 Suo, H., Wan, J., Zou, C., and Liu, J. (2012). Security in the internet of things: a review. *International Conference on Computer Science and Electronics Engineering*, Hangzhou (23-25 March 2012) 3: 648–651.

55 Viswanathan, V. (2019). IT ProPortal. https://www.itproportal.com/features/eight-ways-to-secure-your-data-on-iot-devices/ (accessed 22 November 2019).

56 Williams, P. (2017). It News Africa. https://www.itnewsafrica.com/2017/04/iot-becomes-internet-of-threats/ (accessed 15 June 2019).

57 Zhang, Z. et al. (2014). *IoT Security: Ongoing Challenges and Research Opportunities.* In: *2014 IEEE 7th International Conference on Service-Orientated Computing and Applications, Matsue, Japan.*

6

Sustainable Renewable Energy and Waste Management on Weathering Corporate Pollution

Choo K. Chin[1] and Deng H. Xiang[2]

[1]*University of Northern Iowa, Bachelor of Arts in Computer Information System*
[2]*He Ying Metal Industries Sdn Bhd., Malaysia*

6.1 Introduction

Today, more and more corporations are venturing into environmentally, socially, and economically responsible ways. Many world-renowned companies are making corporation greening a part of their tender process. In Corporate Social Responsibility (CSR), a corporation needs not only to reduce the negative environmental impact but also should leave a positive and lasting legacy for the local community. It is necessary to search for solutions to reduce the environmental burden of business activity at all phases of the production process [1]. Many corporates would like to reduce their carbon footprints on the environment but may not be sure how to do it. It is crucial to involve all of the stakeholders that play a role in the planning or production of the events. If there is no commitment from within the company leadership, efforts may be ineffective and unnoticed. If commitment from within the leadership of a company is engaged, it needs to communicate this fact with all other partners involved to ensure full cooperation from the whole organization [2]. The sooner the company starts this process, the better it is.

Planning ahead is the single most important element in achieving CSR within the corporation. It involves including sustainable development principles and practices at all levels of the organization and aims to ensure that corporate ventures in a responsible manner. It represents the total package of interventions within a corporation that needs to be done in an integrated manner. Corporate Social Responsible practices should start at the inception of every level and should involve all the key role players, such as staff members, clients, management, contractors, and suppliers. Socially responsible entities assume responsibility for

Artificial Intelligence in Industry 4.0 and 5G Technology, First Edition.
Edited by Pandian Vasant, Elias Munapo, J. Joshua Thomas, and Gerhard-Wilhelm Weber.

the ecological ramifications of their activities, strive to conserve energy, reduce pollution and emissions of harmful substances, and attempt to increase the efficiency of using natural resources, thereby alleviating their ecological footprints [3]. One should remember that ever-changing economic growth connected with intense exploitation of natural resources is in overt contraction with the need to preserve these resources for future generations. In fact, every nation can use available resources for the benefit of its people; however, nations are also responsible for their protection and preservation for the coming generations [4].

A review of literature on CSR shows that numerous studies have focused on whole organizations and their impacts on different groups of stake holders. Social responsibility was considered in the context of profitability for the organizations undertaking socially responsible activities [5–7]. A. Glavas reports three main trends in the debate in CSR [8]. The first trend concerns the role of enterprises in society. The main question, often put by default, is whether the companies play a role in societies beyond generating profit. Then the second trend in the debate is about whether CSR is normative, which means that the companies must get involved in socially responsible undertakings, or instrumental, i.e. it is in the company's best interest to get involved in CSR actions. If CSR is normative, companies have a moral obligation toward society to care about its well-being [9]. The third trend of the debate is focused on the impacts of CSR on the company's financial performance. Proving that the company's involvement in CSR can generate additional profit makes other reasons (normative or instrumental) irrelevant as CSR is about doing good for both the company and society [8]. In this study, environmental responsibility can be seen as a fundamental aspect of social responsibility and a mechanism to longevity within a changing event environment. This can be exemplified by clarifying environmental concerns within a corporation.

According to the World Business Council for Sustainable Development, CSR is crucial to sustainable economic development and the well-being of societies [10]. Therefore there is always an urgent need for in-depth studies on the profitability of socially responsible activities towards the silent stakeholder, our world, and the environment. An increase in social awareness is forcing businesses to reduce their environmental burden. By taking environmental responsibilities seriously, business owners are the core to position themselves to communicate the value of environmental sustainability to the local community. Fulfilling environmental responsibilities can help the corporation gain green awareness, can assist in procuring from local government, create a competitive edge, and create greater value for business owners. CSR is about progress, not perfection. If started promptly, every subsequent decision and action of CSR will present opportunities for continuous improvement. Similarly, sustainable activities within corporate will evolve with the times and events will learn from experience and each other.

Both sustainable renewable energy and energy management are very important to the corporation. Throughout this theme issue, practitioners, pioneers, and academics have revealed a change in the way how business owners need to respond to consumer demand; how the corporate responds to political pressure, legislation, the competitive environment, and gives future glimpses into supply chain management and market trends and corroborate each other's work and posit that events need to be cognizant of social impacts, and evaluate these impacts impact upon the "triple bottom line" model. The literature proposes that corporate objectives can differ in their impacts that relate to dimensions of economic, commercial, bio-physical, socio-cultural, political, human resources, and environmental concerns as indicated in [11, 12].

Therefore, this research is designed to fulfill the following objectives:

- To define the characteristics of CSR
- To develop a plan to incorporate CSR
- To explore some of the issues encompassing the management and staging of CSR
- To examine the importance of engaging a range of key stakeholders and consider various ways in which corporations can be socially responsible for their operations

6.2 Literature Review

One of the most frequently quoted definitions explaining the essence of CSR was proposed by Archie B. Carroll [12] who asserted that CSR encompasses the economic, legal, ethical, and philanthropic expectations that society has of organizations at a given point in time [12] (p. 500). Caroll refers to the concept of social responsibility in which the most important task of the company is to generate profit. Therefore, he claims that the company's involvement in sponsoring charity programs or campaigns, when it is incurring financial losses and replacing economic goals, for which businesses are set up with social goals only, is irrational [13] (pp. 315–316). However, it is important that the way the company generates revenues must be honest and a fast-growing profit is not a top priority that overshadows other goals.

The base of Caroll's model is economic responsibility (Figure 6.1a). Caroll argues that enterprises should be profitable in the first place [15] (p. 55). The legal component of this model includes the firm's responsibility, such as the requirement to operate in a manner consistent with the law, and other binding applicable regulations, which determine the frameworks for the firm's operations and reflect the concept of proper business practices. Businesses should obey existing environmental protection regulations, respect consumer rights as well as employee rights, and

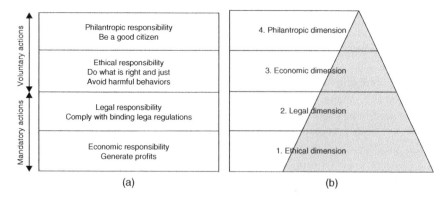

Figure 6.1 Social responsibility in the process of management; a comparison of models proposed by A. B. Carroll (a) and D. Baden (b). Source: Worked out by the authors based on carroll [11] (p. 499); Baden [14] (p. 11).

follow all contractual obligations they have agreed upon [16] (p. 3). Although the law is based on ethical premises, the ethical dimension of social responsibility is about something more, it concerns social expectations, which may not be incorporated into law. Activities performed by enterprises should reflect the spirit of the law [16] (p. 29) and [14] (p. 3). The philanthropic component of social responsibility includes the requirement to devote part of the firm's resources (e.g., financial resources or labor time) to gratify the needs of society and communities.

Denise Baden [17] presents a different view by placing the ethical and legal responsibilities before the economic ones, as she argues that the way enterprises pursue their business goals is very important. In Baden's view, enterprises should accept their ethical responsibility, which means that they should refrain from any harmful activity, and they should take account of social expectations in their operations, that is the right foundation for generating a profit and multiplying values (Figure 6.1b).

In literature from Poland and other countries, we can observe a growing interest in the concept of CSR. This resulted in many research approaches. Some analyses seem to be incomplete which creates problems with the interpretation of proposed CSR definitions. For example, A. Dahlsrud [18] identified at least 37 definitions of CSR. Most approaches, however, emphasize the influence of CSR on the economic, social, and ethical spheres of activity of enterprises [19] (pp. 51–57). Until recently, the concept of social responsibility was studied through the prism of potential benefits for enterprises and related stakeholders [20]. However, at the end of the 20th century, growing contemporary definitions of CSR highlight the increased expectations of enterprises in ethical, ecological, social, and economic spheres (Table 6.1).

Table 6.1 A review of definitions of corporate social responsibility (CSR).

Author(s), Year, Proposed Definition	Spheres Influenced by Enterprises	Reference Areas in Line with the Assumptions of the Conception of Corporate Social Responsibility
P. Hawken (1993)—sustainable development in the aspect of corporate social responsibility (CSR) enables planning specific activities, both economical for the environment and the organization's surrounding	Economic, ethical, social	Corporate governance; relations with employees; natural environment
T. Gossling, C. Vocht (2007)—social responsibility is perceived as the organization's *eff*orts undertaken to achieve a balance between economic, environmental, and social requirements, while being able to contribute to the well-being of a wider community and all stakeholders	Economic, ethical, ecological, social	Corporate governance; relations with employees and external stakeholders; commitment to social issues; natural environment
A.T. Lawrence, J. Weber (2008)—as part of CSR, organizations should assume the responsibility for all activities *aff*ecting the communities and the environment, additionally, they should undertake activities aimed at eliminating various negative *eff*ects resulting from their business operations and, if possible, repair inflicted damage	Economic, ethical, social, ecological	Corporate governance; relations with employees and external stakeholders; commitment to social issues; natural environment
M.P. Miles, L.S. Munilla, J. Darroch (2008)—the concept of corporate social responsibility is understood in terms of the product, process, strategic and innovative aspects; key social, economic and environmental aspects	Economic, ethical, ecological, social	Corporate governance; relations with employees and external stakeholders; commitment to social issues; natural environment
I. Gallego-Alvares, J.M. Prado-Lorenzo, I.M. Garcia-Sanchez (2011)—the concept of corporate social responsibility (CSR) and the conditions of its implementation are contingent upon the size of the enterprise, potential risks, and innovativeness	Economic, ethical, ecological	Corporate governance; relations with employees and external stakeholder

Source: Worked out by the authors based on: [19] (pp. 51–57); [21–25].

Presently, the concept of CSR associated with the care of the environment, justice, social order, and ethical conduct of enterprises, creates a real chance to implement principles of sustainable development at the lowest level – enterprises and entrepreneurs, employees, and local residents [26]. Few studies have developed the conceptual framework for the description of the concept of sustainability tailored from the systemic perspective, the perspective of individual entities, and their role in creating the foundations of sustainable society [27]. There are also studies concerning the perception of various CSR initiatives and programs by different groups of stakeholders, in particular employees, the addressees whose voices are important in the process of decision-making and implementation [28]. It is also possible to evaluate the results of the implementation of the "Green Human Resources Management" (GHRM), as well as an assessment based on an effective scale of measuring CSR in the following three areas: economy, society, and the environment [29].

It is assumed that socially responsible entities bear the consequences of ecological effects of their activities, strive to eliminate pollution, emission of harmful substances, and take steps to maximize the efficiency of using natural resources and minimize the negative impact on the environment [3]. Therefore, for some authors, the concept of CSR is a synonym of the concept of sustainability, translated as sustainable development [30] (p. 725); [31] (p. 81); [32] (p. 36), while other authors, in their definitions of social responsibility, refer directly to the concept of sustainable development [21]. An attempt to tell the difference between sustainability and CSR is a difficult task, as pointed out by I. Montiel in his extensive analysis [33]. The study mentioned in longitudinal of corporate sustainability has covered multiple dimensions [34] (p. 132). However, many authors differentiate between the concepts of sustainability and CSR, pointing how differences in time activities and effects interact between the present and the future.

A. B. Carroll [14] maintains that his model of the pyramid reflects responsibilities towards present stakeholders, as well as the future generations of stakeholders. Other authors distinguish CSR from sustainability, explaining that the concept of sustainability time is crucial. P. Bansal [35] argues that to ensure sustainable development at the organization's level, it is essential to implement the following three principles: environmental integrity, which can be achieved by environmental management, social equity, which can be achieved by CSR, and economic prosperity, which can be achieved by creating values. Krajnakova et al. [36] analyzed trends in the development of CSR in different periods of the economic cycle. They showed that strong socio-cultural ties, favorable cultural and legal climate, as well as scientific progress and technological innovations, exert a positive influence on the development of CSR. Climate changes and lasting ecological crises force enterprises to engage more actively in environmental issues. Rising inflation and growing unemployment rates, and decreasing consumer confidence result in a greater emphasis of businesses put on social and economic solutions. E. Seidel [36] pointed out that, regardless of the fluctuations in the macro-economic environment of both economic and non-economic nature, both businesses and society, at large, are more actively contributing to the promotion of CSR ideas and practices.

The assumption that businesses should consider ethical aspects of their activities in the first place, however, does not exempt them from pursuing economic goals. The impact of business activity on cost accounting is significant, and environmental protection and cost accounting are becoming mutually supportive elements. As E. Seidel noted, environmental costing is a pivotal instrument of pro-ecological management in enterprises and an indispensable component on the way to sustainable management. Unfortunately, giving a general definition of environmental costs is a quite challenging task [37] (p. 360) due to varied approaches and changing understanding of environmental costs over time.

Nowadays, it is commonly recognized that using and protecting the environment are essential parts of business processes that generate costs for enterprises. Environmental protection costs in enterprises can be divided into internal and external costs [38] (p. 55). From the economic viewpoint of the enterprise, the total internal costs of environmental protection are most important. Environmental protection costs are the sum of all investment outlays and current expenditures borne by the enterprise to reduce its environmental burden, incurred by safe storing, preventing, neutralizing, reducing, or eliminating pollutions and/or environmental damage.

Environmental management can yield numerous benefits for enterprises, including costs and resources savings, increased satisfaction and loyalty of customers, and morale of employees [39].

In addition to research on the relationship between socially responsible activity and financial results, particularly important are studies that indicate the relationship between corporate responsibility and its impact on financial results that have been achieved through effective management of the company's intangible resources, such as innovation, human capital, goodwill, and culture [40], value for stakeholders, including consumers [41], and measurement of consumer perception of the company's socially responsible activity in different areas [42].

From the 1990s, the European Community has supported active environmental policy instruments, including Environmental Management Systems (EMSs) [43]. By implementing such systems, enterprises can demonstrate their active approach to different environmental protection issues [44].

6.2.1 Energy Efficiency

Energy is an important daily resource for which demand is growing rapidly. Considering that half of the global population now lives in urban areas, natural resources are dwindling, and climate change is one of the main challenges facing us today, it is appropriate time to change the way we consume energy – and to reduce our impact on the environment. Sustainable energy management is a far-reaching concept that covers all phases of the sector – from fuels and their extraction to energy generation and the systems efficiency, energy distribution, energy consumption (in terms of amount and efficiency), and energy security implications, and application of renewable energy.

When managing energy, consequences need to be considered beyond the local level. Global aspects must be taken into account, such as the environmental, economic, and social results of fuel extraction and transportation from its source to its destination. Sustainable energy management should be conducted as an isolated operation. Many other systems depend on and interact with the energy system. Energy management is one of the layers of community management as a whole; hence, the principles of sustainability should ideally be applied in a cross-sectoral, integrated approach, beyond strictly energy themes. The challenge with energy management is that the amount of energy being consumed is usually not instantly apparent. Unlike waste, which provides a visual cue of its impact on the environment, energy is much harder to quantify and is often taken for granted and treated as an unavoidable cost.

6.2.2 Waste Minimization

The key question for consideration of waste relates to the real impact of different approaches to processing and waste and how does this put consumption into context. Initially, this can be viewed from four perspectives. First, in effect, waste is the disposal of something that has/had value. After all, it was spared. Disposal often has a cost, in that you may have to pay to have it removed and then pay tax on that [45]. An overall context has to be considered for waste. Waste is fundamentally bad in terms of damage to the environment but also erodes economic efficiency. If a manufacturer is wasting products that have been produced sustainably then it is possible to negate the accrued environmental benefit [46]. Recycling waste at a corporation makes sustainable sense but, perhaps, not if that waste creates further planet problems by being transported some way for the recycling process to happen, or, that cheap labor is required to separate the waste causing people issues. From this, the importance of individual responsibility on the consumer side and organizational responsibility on the supply side of a corporation can be seen to affect the overall sustainability of a corporate.

In terms of waste, measuring and reporting may be the key to working with such staff members to reduce this aspect of the impact within the corporation. Statistics revealing the amount of waste diverted from landfills and particularly those which demonstrate a cost-saving for the client can be powerful means of addressing a client's concerns. Gaining the ability to show a client how waste reduction techniques can work in practice can also help to alleviate any anxiety the client may have around the effectiveness or practicality of such changes. With these new waste regulations being brought in, it is expected that waste management will become more complex, and thus associated charges will rise. The extent to which these cost increases are passed down the supply chain may have a significant impact on the attractiveness of waste reduction. If successful, these regulations should increase the amount of life-cycle planning and designing out of waste, and push waste minimization up the agenda for clients.

6.2.3   Water Consumption

Water scarcity is an abstract concept to many and a stark reality for others. It is the result of myriad environmental, political, economic, and social forces [47]. Fresh water makes up a very small fraction of all water on the planet. While nearly 70% of the world is covered by water, only 2.5% of it is fresh. The rest water is saline and ocean-based. Even then, just 1% of our freshwater is easily accessible, with much of it trapped in glaciers and snowfields. In essence, only 0.007% of the planet's water is available to fuel and feed its 7.9 billion people (as of September 2021 – from worldmeters.info) [48]. Water is life, people need water to survive everywhere and every time. However, we humans have proven ourselves to be inefficient water users. Industries that use the water most are fruit and vegetable farming, textiles and garments, meat production, beverage industry, and automotive manufacturing [49]. According to the United Nations, water use has grown at more than twice the rate of population increase in the last century. By 2025, an estimated 1.8 billion people will live in areas plagued by water scarcity, with two thirds of the world's population living in water-stressed regions as a result of use, growth, and climate change. The challenge we now face is how to effectively conserve, manage, and distribute the water we have as we head into the future [49].

6.2.4   Eco-Procurement

Green procurement would seem to be a fairly simple concept. It would involve developing supply chains that delivered goods, of a type and in a way, which minimized the impact on the environment. Minimizing environmental impact would seem, at least initially, to be a reasonable working definition of "green" in this context [50]. A cap on oil consumption by 2025 will impose green procumbent options [51]. This is already evident within "the green procurement code" London, and transparency of supply chain within the food sector, ethical supply chain standards, "fair-trade" and others. Moreover, another [52] report suggests that threshold industry-standard will drive life-cycle analysis reporting through the value chain and initiatives such as the Global Reporting Initiative and standards like Publicly Available Specification (PAS) 2050 will be seen as entry-level management standards for many types of events. The Clinton Climate Initiative acknowledges the opportunity that the carbon economy can provide. It announced that the creation of skills, specific to carbon-neutral technology, will encourage frameworks and models which should be adopted elsewhere and mold influence and move sustainable practice to developing countries and markets [53].

6.2.5 Communication

Communications involve more than legitimacy alone because they also feature an instrumental dimension that deserves further exploration [54]. Finding novelty in sustainability communications does not lie in companies. Recent commitment to social responsibility has started to publish once-implicit objectives and approaches. Citizen companies and responsible management personnel have been around for a while but the publication of formalized commitments has become compulsory in various corporate institutional communications.

6.2.6 Awareness

Creating an eco-friendly culture often involves reinforcing behavior that people already want to adopt, but there is still a need for the appropriate tools and training in order to change. Businesses that cultivate an eco-friendly culture today are often immediately noticeable to clients as unique, even though at times the differences in an eco-friendly culture are imperceptibly small. For the former, an environmentally sound culture is often part of the core business strategy to encourage eco-friendly considerations in every decision that is taken [53]. An eco-friendly strategy should generally lead to cost-effective transformation initiatives that meet or exceed regulatory requirements. Still understanding benefits that are typically qualitative is critical to understanding the total eco-friendly proposition. Eco-friendly value propositions will include benefits to the company environment (building and facilities), benefits to the community, and improvements to the global environment. As for a business to operate toward sustainability, it should start "with the belief that we are part of a larger system – a business ecology – and extends the willingness to examine the larger socio-economic system and how we impact it at the individual, community, and organizational levels, and eventually at the planetary level." This provides the necessity to incorporate the systems approach which can help in identifying and solving environmental problems through a holistic vision towards sustainability [55].

6.2.7 Sustainable and Renewable Energy Development

Sustainable and renewable energy development involves considering environmental, economic, and social objectives when developing and implementing

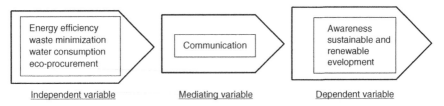

Figure 6.2 Proposed conceptual framework.

eco-friendly public policies and programs (Solar Energy Purchase by Malaysia National Electricity Provider [tenaga nasional berhad (TNB)]). It also involves considering the needs of the present as well as the future generations. Integrated decision-making and a long-term approach to planning are defining characteristics of sustainable and renewable energy development and represent key management challenges as concluded by long-term eco-friendly goals.

6.3 Conceptual Framework

In order to conduct this study, Corporate Social Responsible eco-friendly movement is taken as an independent variable or input which comprises four major constructs: energy efficiency, waste minimization, water consumption, and eco-procurement. On the other hand, awareness and sustainable and renewable energy development are labeled as the dependent variables. Earlier literature has proven the emergence of communication in mediating the link between eco-friendly Corporations and the level of awareness that will contribute towards sustainable and renewable energy development. Thus in this framework, communication is included as a mediator between the independent variable and dependent variables. Therefore, this study proposes the following framework, as illustrated in Figure 6.2.

6.4 Conclusion

The findings of this study will benefit all relevant stakeholders including suppliers, business owners, clients, as well as the local community to generate awareness, develop better CSR practices to improve business services that will contribute toward social transformation, and increase the quality of life. Organizations should take initiatives to encourage sustainable management in all aspects of their operations. We are mainly focusing on the area of corporate management. We propose the following recommendations.

6.4.1 Energy Efficiency

The research would recommend (i) utilizing Energy Star appliances for energy efficiency within corporate buildings, (ii) redesigning or renovating the corporate building to utilize more natural light and ventilation to reduce energy consumption, and (iii) implementing a smart system for corporate temperature, appliances, and lighting control. The corporate management should also monitor, evaluate and accommodate best electricity usage control through smart system implementation on- or off-peak hours. The management team must endeavor to have smart system control to switch off all lights and air-conditioners when no one is in the corporate building.

6.4.2 Waste Minimization

Corporate management should avoid printed handouts and if it is essential, then use recycled or eco-friendly paper only. All staff members should rely mostly on digital content to create a paperless corporate environment. Also, it is recommended to recycle bins within a corporate building to encourage waste separation at the source and reduce waste to landfills. Recently, Effective Microorganisms (EM in water) have been highly popular for home or organic fruits and vegetable plantations. It would be one of the best practices for corporate with the cafeteria to reduce food waste enormously, moreover, it is very eco-friendly to fertilize corporate building greenery landscape.

6.4.3 Water Consumption

Corporations in many sectors of industry use a lot of water. In general, industry uses around 20% of the world's freshwater withdrawals, while in the wealthiest nations, corporate water consumption can be as much as 40% of the total consumption. When it comes to doing something to conserve water usage, many companies have thrown up their hands. According to the current study, the following are a few quick water management practices to conserve water usage:

i. **Meter/Measure/Manage**: Metering and measuring facility water use help analyze saving opportunities. This also assures the equipment is run correctly and maintained properly to help prevent water waste from leaks or malfunctioning mechanical equipment.
ii. **Replace Restroom Fixtures**: Old water fixtures should be replaced with water-efficient standards fixtures.
iii. **Recover Rainwater**: A rooftop rain-water recovery system captures rainwater from the roof and redirects it to a storage tank. This water is used for flushing toilets, supplying cooling towers, and irrigating the landscape.

iv. **Recover Air Handler Condense**: Air conditioning units produce condensate water from the cooling coils. Companies are encouraged to capture this water for recycling purposes.

6.4.4 Eco-Procurement

Corporate management should make sure their purchasing power is fully exercised when to accommodate eco-friendly practices purchase such as energy efficiency and waste reduction products. In issuing and purchasing orders, companies should request products that are eco-friendly and locally manufactured. Also, it should be ensured that products are sustainably sourced. They should give preference to suppliers and contractors that implement eco-friendly practices.

6.4.5 Communication

It is recommended that corporate management develop an eco-friendly plan which can be used to communicate to all relevant parties for corporate activities. They must communicate the eco-friendly interventions implemented within the organization in long run. The research will recommend to corporate management to develop an eco-friendly plan which can be used to communicate to all relevant parties for corporate activities.

6.4.6 Sustainable and Renewable Energy Development

It is encouraged to all business owners' interest in every industry, to make environmental sustainability a management competency and an aspect of organizational excellence. By taking environmental responsibilities seriously, corporations in applying renewable energy (solar, wind, hydro, tide energy, etc.) are uniquely positioned to communicate the value of environmental sustainability to the community. Fulfilling responsibilities. Company managers need training on sustainability and how it applies to their daily activities and influence all staff members to embrace the eco-friendly movement. As sustainability is a difficult subject, training is required to educate managers on the impacts of one company and how they improve sustainability through their work. Following up the training with information and notes for attendees to refer back to, is important in ensuring what is learned is followed up by daily action.

Thus, in many ways, the best way forward may be to adopt some broad form of accreditation such as eco-friendly or to develop individually branded supply chains that will give a unique selling point for the corporation. The benefits from having a formalized, well-articulated eco-friendly strategy vary definitely in industry and individual business, but early adopters can still harness the enormous potential to opportunistically position them with a sustainable eco-friendly strategic advantage. It is essential to develop techniques to make informed decisions, but in a commercial world, the best approach is to adopt a broad set of criteria that may offer environmental benefits both to the business, the consumer, and the local community. This may well be difficult, particularly because many of the benefits of eco-friendly supply chains, such as decreased pollution, increase water quality or decreased production of greenhouse gasses are not effectively quantified economically in the marketplace.

Acknowledgment

The authors gratefully acknowledge the contribution by Dr. Pandian Vasant, modeling evolutionary algorithms simulation and artificial intelligence (MERLIN), Ton Duc Thang (TDTU), Ho Chi Minh City, Vietnam, and Dr. J. Joshua Thomas, UOW Malaysia KDU Penang University College, Malaysia, and their colleagues, family, and friends for the immense support and guidance which have significantly contributed to the quality of this study.

References

1 (a) Bojar, E., Bojar, M., Żelazna, A., and Blicharz, P. (2012). Eco-management in polish companies. *Problems Sustainable Development* 7: 107–113. (b) Lee, M.D.P. et al. (2008). *International Journal of Management Reviews* 10: 53–73.

2 Mazurkiewicz, P. and Grenna, L. (2008). Corporate social responsibility and multistakeholder dialogue. Towards environmental behavioral change. In: *Społeczna odpowiedzialność przedsiębiorstw międzynarodowych*; Discussion Paper, May 2003 (ed. J. Nakonieczna). Warszawa, Poland: World Bank.

3 Ikerd, J. (2008). Sustainable capitalism: a matter of ethics and morality. *Problems Sustainable Development* 3: 13–22.

4 The Economist. The Good Company (A survey of corporate social responsibility). 2005. Available online: https://www.economist.com/sites/default/files/special-reports-pdfs/3555199.pdf (accessed on 01 September 2021).

5 Peloza, J. et al. (2009). *Journal of Management* 35: 1518–1541.

6 Wood, D.J. et al. (2010). *International Journal of Management Reviews* 12: 50–84.

7 Glavas, A. et al. (2016). *Frontiers in Psychology* 7: 2–5.

8 Donaldson, T., Preston, L.E. et al. (1995). *The Academy of Management Review* 20: 65–91.

9 Garde-Sanchez, R., López-Pérez, M.V., López-Hernández, A.M. et al. (2018). *Sustainability* 10: 2403.

10 Adamczyk, J. and Nitkiewicz, T. (2007). *Programowanie zrównoważonego rozwoju przedsiębiorstw.* Warsaw, Poland: Polskie Wydawnictwo Ekonomiczne.

11 Carroll, A.B. et al. (1979). *The Academy of Management Review* 497–505: [CrossRef].

12 Bojar, E. and Żelazna, A. (2008). Przejawy społecznej odpowiedzialności przedsiębiorstw względem środowiskanaturalnego. In: *Przedsiębiorstwo wobec wyzwań globalnych* (ed. A. Herman and K. Poznańska), 133–136. Warszawa, Poland: Szkoła Główna Handlowa w Warszawie—Oficyna Wydawnicza.

13 Zaleśna, A. (2019). *Kompetencje zarządzających w przedsiębiorstwach społecznie odpowiedzianych.* Warszawa, Poland: Wyd. Difin.

14 Baden, D. (2016). A reconstruction of Carroll's pyramid of corporate social responsibility for the 21st century. *International Journal of Corporate Social Responsibility* 1: 1–15.

15 Rybak, M. (2004). *Etyka menedżera—Społeczna odpowiedzialność przedsiębiorstw.* Warszawa, Poland: Wydawnictwo PWN.

16 Carroll, A.B. (2016). Pyramid of CSR: taking another look. *International Journal of Corporate Social Responsibility* 1: 1–8.

17 Dahlsrud, A. et al. (2008). *Corporate Social Responsibility and Environmental Management* 15: 1–13.

18 Ratajczak, M. (2020). *Społeczna odpowiedzialność mikro, małych i średnich przedsiębiorstw w sektorze agrobiznesu. Podejście modelowe.* Warszawa, Poland: Wyd SGGW.

19 Wiszczun, E. (2013). Koncepcja społecznej odpowiedzialności w świetle realizacji usług polityki społecznej. In: *Etyczny wymiar odpowiedzialnego biznesu i konsumeryzmu na początku XXI wieku* (ed. L. Karczewski and H. Kretek), 8–19. Racibórz, Poland: Państwowa Wyższa Szkoła Zawodowa w Raciborzu.

20 Hawken, P. (1993). *The Ecology of Commerce.* New York, NY, USA: Harper Collins.

21 Gossling, T., Vocht, C. et al. (2007). *Journal of Business Ethics* 74: 363–372.

22 Lawrence, A.T. and Weber, J. (2008). *Business and Society. Stakeholders, Ethics, Public Policy.* New York, NY, USA: Mc-Graw-Hill Irwin.

23 Miles, M.P., Munilla, L.S., and Darroch, J. (2008). Sustainable corporate entrepreneurship. *The International Entrepreneurship and Management Journal* 5: 65–76.

24 Gallego-Alvarez, I., Prado-Lorenzo, J.M., and Garcia-Sanchez, I.M. (2011). Corporate social responsibility and innovation: a resource based theory. *Management Decision* 49: 1709–1727.

25 Bojar, E. and Żelazna, A. (2013). Realizowanie zasad społecznej odpowiedzialności poprzez zrównoważoną produkcję i konsumpcję. In:

Paradygmat sieciowy, wyzwania dla teorii i praktyki zarządzania (ed. A. Karbownik). Gliwice, Poland: Wydawnictwo Politechniki Ś ląskiej.

26 Formisano, V., Quattrociocchi, B., Fedele, M. et al. (2018). *Sustainability* 10: 725.

27 El Akremi, A., Gond, J.P., Swaen, V. et al. (2018). *Journal of Management* 44: 619–657.

28 Cheema, S. and Javed, F. (2017). The effects of corporate social responsibility toward green human resource management: the mediating role of sustainable environment. *Cogent. Business and Management* 4: 1–8.

29 Strand, R. et al. (2013). *Journal of Business Ethics* 112: 721–734.

30 Porter, M.E., Kramer, R.K. et al. (2006). *Harvard Business Review* 78–93. https://hbr.org/2006/12/strate gy-and-society-the-link-between-competitive-advantage-and-corporate-social-responsibility (accessed on 01 September 2021).

31 Pfeffer, J. (2010). Building sustainable organizations: the human factor. *Academy of Management Perspectives* 24: 34–45.

32 Montiel, I. (2008). Corporate social responsibility and corporate sustainability: separate pasts, common futures. *Organization and Environment* 21: 245–269.

33 Dyllick, T., Hockerts, K. et al. (2002). *Business Strategy and the Environment* 11: 130–141.

34 Bansal, T. (2005). Evolving sustainably: a longitudinal study of corporate sustainable development. *Strategic Management Journal* 26: 197–218.

35 Krajnakova, E., Navickas, V., Kontautiene, R. et al. (2018). *Oeconomia Copernicana* 9: 477–492.

36 Seidel, E. (2005). Rachunek kosztów środowiskowych. In: *Międzynarodowe zarządzanie środowiskiem, Volume III: Operacyjne zarządzanie środowiskiem w aspekcie międzynarodowym i interdyscyplinarnym* (ed. M. Kramer, H. Strebel and L. Buzek), 1–18. Warszawa, Poland: C.H. Beck.

37 Kobyłko, G. (ed.) (2007). *Proekologiczne zarządzanie przedsiębiorstwem.* Wrocław, Poland: Wyd. Akademii Ekonomicznej im. Oskara Langego we Wrocławiu.

38 Tran, K.T., Nguyen, P.V. et al. (2020). *Sustainability* 12: 1044.

39 Surroca, J., Tribó, J.A., Waddock, S. et al. (2010). *Strategic Management Journal* 31: 463–490.

40 Peloza, J., Shang, J. et al. (2011). *Journal of the Academy of Marketing Science* 39: 117–135.

41 Öberseder, M., Schlegelmilch, B.B., Murphy, P.E. et al. (2014). *Journal of Business Ethics* 124: 101–115.

42 Neugebauer, F. et al. (2012). *Journal of Cleaner Production* 37: 249–256. https://pureadmin.qub.ac.uk/ws/portalfiles/portal/2532045/EMAS_a nd_ISO_14001_in_the_German_industry_complements_or_substitutes.pdf (sccessed on 02 September 2021).

43 Preziosi, M., Merli, R., D'Amico, M. et al. (2016). *Sustainability* 8: 191.

44 Musgrave, J. et al. (2011a). *Worldwide Hospitality and Tourism Themes* 3 (3): 1–5.

45 Musgrave, J. (2011b). Moving towards responsible events management. *Worldwide Hospitality and Tourism Themes* 3 (3): 258–274.

46 Copyright © 1996-2015 National Geographic Society nationalgeographic.com/environment/article/freshwater-crisis.

47 Copyright © Worldometer All Rights Reserved worldometers.info/world-population (accessed on 8 September 2021).

48 Copyright © 2021 Thomas Publishing Company. All Rights Reserved. thomasnet.com/insights/which-industries-use-the-most-water (accessed on 10 September 2021).

49 Beer, S. and Lemmer, C. (2011). A critical review of 'green' procurement: life cycle analysis of food products within the supply chain. *Worldwide Hospitality and Tourism Themes* 3 (3): 229–244.

50 Mysen, T. (2012). Sustainability as corporate mission and strategy. *European Business Review* 24 (6): 496–509.

51 Olson, E.G. (2008). Creating an enterprise-level "green" strategy. *Journal of Business Strategy* 29 (2): 22–30.

52 Paterson, M. and Ward, S. (2011). Roundtable discussion: applying sustainability legislation to events. *Worldwide Hospitality and Tourism Themes* 3 (3): 203–209.

53 Esquer-Peralta, J., Velazquez, L., and Munguia, N. (2008). Perceptions of core elements for sustainability management systems (SMS). *Management Decision* 46 (7): 1027–1038.

54 Ritchie, J.R.B. and Smith, B.H. (1991). The impact of a mega-event on host region awareness: a longitudinal study. *Journal of Travel Research* 30 (3): 3–10.

55 Moldavska, A. and Welo, T. (2019). A holistic approach to corporate sustainability assessment: Incorporating sustainable development goals into sustainable manufacturing performance evaluation. *Journal of Manufacturing Systems* 50: 53–68.

7

Adam Adaptive Optimization Method for Neural Network Models Regression in Image Recognition Tasks

Denis Y. Nartsev[1], Alexander N. Gneushev[1,2], and Ivan A. Matveev[2]

[1]*Moscow Institute of Physics and Technology, Department of Control and Applied Mathematics, Institutskiy per., 9, Dolgoprudny, Moscow Region 141701, Russia*
[2]*Federal Research Center "Computer Science and Control" of Russian Academy of Sciences, Dorodnicyn Computing Center, Vavilov str., 40, Moscow 119333, Russia*

7.1 Introduction

Currently, as the most effective approach for solving complex and poorly formalized problems, building models in the field of intelligent systems is the usage of neural network models and machine learning. Such models are applied to the tasks of automatic localization, segmentation, and recognition of objects, identifying a person by the images of a face, iris, hand, etc. In intelligent image analysis systems, image preprocessing is important for the stable extraction of object features, as well as for filtering those images where reliable object recognition cannot be achieved with the required reliability.

In particular, an important step for the task of identifying a person from a face image is preliminary image alignment. To do this, one needs to assess the position of the face in the image. Approaches to solving this problem can be divided into three types: cascade regression of the form [1], statistical models of the form [2–6], and neural network regression models [7, 8], evaluating the position of the face landmarks and the angle of rotation in the image plane.

Image quality indicators are determined by the measure of their usefulness for the problem of object recognition. For example, for personality recognition by the iris of the eye, filtering eye images by quality allows not only to build a reliable training base of images but also to reduce the level of identification errors during the operation of the system [9]. Approaches to assessing the quality of images can be divided into absolute and relative. Relative indicators characterize the severity of features of a particular class of objects [10–12] and can be used only for the

Artificial Intelligence in Industry 4.0 and 5G Technology, First Edition.
Edited by Pandian Vasant, Elias Munapo, J. Joshua Thomas, and Gerhard-Wilhelm Weber.

relative ranking of images. Absolute indicators do not depend on the structure of the objects in an image. To assess them, neural network regression models are used [13].

The usage of neural network models in the considered tasks is a versatile and effective solution. To build a working system the model needs to be trained. The training problem is usually formulated using the optimization of some quality functional, the functional of the error level of the outputs of the neural network regression model by its parameters or weights on the training dataset.

To optimize the quality functional the iterative gradient methods and their generalizations are commonly used. For example, to overcome local optima, the stochastic gradient descent method (SGDM) with moment [14] uses gradient filtration along with the iterations, which allows accumulating inertia of movement along with the functional error relief. Additionally, the L2-regularization method is used, which minimizes the norm of the model weights to avoid overfitting. The approach based on the estimation of the moving average value of the gradient during iterations is used to adapt the learning rate in Nesterov accelerated gradient method [15].

In the process of training, the gradient of the loss function for some weights of the model can become significantly less than for the rest of the weights, thus optimization stops for some of the components. In adaptive methods, the step of iterative descent is adjusted to the complexity of the loss function relief. It is increased for those components along which the average gradient is less than along others. This approach compensates for the heterogeneity of the training sample.

In the Adagrad algorithm [16], to adapt the convergence rate, the gradient descent step normalization is proposed with the accumulated sum of the loss function gradient components squares. A serious limitation of this approach is a significant decrease in the learning rate with an increase in the accumulated normalizing amount. To solve this problem, RMSprop [17] and Adadelta [18] methods use the mean estimate of the gradient components squares instead of their sum. In Adam's algorithm [19], in addition to the gradient descent step normalizing, to adapt the convergence rate exponential gradient averaging is used to overcome local optima, similar to the SGDM method.

Comparison of SGDM method and methods with adaptive step change, presented in [20], in various machine learning problems shows that the use of an adaptive step reduces the generalizing ability of the trained model. Some of the problems with these methods can be associated with the implementation of the corresponding step adaptation procedure. For example, in [21], a modification of the Adam algorithm was investigated. It was found that the addition to the gradient of the loss function associated with the weights L2-regularization affects the results of calculating its average characteristics. For the AdamW method [21] the authors propose to filter the gradient without this correction and add it directly to the iteration step.

The convergence of methods using different gradient moments of the loss function is very sensitive to the parameter initialization, and it depends upon the values of the corresponding moments at the beginning of training. For the initial estimation of the gradient moments, various approaches are used, for example, the Warmup method [22] in which the gradient descent step is, as a rule, a linearly increasing function of the iteration number. In the RAdam method [23] a nonlinear dependence of the gradient descent step normalization on the iteration number is specified to obtain stable average gradient estimates at the beginning of training.

We compare optimization methods, such as SGDM, Adam and their various modifications for training neural network models that solve some regression problems for image preprocessing. In particular, we estimate the face rotation angle for subsequent alignment in the problem of identification of the person and the estimation of the eye blurring degree for the problem of identification by the iris.

To solve these tasks, we consider the original images of objects as distorted by some transformation with a parameter that should be estimated. We consider the approach of direct estimation of these parameters by solving the image regression problem by training neural network models. For the problem of assessing the quality of the eye image, the estimated parameter is the root-mean-square deviation in the Gaussian low-frequency filtering model, and for the problem of assessing the face rotation, the estimated parameter is the angle of rotation. The corresponding regression models are constructed by training with the considered optimization methods on training samples obtained by modeling the corresponding distortion functions. Images obtained by modeling the corresponding distortions are used as test samples, on which the quality of the constructed models is determined for each method.

7.2 Problem Statement

The set of reference images is denoted as $\widetilde{I} \subset R^{CxHxV}$, where C denotes the number of color channels; H, W denote image height and width, respectively. The model of reference images distortion is defined as the transformation of images depending on the parameter $y \in Y$: $M_y : \widetilde{I} \to I, I \subset R^{CxHxV}$, where Y – given set of possible parameter values. A training set of distorted images is formed as

$$S = \{I_i\}_{i=1}^{N}, I_i = M_{y_i}(\widetilde{I}_i), \ \widetilde{I}_i \in \widetilde{I}, \ y_i \in Y$$

where N is the number of images; $\{y_i\}_{i=1}^{N}$ – set of parameter values, forming the markup for the training set S. There is considered a task of finding a neural network model $h(I \mid \boldsymbol{\theta})$ for the estimation of transformation M_y parameters by solving the regression problem on a training set of images S depending on the weights $\boldsymbol{\theta}$:

$$h(I_i \mid \boldsymbol{\theta}) : I_i \rightarrow y_i, I_i \in S, i = 1, \dots, N.$$

This problem is solved by minimizing the predictive loss function of the model on the training data by optimizing weights. The expression for the general loss function can be represented as

$$F(\boldsymbol{\theta}) = \frac{1}{N} \sum_{i=1}^{N} L(h(I_i \mid \boldsymbol{\theta}) - y_i), \tag{7.1}$$

where $L : R \rightarrow R$ – a convex function with a single global minimum at zero. Then, the problem of training the regression model has the form

$$\min_{\boldsymbol{\theta}} F(\boldsymbol{\theta}). \tag{7.2}$$

We will use the mean absolute error (MAE) function $L(x)$:

$$L_{\text{MAE}}(x) = |x|. \tag{7.3}$$

The article [24] reports that models trained using the WingLoss function achieved better results compared to models trained with other loss functions in problems of assessing the positions of key points on the face. The form of the WingLoss function with parameters p and ε is used for learning strategy, in which small and medium level errors have a greater weight:

$$L_{\text{WingLoss}}(x \mid w, \varepsilon, p) = \begin{cases} pw \ln(1 + |x|/\varepsilon), & |x| < w, \\ p(|x| - C(w, \varepsilon)), & |x| \geq w, \end{cases}$$

$$C(w, \varepsilon) = w - w \ln(1 + w/\varepsilon). \tag{7.4}$$

Consider the following tasks – building and training a neural network regression model $h(I \mid \boldsymbol{\theta})$ for two different image distortion functions M_y.

For the distorting transformation M_y in the task of assessing the blurring level of an eye image (Figure 7.1) we will use the Gaussian blur model [13], low-frequency filtering with Gaussian kernel with standard deviation σ, so y depends on σ:

$$I(m, n) = \frac{1}{2\pi\sigma} \sum_{u=-X}^{X} \sum_{v=-X}^{X} e^{\frac{-(u^2 + v^2)}{2\sigma^2}} \tilde{I}(m + u, n + u), \tag{7.5}$$

where $X = [3\sigma]$ – integer value of the radius of the Gaussian filter kernel size $(2X + 1) \times (2X + 1)$.

For the task of estimating the angle face orientation in an image as a transformation, M_y we will use a rotation by an angle α with bilinear interpolation.

Figure 7.1 Scheme of neural network model training to assess the blurring level of an eye image.

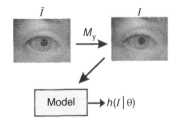

7.3 Modifications of the Adam Optimization Method for Training a Regression Model

Let's assume that the loss function $F(\theta)$ is differentiable. Then to solve the minimization problem (7.2) an iterative gradient descent method with exponential smoothing along iterations with the parameter can be used $0 < \beta < 1$ (gradient descent with an accumulation of moment):

$$\mathbf{m}_t = \beta \mathbf{m}_{t-1} + (1 - \beta)\nabla F(\theta_{t-1}),$$

$$\theta_t = \theta_{t-1} - \lambda_t \mathbf{m}_t.$$

where θ_0 – initial initialization of the model weights, determined at the beginning of the training process; λ_t – step on tth iteration of algorithm; $\nabla F(\theta)$ – gradient function $F(\theta)$ by model weights θ; $\mathbf{m}_t$ – accumulated moment $\nabla F(\theta)$. For parameter λ_t there is usually specified a stepwise decreasing of the iteration number function which is determined by the training schedule table.

This method makes it possible to overcome local minima for complex loss functions of large dimensions since even with a small value of the gradient at some optimization step, the value of the accumulated moment remains significant and allows the optimization process to continue.

The trained model $h(I \mid \theta)$ shows generalizing ability if the errors of its prediction on the test data $I \notin S$ are small enough or slightly exceed the errors on the training data, otherwise, the model becomes overfitted. In order to avoid overfitting, the model regularization method is applied [25], which consists of adding to the functional the error L_2 – norm of the weights:

$$F(\theta) = \frac{1}{n}\sum_{i=1}^{n} F_i(\theta) + \frac{R}{2}\|\theta\|_2^2,$$

where $F_i(\theta) = L(h(I_i \mid \theta) - y_i)$, R – regularization factor. An additional term $R\theta$ arises in the gradient $\nabla F(\theta)$, leading to a decrease in the weights norm.

To find the exact gradient $\nabla F(\theta)$ value, it is required to calculate the gradients of all functions F_i. If their number is large, then the SGDM is used for optimization, in which at each iteration step the value of the true gradient of the function $\nabla F(\theta)$

is approximated by the sum $\nabla F_i(\theta)$ at the point θ for a random subset of training examples (mini-batch), different at each step (Algorithm 7.1).

Algorithm 7.1   In the SGDM method, for vector values, operations are performed element by element.

---

1. $\mathbf{m}_0 = 0$, $t = 0$, parameters of the algorithm: β, R.
   Cycle. Repeat steps 2–5 until the stop criterion is met.
2. $t = t + 1$.
3. $\mathbf{g}_t = \nabla F_t(\theta_{t-1}) + R\theta_{t-1}$.
4. $\mathbf{m}_t = \beta \mathbf{m}_{t-1} + (1 - \beta)\mathbf{g}_t$.
5. $\theta_t = \theta_{t-1} - \lambda_t \mathbf{m}_t$.
6. The algorithm terminates.

Despite the smoothing of the gradient and the moment accumulation in the SGDM method, some of the weights components can be stabilized and stop changing during the optimization process, so the gradient descent along them stops. In the Adam method [19], to solve this problem, it is proposed to estimate the gradient spread along the descent trajectory and normalize the step of the next iteration with this indicator, so that the more gently sloping the loss function's relief, the larger step will be used for the next iteration. Using as such indicator the average estimate of the gradient components squares characterizing the relief curvature variation of the loss function, the Adam method can be represented in the form of Algorithm 7.2.

Algorithm 7.2   Adam's method [19], for vector values, operations are performed element by element.

---

1. $\mathbf{m}_0 = 0$, $\mathbf{v}_0 = 0$, $t = 0$, parameters of the algorithm: β_1, β_2, R.
   Cycle. Repeat steps 2–8 until the stop criterion is met.
2. $t = t + 1$.
3. $\mathbf{g}_t = \nabla F_t(\theta_{t-1}) + R\theta_{t-1}$.
4. $\mathbf{m}_t = \beta_1 \mathbf{m}_{t-1} + (1 - \beta_1)\mathbf{g}_t$.
5. $\mathbf{v}_t = \beta_2 \mathbf{v}_{t-1} + (1 - \beta_2)\mathbf{g}_t^2$.
6. $\mathbf{m}_t^* = \mathbf{m}_t / (1 - \beta_1^t)$.
7. $\mathbf{v}_t^* = \mathbf{v}_t / (1 - \beta_2^t)$.
8. $\theta_t = \theta_{t-1} - \lambda_t (\mathbf{m}_t^* / \sqrt{\mathbf{v}_t^*})$.
9. The algorithm terminates.

In practice, the SGDM method achieves better results than Adam if the schedule for decreasing the step in the process of iterations is correctly selected for the training set [26]. However, in [21], it is indicated that the unsatisfactory adaptation of the step in the Adam algorithm occurs due to the incorrect implementation of the estimate of the mean values of the gradient distribution along the descent. The addition to the gradient of the loss function associated with the weights L2-regularization distorts this estimate. A modified AdamW method is proposed, in which the regularization addition is transferred to the expression for the iteration step, and step 3 of Algorithm 2 takes the form $\mathbf{g}_t = \nabla F_t(\boldsymbol{\theta}_{t-1})$, step 8 is changed as follows:

$$\boldsymbol{\theta}_t = \boldsymbol{\theta}_{t-1} - \lambda_t \left(\mathbf{m}_t^* / \sqrt{\mathbf{v}_t^*} + R\boldsymbol{\theta}_{t-1} \right). \tag{7.6}$$

Methods such as Adam and AdamW which use the statistical estimation of gradients along a path have the problem of initializing the average estimates. The inadequacy of their values due to insufficient initial statistics at the beginning of training and, accordingly, incorrect normalization of the iteration step has a negative impact on the learning process. The usual solution to this problem is to use the preparatory stage of training using the Warmup method [22], in which the iteration step is initialized to a small value; therefore, it is increased in accordance with a given function during a specified number of iterations. Thus the average gradient values are initialized in the vicinity of the initial values of the model weights.

However, in practice, the result of the Warmup method depends significantly on the correct choice of the increasing step function type and the duration of the preparatory stage. In the RAdam method [23], the preparatory stage is not used, and the scale coefficient for the adaptive step is found, which has a non-linear dependence on the iteration number t:

$$r_t = \sqrt{\frac{(\rho_t - 4)(\rho_t - 2)\rho_\infty}{(\rho_\infty - 4)(\rho_\infty - 2)\rho_t}},$$

where $\rho_\infty = 2/(1 - \beta_2) - 1$, $\rho_t = \rho_\infty - 2t\beta_2^t / \left(1 - \beta_2^t\right)$. Coefficient r_t stabilizes gradient step normalization $1/\sqrt{\mathbf{v}_t}$. If conditions are met $0 < \beta_2 < 1$, $\rho_t > 4$, then the parameter r_t is a real number, monotonically increased by t and it tends to unity in the limit $t \to \infty$ [23]. Thus, to modify the AdamW method and obtain the RAdam algorithm, formula (7.6) takes the following form:

$$\boldsymbol{\theta}_t = \begin{cases} \boldsymbol{\theta}_{t-1} - \lambda_t \left(\mathbf{m}_t^* r_t / \sqrt{\mathbf{v}_t^*} + R\boldsymbol{\theta}_{t-1} \right), & \text{if } \rho_t > 4, \\ \boldsymbol{\theta}_{t-1} - \lambda_t \mathbf{m}_t^*, & \text{otherwise.} \end{cases} \tag{7.7}$$

Note that for the RAdam method the scale factor r_t is multiplied only by the normalized mean gradient of the loss function. Thus, the method does not take into account that the weight of the regularization term turns out to be more significant than the step along the gradient of the loss function. The regularization term defines the scale of the entire iteration step that remains unstabilized in the vicinity of the initially initialized weights. This fact makes it difficult to obtain an adequate estimate of the initial mean characteristics of the gradient and returns to the problem of the AdamW method. Moreover, this problem leads to an underestimation of the weights norm and strengthening of restrictions for the optimization process. So it leads to a deterioration in the convergence of the method. To solve this problem, the factor r_t is used to scale the entire iteration step, not only the normalized mean gradient component. Thus the regularization term is stabilized. Then formula (7.7) takes the form and defines the RAdamW method (see Algorithm 7.3):

$$\boldsymbol{\theta}_t = \begin{cases} \boldsymbol{\theta}_{t-1} - \lambda_t r_t \left(\mathbf{m}_t^* / \sqrt{\mathbf{v}_t^*} + R\boldsymbol{\theta}_{t-1} \right), & \text{if } \rho_t > 4, \\ \boldsymbol{\theta}_{t-1} - \lambda_t \mathbf{m}_t^*, & \text{otherwise.} \end{cases}$$

Algorithm 7.3   In the RAdamW method, for vector values, operations are performed element by element.

---

1. $\mathbf{m}_0 = 0$, $\mathbf{v}_0 = 0$, $t = 0$, parameters of the algorithm: β_1, β_2, R.
   Cycle. Repeat steps 2-11 until the stop criterion is met.
2. $t = t + 1$.
3. $\mathbf{g}_t = \nabla F_t(\boldsymbol{\theta}_{t-1})$.
4. $\mathbf{m}_t = \beta_1 \mathbf{m}_{t-1} + (1 - \beta_1)\mathbf{g}_t$.
5. $\mathbf{v}_t = \beta_2 \mathbf{v}_{t-1} + (1 - \beta_2)\mathbf{g}_t^2$.
6. $\mathbf{m}_t^* = \mathbf{m}_t / \left(1 - \beta_1^t\right)$.
7. $\mathbf{v}_t^* = \mathbf{v}_t / \left(1 - \beta_2^t\right)$
8. $\rho_\infty = 2/(1 - \beta_2) - 1$.
9. $\rho_t = \rho_\infty - 2t\beta_2^t / \left(1 - \beta_2^t\right)$.

10. $r_t = \sqrt{\dfrac{(\rho_t - 4)(\rho_t - 2)\rho_\infty}{(\rho_\infty - 4)(\rho_\infty - 2)\rho_t}}$.

11. $\boldsymbol{\theta}_t = \begin{cases} \boldsymbol{\theta}_{t-1} - \lambda_t r_t \left(\mathbf{m}_t^* / \sqrt{\mathbf{v}_t^*} + R\boldsymbol{\theta}_{t-1} \right), & \text{if } \rho_t > 4, \\ \boldsymbol{\theta}_{t-1} - \lambda_t \mathbf{m}_t^*, & \text{otherwise.} \end{cases}$
12. The algorithm terminates.

7.4 Computational Experiments

One of the requirements for recognition systems is real-time operation. In this case, the subtask of image preprocessing, being auxiliary, should consume a minimum of computational resources. Therefore, to solve the set tasks, we chose a relatively simple for modern computing platforms and widely used neural network model of the ResNet architecture [27] with a pre-activated layer (Full-pre-activation) [28] and 18 convolutional layers – ResNet18. We train the model with the formed training sample by solving the problem (7.2). For the optimization methods Adam, AdamW, RAdam and RAdamW the parameters $\beta_1 = 0.9$, $\beta_2 = 0.99$ are used and $\beta = 0.9$ is used for SGD.

7.4.1 Model for Evaluating the Eye Image Blurring Degree

Samples for the model training and testing are formed from open databases of eye images: BATH [29] и CASIA [30] (Figure 7.2). For each image, for distorting conversion M_y the value of the blurring filter kernel parameter $\sigma \in [0, 6]$ is chosen randomly, it determines the estimated parameter of the model $y = (\sigma - 3)/6$.

The size of the training sample based on the BATH database is 29,749 images. The training takes 130 epochs. The schedule of decreasing the iteration step with a multiplier of 0.1 by 50 and 100 epochs are used.

The size of the training sample based on the CASIA database is 5600 images. The training takes 2500 epochs. The schedule of decreasing the iteration step with a multiplier of 0.1 by 900 and at 1800 epochs is used.

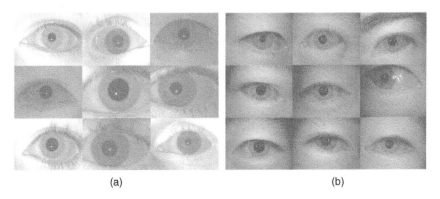

(a)　　　　　　　　　　　　　　　　(b)

Figure 7.2 Sample images from databases: (a) BATH; (b) CASIA.

For optimization (7.2), the MAE function (7.3) is used. For the Warmup method with a linear function, the number of preparatory learning epochs based on BATH is equal to two, for the CASIA base 10 epochs are set, the iteration step changed from 10^{-9} to 10^{-2} for the SGD method and to 10^{-4} for the Adam and AdamW methods. For optimization by the RAdam and RAdamW methods, the Warmup algorithm is not applied, the initial training step is equal 10^{-3}. The coefficient of L_2-regularization R for the Adam and AdamW methods for BATH-based training is equal 0.01; in other cases, it is set by a value 0.1. The mini-batch size is 128. These parameters show the best results in the tests.

The test sample based on the BATH database contains 2239 images. The test sample based on the CASIA database contains 2100 images. The accuracy of the trained models is assessed for each training epoch.

In Figure 7.3, the dependences of the MAE of the model on the number of the learning epoch by the Adam and AdamW methods on test bases are shown. The notation "wd" indicates the usage of L_2 – regularization in the method, the designation "Warmup" – the usage of the Warmup method. It can be seen from the graphs that when training on both bases by the AdamW method, the usage of the Warmup algorithm in addition to L_2 – regularization significantly reduces the model error. In contrast, for the Adam method, L_2 – regularization increases the error of the CASIA-based model. In further tests, algorithms showing the best accuracy are used for the considered methods.

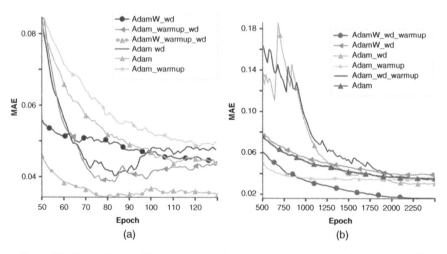

Figure 7.3 Dependence of the average model error on the epoch number when teaching by the Adam and AdamW methods: (a) accuracy based on BATH; (b) accuracy based on CASIA.

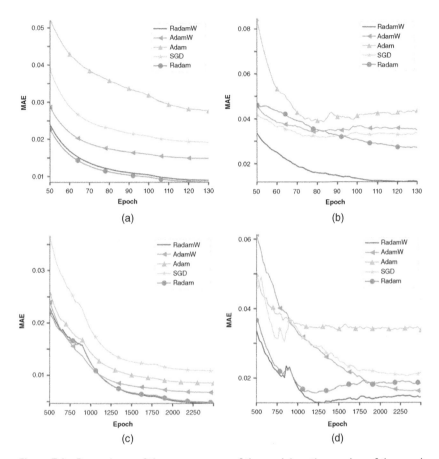

Figure 7.4 Dependence of the average error of the model on the number of the epoch when teaching by the considered methods: (a) for the BATH-based training; (b) for the testing on BATH test part with BATH-based training; (c) for the CASIA-based training; (d) for the testing on CASIA test part with CASIA-based training.

The dependences of MAE on the epoch number of the models trained by the SGDM, Adam, AdamW, RAdam, and RAdamW methods are shown in Figure 7.4. Training by the AdamW method converges faster than using the Adam and SGDM methods, while the model achieves half the error on BATH and 1.2 times the error on CASIA compared to Adam on the training set.

The test results and the accuracy of the corresponding models are presented in Table 7.1. On all test bases, the AdamW-trained model has comparable or less

Table 7.1 Accuracy of trained regression models of the eye images blurring degree estimation.

Algorithm	Training base	MAE on BATH	MAE on CASIA
SGD	BATH	0.034	0.052
Adam	BATH	0.039	0.068
AdamW	BATH	0.036	0.040
RAdam	BATH	0.028	0.034
RAdamW	**BATH**	**0.015**	**0.022**
SGD	CASIA	0.018	0.022
Adam	CASIA	0.039	0.036
AdamW	CASIA	0.017	0.016
RAdam	CASIA	0.013	0.016
RAdamW	**CASIA**	**0.012**	**0.014**

error than the SGDM-trained model and outperforms the Adam-trained models. However, the best result is obtained by the RAdam and RAdamW methods. The model trained by the RAdamW method on the BATH test base shows 1.8 times less error compared to training RAdam and 2.4 times less error compared to training AdamW.

In Figure 7.5 the distributions of the number of the example by the deviation of the trained model's predictions are shown. The model trained by the RAdamW method has an unbiased and narrow error distribution on all test samples. Estimates of models trained by other methods are biased and with greater variance.

7.4.2 Facial Rotation Angle Estimation Model

The training set consisted of 400,000 prepared face images from the DeepGlint database [31] (Figure 7.6). For each image, the value of the rotation angle $\alpha \in [-\pi/4, \pi/4]$ for the transformation M_y is chosen at random. The estimated parameter of the model is the normalized value $y = \frac{4\alpha}{\pi}$. The size of the mini-batch for training is 512, the iteration step is 0.01 with a 10-fold decrease at 100 and 200 epochs. For optimization (7.2), the WingLoss loss function (7.4) uses parameters $w = 0.03$, $\varepsilon = 0.03$, $p = 1$. The test set contains 35,000 face images.

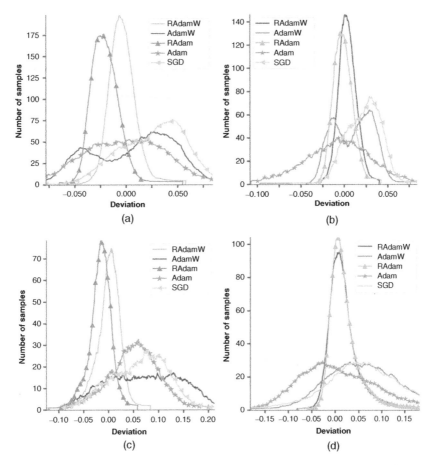

Figure 7.5 Distribution of the number of the example by the deviation of the trained model's predictions: (a) on the BATH test set when training on BATH; (b) on the CASIA test set when training on CASIA; (c) on the CASIA test set when training on BATH; (d) on the BATH test set when training on CASIA.

The accuracy of the model is estimated by the level of the mean square error (RMSE) of the model prediction and the true values of the angles. Table 7.2 shows the values of the standard deviation of the models trained by various optimization methods. The results show that the RAdamW method allows training the model with the lowest error compared to all other methods considered.

Figure 7.6   Examples of images from the DeepGlint database.

Table 7.2   The accuracy of a model trained by various optimization methods for the problem of determining the face rotation angle in an image.

Algorithm	R	RMSE (°)
Adam	0.1	2.13
Adam	0.0	2.10
AdamW	0.1	1.94
RAdam	0.1	1.92
RAdamW	**0.1**	**1.16**

7.5   Conclusion

There are considered adaptive methods of training a regression neural network model for direct parameters estimation in some image preprocessing problems. The criterion for the quality of the optimization method was the level of prediction errors for trained models on test samples. The results of the experiments demonstrate the advantage of the RAdam and RAdamW methods among the SGDM, Adam and AdamW algorithms for all the considered problems. Modification of RAdamW allows to reduce the average error of the trained model by more than 1.5 times in comparison with the RAdam method for assessing the level of eye image blurring. For the face rotation angle estimation problem the RAdamW method reduces the prediction error of the trained model by 65% compared to the RAdam method, which turned out to be less accurate than the AdamW algorithm. In general, modifications of the Adam method allow training a more accurate neural network regression model with less effort in selecting training parameters than the SGDM method.

Acknowledgments

The work is supported by the Russian Foundation of Basic Research, grant no. 19-07-01231.

References

1 Xiong X., De la Torre F. Supervised descent method and its applications to face alignment, *Proceedings of the IEEE Conference on Computer Vision and Pattern Recognition*, IEEE (Institute of Electrical and Electronic Engineers), 2013, pp. 532–539, https://doi.org/10.1109/CVPR.2013.75, https://ieeexplore.ieee .org/document/6618919.

2 Zhang, Z., Luo, P., Loy, C.C., and Tang, X. (2016). Learning deep representation for face alignment with auxiliary attributes. *IEEE Transactions on Pattern Analysis and Machine Intelligence* 38 (5): 918–930.

3 Gneushev, A.N. (2007). Construction and optimization of a texture–geometric model of a face image in the space of basic gabor functions. *Journal of Computer and Systems Sciences International* 46 (3): 418–428.

4 Novik, V., Matveev, I.A., and Litvinchev, I. (2020). Enhancing iris template matching with the optimal path method. *Wireless Networks* 26 (7): 4861–4868.

5 Matveev, I. (2010). Detection of iris in image by interrelated maxima of brightness gradient projections. *Applied and Computational Mathematics.* 9 (2): 252–257.

6 Desyatchikov, A.A., Kovkov, D.V., Lobantsov, V.V. et al. (2006). A system of algorithms for stable human recognition. *Journal of Computer and Systems Sciences International.* 45 (6): 958–969.

7 Sun Y., Wang X., Tang X.. Deep convolutional network cascade for facial point detection, *Proceedings of the IEEE Conference on Computer Vision and Pattern Recognition*, IEEE (Institute of Electrical and Electronic Engineers), 2013, pp. 3476–3483, https://doi.org/10.1109/CVPR.2013.446, https://ieeexplore.ieee .org/document/6619290.

8 Bulat A., Tzimiropoulos G. Binarized convolutional landmark localizers for human pose estimation and face alignment with limited resources, *Proceedings of the IEEE International Conference on Computer Vision (ICCV)*, IEEE (Institute of Electrical and Electronic Engineers), 2017, pp. 3726–3734, https://doi .org/10.1109/ICCV.2017.400, https://ieeexplore.ieee.org/document/8237662.

9 Gneushev, A.N., Kovkov, D.V., Matveev, I.A., and Novik, V.P. (2015). Optimizing the selection of a biometric template from a sequence. *Journal of Computer and Systems Sciences International* 54 (3): 399–405.

10 Monich, Y., Starovoitov, V. et al. (2008). *Artificial Intelligence* 4: 376–386. (in Russian).

11 Matveev, I.A., Novik, V., and Litvinchev, I. Influence of degrading factors on the optimal spatial and spectral features of biometric templates. *Journal of Computational Science* 2018, 25: 419–424.

12 Lobantsov, V.V., Matveev, I.A., and Murynin, A.B. (2011). Method of multimodal biometric data analysis for optimal efficiency evaluation of recognition algorithms and systems. *Pattern Recognition and Image Analysis.* 21 (2): 494–497.

13 Bosse S., Maniry D., Wiegand T. A deep neural network for image quality assessment, *Proceedings of the IEEE International Conference on Image Processing*, IEEE (Institute of Electrical and Electronic Engineers), 2016, pp. 3773-3777, https://doi.org/10.1109/ICIP.2016.7533065, https://ieeexplore.ieee .org/document/7533065.

14 Polyak, B.T. (1964). Some methods of speeding up the convergence of iteration methods. *USSR Computational Mathematics and Mathematical Physics* 4 (5): 1–17. (in Russian).

15 Nesterov, Y. (1983). A method for unconstrained convex minimization problem with the rate of convergence $o(1/k^2)$. *Proceedings of the USSR Academy of Sciences* 269 (3): 543–547. (in Russian).

16 Duchi, J., Hazan, E., and Singer, Y. (2010). Adaptive subgradient methods for online learning and stochastic optimization. *Journal of Machine Learning Research* 12: 2121–2159.

17 Mukkamala, M.C. and Hein, M. (2017). Variants of RMSProp and Adagrad with logarithmic regret bounds. In: *Proceedings of the 34th International Conference on Machine Learning, Proceedings of Machine Learning Research*, vol. 70, 2545–2553. https://dl.acm.org/doi/10.5555/3305890.3305944.

18 Zeiler M.D. (2012) ADADELTA: An Adaptive Learning Rate Method, arXiv preprint arXiv: 1212.5701.

19 Diederik P. Kingma, Jimmy Ba. (2014) Adam: a method for stochastic optimization, *In Conference Track Proceedings of 3rd International Conference on Learning Representations (ICLR), San Diego*, 2015, arXiv preprint arXiv:1412.6980, 1–15, https://dblp.org/rec/conf/iclr/2015.html?view=bibtex; https://www.scimagojr.com/journalsearch.php?q=21100924741&tip=sid& clean=0.

20 Wilson, A.C., Roelofs, R., Stern, M. et al. (2017). The marginal value of adaptive gradient methods in machine learning. *Advances in Neural Information Processing Systems* 30: 4148–4158.

21 Loshchilov, I. and Hutter, F. (2019). Decoupled weight decay regularization. In: *Proceedings of 7th International Conference on Learning Representations (ICLR), New Orleans*, 1–19, available at: https://openreview.net/pdf?id=Bkg6RiCqY7. https://dblp.org/rec/conf/iclr/LoshchilovH19.html?view=bibtex.

22 Vaswani, A., Shazeer, N., and Uszkoreit, J. (2017). Attention is all you need. *Advances in Neural Information Processing Systems* 30: 5998–6008.

23 Liu, L., Jiang, H., He, P. et al. (2020). On the variance of the adaptive learning rate and beyond. In: *Proceedings of 8th International Conference on Learning Representations, Addis Ababa*, 1–14, available at: https://openreview.net/pdf?id=rkgz2aEKDr. https://dblp.org/rec/conf/iclr/LiuJHCLG020.html?view=bibtex.

24 Feng Z., Kittler J., Awais M., Huber P., Wu X. (2018) Wing loss for robust facial landmark localisation with convolutional neural networks, *In Proceedings of the IEEE/CVF Conference on Computer Vision and Pattern Recognition*, IEEE (Institute of Electrical and Electronic Engineers), pp. 2235–2245, https://doi.org/10.1109/CVPR.2018.00238, https://ieeexplore.ieee.org/document/8578336.

25 Girosi, F., Jones, M., and Poggio, T. (1995). Regularization theory and neural networks archtectures. *Neural Computation* 7 (2): 19–269.

26 Krizhevsky, A. (2009). Learning Multiple Layers of Features from Tiny Images, Technical Report, University of Toronto, pp. 1-58.

27 He K., Zhang X., Ren S., Sun J. Deep residual learning for image recognition, *Proceedings of the IEEE Conference on Computer Vision and Pattern Recognition (CVPR)*, IEEE (Institute of Electrical and Electronic Engineers), 2016, pp. 770–778, https://doi.org/10.1109/CVPR.2016.90, https://ieeexplore.ieee.org/document/7780459.

28 He K., Zhang X., Ren S., Sun J. Identity mappings in deep residual networks, Leibe B., Matas J., Sebe N., Welling M. *Computer Vision – ECCV 2016. ECCV 2016. Lecture Notes in Computer Science*, Springer, Cham, vol 9908, 2016, pp 630–645, https://doi.org/10.1007/978-3-319-46493-0_38, https://link.springer.com/chapter/10.1007/978-3-319-46493-0_38.

29 Iris Image Database [Electronic resource], University of Bath: http://www.bath.ac.uk/eleceng/research/sipg/irisweb/ (accessed 2005).

30 Iris Image Database, Version 3 [Electronic resource], Chinese Academy of Sciences Institute of Automation: http://www.cbsr.ia.ac.cn/IrisDatabase.htm (accessed 2005).

31 Trillionpairs [Electronic resource], DeepGlint: http://trillionpairs.deepglint.com (accessed 2018).

8

Application of Integer Programming in Allocating Energy Resources in Rural Africa

Elias Manopo

Department of Statistics and Operations Research, School of Economic Sciences, North West University, Mahikeng, South Africa

8.1 Introduction

The quadratic assignment problem (QAP) can be defined as the problem whereby a set of facilities are allocated to a set of locations in such a way that the cost is a function of the distance and flow between the facilities. In this problem, the costs are associated with a facility being placed at a certain location. The objective is to minimize the assignment of each facility to a location. To date, the QAP has been believed to be very difficult, and heuristics were believed to be the methods of choice as given in Munapo [1]. Various heuristics have been developed, for example, Hahn and Grant [2], Ramakrishnan et al. [3], and Drezner [4]. For more developments in solving QAP, you may see Mohamed et al. [5], Adams and Johnson [6], Cela [7], Nagarajan and Sviridenko [8], Rego et al. [9], Xia [10] and Yang et al. [11].

8.1.1 Applications of the QAP

In addition to allocating resources and having importance in decision fame work, the QAP can also be used in numerical analysis, dartboard construction, archaeology, statistical analysis, reaction chemistry, economic problem modeling, hospital lay-out, backboard wiring problem, and campus planning model. These are briefly explained as follows.

Numerical analysis. The QAP can be used to generate initial permutations for the least-squares unidimensional scaling of symmetric proximity matrices. This method is used in numerical analysis.

Artificial Intelligence in Industry 4.0 and 5G Technology, First Edition.
Edited by Pandian Vasant, Elias Munapo, J. Joshua Thomas, and Gerhard-Wilhelm Weber.

Dartboard construction. A dart game is a game played by throwing darts at a circular board with numbered spaces. The problem of locating the numbers around a dartboard optimally can be formulated as a QAP. The same problem can also be modeled as a traveling salesman problem.

Chemical reactions. In some chemical reactions, some chemical processes and complexes can be modeled as a QAP. The solution to this problem can be modeled as a QAP, and its optimal solution gives the optimal chemical process or optimal chemical reaction.

Computer backboard wiring. The objective of computer backboard wiring is to minimize the total length of wiring used for the interconnection of components. A backboard is made of electronic components which require wiring. If the wiring is minimized, we save wiring costs and an improved computational time.

Archaeology. In archaeology, various forms of data can be collected from all angles of archeology. There are also so many ways that are used to analyze these forms of data. Qualitative, quantitative, and mixed methods are some of these ways. In addition, various types of these data forms can be expressed as mathematical or economic models. The QAP is one such economic model.

Hospital Layout. This problem comprises of assigning locations or departments to facilities or clinics in a way that minimizes the total distance that is traveled by patients. Some important factors such as human and economic factors are considered when planning a hospital.

8.2   Quadratic Assignment Problem Formulation

There are so many mathematical formulations for QAP. In this paper, we use the linear form proposed by Munapo in 2012 [1]. This linear form is an extension of the formulation introduced by Koopmans and Beckmann in 1957 [12]. In this formulation, we assume that new buildings are to be placed on a piece of land and n sites have been identified as sites for the buildings. We also assume that each building has a special function.

8.2.1   Koopmans–Beckmann Formulation

Let

a_{ij} be the walking distance between sites i and j.
b_{kl} be the number of people per week who circulate between buildings k and l.

Then, the Koopmans–Beckmann formulation of the QAP is given as follows:

$$
\begin{aligned}
\textit{Maximize} \quad & Z = \sum_{i=1}^{n} \sum_{j=1}^{n} \sum_{k=1}^{n} \sum_{l=1}^{n} a_{ij} b_{kl} x_{ik} x_{jl} + \sum_{i=1}^{n} \sum_{k=1}^{n} c_{ik} x_{ik}, \\
\textit{Such that :} \quad & \sum_{i=1}^{n} x_{ij} = 1, \quad 1 \leq j \leq n, \\
& \sum_{j=1}^{n} x_{ij} = 1, \quad 1 \leq i \leq n, \\
& x_{ij} \in \{0,1\}, \quad 1 \leq i \leq n, \quad 1 \leq j \leq n.
\end{aligned}
$$

$$(8.1)$$

In this formula, there are n^2 variables and $2n$ constraints [12].

8.3 Current Linearization Technique

The current technique can linearize the Koopmans–Beckmann model to the form given in (8.2).

$$
\begin{aligned}
\textit{Maximize} \quad & Z = \sum_{i=1}^{n} \sum_{j=1}^{n} \sum_{k=1}^{n} \sum_{l=1}^{n} a_{ij} b_{kl} y_{ijkl}, \\
\textit{Such that :} \quad & \sum_{i=1}^{n} \sum_{j=1}^{n} \sum_{k=1}^{n} \sum_{l=1}^{n} y_{ijkl} = n^2, \\
& x_{ik} + x_{jl} \geq 2 y_{ijkl}, \forall i, j, k, l, \\
& y_{ij} \geq x_{ik} + x_{jl} - 1.
\end{aligned}
$$

$$(8.2)$$

Solving this, linearized QAP model becomes very difficult as n increases in size. This linearized model has $(n^4 + n^2)$ variables and $O(n^4)$ constraints. This is very difficult to manage as n becomes large.

8.3.1 The General Quadratic Binary Problem

The Koopmans–Beckmann is a special case of a quadratic binary problem.
    Let a general case of the quadratic binary problem be represented in (8.3a).

$$
\begin{aligned}
\textit{Maximize} \quad & Z = \sum_{i=1}^{n} \sum_{j=1}^{n} c_{ij}^{0} x_i x_j + \sum_{k}^{n} c_{k}^{1} x_k, \\
\textit{Such that :} \quad & a_{11} x_1 + a_{12} x_2 + \cdots + a_{1n} x_n \leq b_1, \\
& a_{21} x_1 + a_{22} x_2 + \cdots + a_{2n} x_n \leq b_2, \\
& \qquad\qquad \cdots \\
& a_{m1} x_1 + a_{m2} x_2 + \cdots + a_{mn} x_n \leq b_m.
\end{aligned}
$$

$$(8.3a)$$

where

a_{ij}, b_i, c_{ij}^{0} and c_{k}^{1} are constants, $1 \leq i \leq m, \quad 1 \leq j \leq n$,
$x_i, x_j, x_k \in \{0,1\}, 1 \leq i \leq n, \ 1 \leq j \leq n, \ 1 \leq k \leq n$.

Budgetary constraints are given in (8.3b).

$$a_{11}x_1 + a_{12}x_2 + \cdots + a_{1n}x_n \le b_1,$$
$$a_{21}x_1 + a_{22}x_2 + \cdots + a_{2n}x_n \le b_2,$$
$$\cdots$$
$$a_{m1}x_1 + a_{m2}x_2 + \cdots + a_{mn}x_n \le b_m. \tag{8.3b}$$

The budgetary constraints give the amounts that are available for the construction of the gas fuel site. These constraints also give the minimum amount of money that is required for the fuel gas sites. These constraints depend on a country.

In this case, there are n cites. The total distance traveled by people around the site is given as the coefficient of that site or combined sites.

This model has application in the allocation of gas stations in rural African villages. Most African governments are now banning the use of firewood as a measure to curb deforestation. People in these areas are being asked to use gas stoves and there is a need to make the fuel gas available to these remote villages. The problem of locating the fuel gas stations in rural areas in such a way that the total distance traveled by the villagers is minimized can be modeled as a QAP. The QAP is given in (8.3a). In this model, the optimal solution gives the minimal total distance traveled by all the villagers to get fuel gas. In developing this model, paths giving the shortest distance to these fuel stations are used. People walk or use bicycles or donkeys on these shortest paths, and it may not be possible with cars. People with cars use the long winding road to access these fuel gas stations. In addition to shortest paths, village or area populations are used in the modeling process. The coefficient of a variable or product of variables gives the average total distance traveled around a site or two sites.

The variables $x_i x_j$ where $i = j$.

If $i = j$ then $x_i^2 = x_j^2$. For binary integer variables, we have the following.

$$x_i(x_i - 1) = 0,$$
$$x_i^2 - x_i = 0,$$
$$x_i = x_i^2. \tag{8.4}$$

Thus x_i^2 can be replaced by x_i in the objective function. Similarly, x_j^2 can also be replaced by x_j in the objective function. Note that this substitution on its own does not change the number of variables in the problem.

The variables $x_i x_j$ where $i \ne j$.

If $i \ne j$ then in the worst case there are $\frac{n(n-1)}{2}$ combinations of such variables in the objective function.

Justification: Suppose that we have:

Two variables. (x_1 and x_2) then in the worst case we can have x_1x_2 as the only possible combination of variables.

Three variables. (x_1x_2 and x_3) then in the worst case we can have x_1x_2, x_1x_3 and x_2x_3 as the possible combinations of variables. Thus these three variables give 3 possible combinations.

n variables. ($x_1, x_2, \ldots, x_{n-1}$ and x_n) then in the worst case we can have $x_1x_2, x_1x_3,$ $\ldots, x_1x_n, x_2x_3, x_2x_4, \ldots, x_2x_n, \ldots, x_{n-1}x_n$ as the possible combinations. The n variables give $(n-1) + (n-2) + \cdots + 1 = \sum_1^{n-1} t = \frac{n(n-1)}{2}$ possible combinations.

8.3.2 Linearizing the Quadratic Binary Problem

The variable combinations x_ix_j where $i \neq j$ must be removed in order to make the objective function linear. This is done by using the following substitution.

8.3.2.1 Variable Substitution

Let

$$x_ix_j = \delta_r, \tag{8.5}$$

where δ_r is also a binary variable and $r = 1, 2, \ldots, \frac{n(n-1)}{n}$.

Such that:

$$\left.\begin{array}{c} x_i + x_j = 2\delta_r + \overline{\delta}_r, \\ \delta_r + \overline{\delta}_r \leq 1, \\ \delta_r, \overline{\delta}_r \in \{0,1\} \text{ and } r = 1, 2, \ldots, \frac{n(n-1)}{n}. \end{array}\right\} \tag{8.6}$$

8.3.2.2 Justification

We have to show that the solution space $\Omega(x_ix_j) = \{0, 1\}$ is also the solution space for $\Omega(\delta_r)$, every point in $\Omega(x_ix_j)$ has a corresponding point in $\Omega(\delta_r)$ and that $x_ix_j = \delta_r$ for all corresponding points.

Solution Space for x_ix_j *i.e.* $\Omega(x_ix_j)$

$$\left.\begin{array}{l} x_i = 0 \text{ and } x_j = 0, \\ x_i = 1 \text{ and } x_j = 0, \\ x_i = 0 \text{ and } x_j = 1, \end{array}\right\} x_ix_j = 0.$$

$$x_i = 1 \text{ and } x_j = 1, \Rightarrow x_ix_j = 1.$$

$$\therefore \Omega(x_ix_j) = \{0,1\}.$$

Solution Space for δ_r i.e. $\Omega(\delta_r)$

$$\delta_r = 1 \text{ and } \bar{\delta}_r = 0 \Rightarrow x_i + x_j = 2 \Rightarrow x_i = x_j = 1 \Rightarrow x_i x_j = 1.$$

$$\delta_r = 0 \text{ and } \bar{\delta}_r = 1 \Rightarrow x_i + x_j = 1,$$

$$\Rightarrow \begin{cases} \text{Either} & x_i = 1 \text{ and } x_j = 0 \Rightarrow x_i x_j = 0. \\ \text{Or} & x_i = 0 \text{ and } x_j = 1 \Rightarrow x_i x_j = 0. \end{cases}$$

$$\delta_r = 0 \text{ and } \bar{\delta}_r = 0 \Rightarrow x_i + x_j = 0 \Rightarrow x_i = x_j = 0 \Rightarrow x_i x_j = 0.$$

$$\therefore \Omega(\delta_r) = \{0,1\}.$$

Corresponding Points

Point in $\Omega(x_i x_j)$	Corresponding point in $\Omega(\delta_r)$
$x_i = 0$ and $x_j = 0$,	$\delta_r = 0$ and $\bar{\delta}_r = 0$,
$x_i = 1$ and $x_j = 0$,	$\delta_r = 0$ and $\bar{\delta}_r = 1$,
$x_i = 0$ and $x_j = 1$,	$\delta_r = 0$ and $\bar{\delta}_r = 1$,
$x_i = 1$ and $x_j = 1$,	$\delta_r = 1$ and $\bar{\delta}_r = 0$.

8.3.3 Number of Variables and Constraints in the Linearized Model

Two extra variables are added to every product of variables $x_i x_j$ where $i \neq j$, that appear in the objective function. In the general case of the quadratic binary problem, there are $\frac{n(n-1)}{2}$ products as shown in Section 8.2. Thus there are

$$2 \times \frac{n}{2}(n-1) = n(n-1) \text{ new variables.} \tag{8.7}$$

This gives a total of $n(n-1)$ new variables $+ n$ original variables $= n^2$ variables.

Also two extra constraints are added for every product of variables $x_i x_j$ where $i \neq j$ that appears in the objective function. The total number of new constraints is given by (8.8).

$$2 \times \frac{n}{2}(n-1) = n(n-1) \text{ constraints.} \tag{8.8}$$

The total number constraints $(\overline{m})$ is given by,

$$\overline{m} = m \text{ original constraints} + n(n-1) \text{ original constraints}$$
$$= (n^2 + m - n) \text{ variables.} \tag{8.9}$$

8.3.4 Linearized Quadratic Binary Problem

Then linearized model becomes as given in (8.10).

$$
\left.
\begin{aligned}
\text{Minimize} \quad & Z = \sum_{r=1}^{\frac{n}{2}(n-1)} \overline{c}_r^0 \delta_r + \sum_i^n \overline{c}_k^{-1} x_k, \\
\text{Such that :} \quad & a_{11}x_1 + a_{12}x_2 + \cdots + a_{1n}x_n \leq b_1, \\
& a_{21}x_1 + a_{22}x_2 + \cdots + a_{2n}x_n \leq b_2, \\
& \qquad\qquad\qquad \cdots \\
& a_{m1}x_1 + a_{m2}x_2 + \cdots + a_{mn}x_n \leq b_m,
\end{aligned}
\right\}
\tag{8.10}
$$

$$x_i + x_j = 2\delta_r + \overline{\delta}_r, \forall i \neq j,$$

$$\delta_r + \overline{\delta}_r \leq 1, \forall i \neq j,$$

$$x_i, x_j, x_k \in \{0,1\}, \quad 1 \leq i \leq n, \quad 1 \leq j \leq n, \quad 1 \leq k \leq n,$$

$$\delta_r, \overline{\delta}_r \in \{0,1\} \quad \text{and} \quad r = 1,2,\ldots,\frac{n(n-1)}{n}.$$

8.3.5 Reducing the Number of Extra Constraints in the Linear Model

As n becomes large, solving a linear model with $n(n-1)$ extra constraints increases in complexity. The number of extra constraints can be reduced by half. The following constraints can be combined into one.

$$x_i + x_j = 2\delta_r + \overline{\delta}_r.$$

$$\delta_r + \overline{\delta}_r \leq 1.$$

The first constraint can be expressed as given in (8.11, 8.12).

$$x_i + x_j = \delta_r + \delta_r + \overline{\delta}_r. \tag{8.11}$$

$$x_i + x_j - \delta_r = \delta_r + \overline{\delta}_r. \tag{8.12}$$

Since $\delta_r + \overline{\delta}_r$ cannot exceed one as given in (8.13),

$$x_i + x_j - \delta_r \leq 1. \tag{8.13}$$

This reduces the number of extra constraints and variables to $\frac{n(n-1)}{2}$.

8.3.6 The General Binary Linear (BLP) Model

Let any BLP model be represented by (8.14).
  Minimize CX^T,

$$AX^T \geq B^T, X^T \leq I^T, X^T \geq 0, \text{Where } I = \begin{pmatrix} 1 & 1 & \cdots & 1 \end{pmatrix},$$

$$A = \begin{pmatrix} a_{11} & \cdots & a_{1n} \\ \cdots & \cdots & \cdots \\ a_{m1} & \cdots & a_{mn} \end{pmatrix}, B = \begin{pmatrix} b_1 & b_2 & \cdots & b_m \end{pmatrix}., \tag{8.14}$$

$$C = (c_1, c_2, \ldots, c_n), X = \begin{pmatrix} x_1 & x_2 & \cdots & x_n \end{pmatrix}. C^T \geq 0. \text{ If } c_j > 0 \text{ then } c_j \geq 1. \tag{8.15}$$

8.3.6.1 Convex Quadratic Programming Model

Let a quadratic programming problem be represented by (8.15).

$$\text{Minimize } f(X) = CX^T + \frac{1}{2}XQX^T,$$

$$\text{Such that } : AX^T \geq B^T,$$

$$X^T \leq I^T,$$

$$X^T \geq 0. \tag{8.16}$$

where $Q = \begin{pmatrix} q_{11} & \cdots & q_{1n} \\ \cdots & \cdots & \cdots \\ q_{n1} & \cdots & q_{nn} \end{pmatrix}.$

We assume that:

(i) matrix Q is symmetric and positive definite,
(ii) function $f(X)$ is strictly convex,
(iii) since constraints are linear, the solution space is convex,
(iv) any maximization quadratic problem can be changed into a minimization and vice versa.

If $f(X)$ is strictly convex for all points in the convex region, the local minimum is also the global minimum of the quadratic problem [13].

8.3.6.2 Transforming Binary Linear Programming (BLP) Into a Convex/Concave Quadratic Programming Problem

Let $S = \begin{pmatrix} s_1 & s_2 & \cdots & s_n \end{pmatrix}.$
Such that: $X^T + S^T = I^T$ and $S^T \geq 0.$
The convex quadratic objective function then becomes as given in (8.17).

$$f(\overline{X}) = \ell_1 \left(c_1 x_1^2 + c_2 x_2^2 + \cdots + c_n x_n^2 \right) + \left(s_1^2 + s_2^2 + \cdots + s_n^2 \right)$$
$$+ \ell_2 (x_1 s_1 + x_2 s_2 + \cdots + x_n s_n). \tag{8.17}$$

In matrix form, it simplifies to (8.18).

$$f(\overline{X}) = \ell_1 CXX^T + SS^T + \ell_2 XS^T. \tag{8.18}$$

The BLP becomes the convex quadratic problem given in (8.19).

$$\text{Minimize } f(\overline{X}) = \ell_1 CXX^T + SS^T + \ell_2 XS^T,$$

Such that : $AX^T \geq B^T$,

$$X^T + S^T = I^T. \tag{8.19}$$

where ℓ_1 and ℓ_2 are very large in terms of their sizes compared to any of the coefficients in the objective function. For this to work, the following must be satisfied.

$$\ell_1 << \ell_2. \tag{8.20}$$

$$\ell_1 = 1000(|c_1| + |c_2| + \cdots + |c_n|). \tag{8.21}$$

$$\ell_2 = 1\,000\,000(|c_1| + |c_2| + \cdots + |c_n|). \tag{8.22}$$

Note that the weights of $1\,000$ and $1\,000\,000$ depend on the sizes of the coefficients of the problem. For some small binary linear problems with small coefficients, weights of 10 and 1000 can be used effectively.

Enforcer $\ell_2(x_1 s_1 + x_2 s_2 + \cdots + x_n s_n)$.

Since this is a minimization quadratic objective function, the objective function will be minimal when (8.23) is satisfied.

$$\ell_2(x_1 s_1 + x_2 s_2 + \cdots + x_n s_n) = 0. \tag{8.23}$$

$$\text{i.e.} x_1 s_1 + x_2 s_2 + \cdots + x_n s_n = 0. \tag{8.24}$$

This is possible when either $x_j = 0$ or $s_j = 0$. Munapo [14] termed the expression in (8.23) an enforcer since it forces the variables to assume only binary values.

8.3.6.3 Equivalence

$$c_1 x_1 + c_2 x_2 + \cdots + c_n x_n = c_1 x_1^2 + c_2 x_2^2 + \cdots + c_n x_n^2. \tag{8.25}$$

The two quantities are equal if $x_j = 0$ or $x_j = 1$.

$$s_1 + s_2 + \cdots + s_n = s_1^2 + s_2^2 + \cdots + s_n^2. \tag{8.26}$$

Similarly the two quantities are equal if $s_j = 0$ or $s_j = 1$.

Convexity of $f(\overline{X})$ Since $f(\overline{X}) = \ell_1 \left(c_1 x_1^2 + c_2 x_2^2 + \cdots + c_n x_n^2\right) + (s_1^2 + s_2^2 + \cdots + s_n^2) + \ell_2(x_1 s_1 + x_2 s_2 + \cdots + x_n s_n)$, then this function $f(\overline{X}) = f(x_1, x_2, \ldots, x_n, s_1, s_2, \ldots, s_n)$ is convex if and only if it has second-order partial derivatives for each point $\overline{X} = (x_1, x_2, \ldots, x_n, s_1, s_2, \ldots, s_n) \in S$ and for each $\overline{X}' \in S$, all principal minors of the Hessian matrix are none negative.

Proof

In this case

$$f(\overline{X}) = \ell_1 \left(c_1 x_1^2 + c_2 x_2^2 + \cdots + c_n x_n^2\right) + \left(s_1^2 + s_2^2 + \cdots + s_n^2\right)$$
$$+ \ell_2(x_1 s_1 + x_2 s_2 + \cdots + x_n s_n)$$

This has continuous second-order partial derivatives and the $2n$ by $2n$ Hessian matrix is given by (8.27).

$$H(x_1, x_2, \ldots, x_n, s_1, s_2, \ldots, s_n) = \begin{bmatrix} 2\ell_1 c_1 & 0 & \cdots & 0 & 0 & 0 & \cdots & 0 \\ 0 & 2\ell_1 c_2 & \cdots & 0 & 0 & 0 & \cdots & 0 \\ & & \cdots & & & & & \\ 0 & 0 & \cdots & 2\ell_1 c_n & 0 & 0 & \cdots & 0 \\ 0 & 0 & \cdots & 0 & 2 & 0 & \cdots & 0 \\ 0 & 0 & \cdots & 0 & 0 & 2 & \cdots & 0 \\ & & \cdots & & & & & \\ 0 & 0 & \cdots & 0 & 0 & 0 & \cdots & 2 \end{bmatrix}$$

(8.27)

Since all principal minors of $H(x_1, x_2, \ldots, x_n, s_1, s_2, \ldots, s_n)$ are non-negative, $f(x_1, x_2, \ldots, x_n, s_1, s_2, \ldots, s_n)$ is convex. See [14] for more on convex functions.

Note that $\overline{X} H \overline{X}^T \geq 0, \forall \overline{X}^T \geq 0$. Thus the matrix H is symmetric and positive definite.

The binary solution that minimizes $f(\overline{X})$ also minimizes $c_1 x_1 + c_2 x_2 + \cdots + c_n x_n$

From $f(\overline{X}) = \ell_1 \left(c_1 x_1^2 + c_2 x_2^2 + \cdots + c_n x_n^2 \right) + \left(s_1^2 + s_2^2 + \cdots + s_n^2 \right) + \ell_2 (x_1 s_1 + x_2 s_2 + \cdots + x_n s_n)$, ℓ_2 is very large and $\ell_1 < \, < \ell_2$ then $\ell_2 (x_1 s_1 + x_2 s_2 + \cdots + x_n s_n) = 0$. This is the same as just, minimize $c_1 x_1^2 + c_2 x_2^2 + \cdots + c_n x_n^2$.

This reduces to, minimize $c_1 x_1 + c_2 x_2 + \cdots + c_n x_n$.

8.4 Algorithm

The procedure is summarized in the following steps.

Step 1: Linearize the QAP to a binary linear problem

Step 2: Convert binary linear problem into a convex quadratic programming problem

Step 3: Use an interior point algorithm to solve the convex quadratic programming problem

Numerical illustration (This is a QAP for locating fuel gas stations to four possible sites)

$$\left. \begin{array}{ll} \text{Minimize} & z = 15x_1 + 11x_2 + 12x_3 + 9x_4 + 4x_1 x_2 + 5x_1 x_3 + 7x_1 x_4 \\ & + 7x_2 x_3 + 11x_2 x_4 + 10x_3 x_4 + 12x_1^2 + 14x_2^2 + 13x_3^2 + 11x_4^2, \\ \text{Such that :} & 12x_1 + 20x_2 + 18x_3 + 19x_4 \geq 49, \\ & 22x_1 + 16x_2 + 20x_3 + 25x_4 \geq 54, \\ & 19x_1 + 23x_2 + 16x_3 + 21x_4 \geq 51, \\ & x_j \in \{0,1\}, j = 1, 2, 3, 4. \end{array} \right\}$$

(8.28)

There are four possible sites for fuel gas stations for this problem. The optimal solution to (8.28) gives the minimum total distance that can be traveled by the villagers to obtain fuel gas.

In this case, the three budgetary constraints give the minimum amounts of money required to open the gas sites in terms of thousands of SA rands.

8.4.1 Making the Model Linear

Using $x_j = x_j^2$ and $x_i x_j = \delta_r$, the linear model in the numerical illustration becomes as given in (8.29).

$$
\left.
\begin{aligned}
&\text{Minimize} \\
&z = 27x_1 + 25x_2 + 25x_3 + 19x_4 + 4\delta_1 + 5\delta_2 + 7\delta_3 + 7\delta_4 + 11\delta_5 + 10\delta_6, \\
&\quad \text{Such that}: 12x_1 + 20x_2 + 18x_3 + 19x_4 \geq 49, \\
&\qquad 22x_1 + 16x_2 + 20x_3 + 25x_4 \geq 54, \\
&\qquad 19x_1 + 23x_2 + 16x_3 + 21x_4 \geq 51, \\
&\quad x_1 + x_2 - \delta_1 \leq 1, x_1 + x_3 - \delta_2 \leq 1, x_1 + x_4 - \delta_3 \leq 1, \\
&\qquad x_2 + x_3 - \delta_4 \leq 1, x_2 + x_4 - \delta_5 \leq 1, \\
&\qquad x_3 + x_4 - \delta_6 \leq 1.
\end{aligned}
\right\}
$$
(8.29)

where $\delta_i \in \{0, 1\}$, $i = 1, 2, 3, 4, 5, 6$. This formulation was proposed by Munapo [1]. Changing the problem into a convex quadratic problem

$$
\left.
\begin{aligned}
&\text{Minimize} \\
&f(\overline{X}) = \ell_1 \left(27x_1 + 25x_2 + 25x_3 + 19x_4 + 4\delta_1 + 5\delta_2 + 7\delta_3 + 7\delta_4 + 11\delta_5 + 10\delta_6 \right. \\
&\quad + (s_1 + s_2 + s_3 + s_4 + \overline{\delta}_1 + \overline{\delta}_2 + \overline{\delta}_3 + \overline{\delta}_4 + \overline{\delta}_5 + \overline{\delta}_6) + \\
&\quad \ell_2(x_1 s_1 + x_2 s_2 + x_3 s_3 + x_4 s_4 + \delta_1 \overline{\delta}_1 + \delta_2 \overline{\delta}_2 + \delta_3 \overline{\delta}_3 + \delta_4 \overline{\delta}_4 + \delta_5 \overline{\delta}_5 + \delta_6 \overline{\delta}_6), \\
&\quad \text{Such that}: 12x_1 + 20x_2 + 18x_3 + 19x_4 \geq 49, \\
&\qquad 22x_1 + 16x_2 + 20x_3 + 25x_4 \geq 54, \\
&\qquad 19x_1 + 23x_2 + 16x_3 + 21x_4 \geq 51, \\
&\quad x_1 + x_2 - \delta_1 \leq 1, x_1 + x_3 - \delta_2 \leq 1, x_1 + x_4 - \delta_3 \leq 1, \\
&\qquad x_2 + x_3 - \delta_4 \leq 1, x_2 + x_4 - \delta_5 \leq 1, \\
&\quad x_3 + x_4 - \delta_6 \leq 1, x_1 + s_1 = 1, x_2 + s_2 = 1, \\
&\quad x_3 + s_3 = 1, x_4 + s_4 = 1, \delta_1 + \overline{\delta}_1 = 1, \delta_2 + \overline{\delta}_2 = 1, \\
&\qquad \delta_3 + \overline{\delta}_3 = 1, \delta_4 + \overline{\delta}_4 = 1, \delta_5 + \overline{\delta}_5 = 1, \\
&\qquad \delta_6 + \overline{\delta}_6 = 1,
\end{aligned}
\right\}
$$
(8.30)

where $s_i \in \{0, 1\}$, $i = 1, 2, 3, 4, 5, 6$ and $\overline{\delta}_i \in \{0,1\}$, $i = 1,2, 3,4, 5,6$.

Using $\ell_1 = 1000(27 + 23 + 25 + 19 + 4 + 5 + 7 + 7 + 11 + 10) = 138\,000$ and $\ell_2 = 1\,000\,000(27 + 23 + 25 + 19 + 4 + 5 + 7 + 7 + 11 + 10) = 138\,000\,000$. The optimal solution by the interior point algorithm becomes as given in (8.31).

$$x_1 = x_2 = x_3 = s_4 = \delta_1 = \delta_2 = \delta_4 = \overline{\delta}_3 = \overline{\delta}_5 = \overline{\delta}_6 = 1,$$
$$s_1 = s_2 = s_3 = x_4 = \delta_3 = \delta_5 = \delta_6 = \overline{\delta}_1 = \overline{\delta}_2 = \overline{\delta}_4 = 0. \tag{8.31}$$

In terms of the original problem, opening fuel gas stations at sites 1, 2, and 3 will minimize the total distance traveled by all the villagers to obtain fuel gas.

8.5 Conclusions

The reason for converting the BLP into a convex quadratic programming model is to use the available efficient interior point algorithms which can solve convex quadratic problems in polynomial time. If the QAP which is NP-hard can be converted into a convex quadratic problem which can be solved in polynomial time then P=NP. This implies that the QAP is not NP-hard. In the future, we will consider increasing the complexity of the problem by making multi-objective/multi-level/variables with uncertainty (i.e. stochastic or fuzzy applications). In addition, when people visit a site, they do that with the intention of doing more than one thing, i.e. getting the gas and going to the grinding mill and/or visiting a clinic. This is not easy to factor this additional task in the nonlinear objective function of the QAP. Efforts will be made in the future to incorporate this additional task into single or multi-objective quadratic assignment model which may be stochastic or fuzzy.

References

1 Munapo, E. (2012). Reducing the number of new constraints and variables in a linearised quadratic assignment problem. *African Journal of Agricultural Research* 2 (29): 3147–3152.

2 Hahn, P. and Grant, T. (2008). An algorithm for the generalized quadratic assignment problem. *Computational Optimization and Applications* 40 (3): 351–372.

3 Ramakrishnan, K., Resende, M. et al. (2002). Tight QAP bounds via linear programming. In: *Combinatorial and Global Optimization* (ed. P. Pardalos, A. Migdalas and R. Burkard), 297–303. Singapore: World Scientific Publishing.

4 Drezner, Z. (2008). Extensive experiments with hybrid genetic algorithms for the solution of the quadratic assignment problem. *Computers & Operations Research* 35 (3): 717–736.

5 Mohamed Manogaran, A.-B., Rashad, G., Zaied, H., and Nasser, A. (2018). A comprehensive review of quadratic assignment problem: variants, hybrids and

applications. *Journal of Ambient Intelligence and Humanized Computing* 9 (5): 1–24. https://doi.org/10.1007/s12652-018-0917-x.

6 Adams, W.P. and Johnson, T.A. (1994). Improved linear programming-based lower bounds for the quadratic assignment problem. In: *Quadratic Assignment and Related Problems*, Volume 16 of DIMACS Series in Discrete Mathematics and Theoretical Computer Science, vol. 16 (ed. P.M. Pardalos and H. Wolkowicz), 43–75. AMS.

7 Cela, E. (1998). The quadratic assignment problem: theory and algorithms. In: *Combinatorial Optimization*. Dordrecht: Kluwer Academic Publishers.

8 Nagarajan, V. and Sviridenko, M. (2009). On the maximum quadratic assignment problem. *Mathematics of Operations Research* 34 (4): 859–868.

9 Rego, C., James, T., and Fred, G.F. (2010). An ejection chain algorithm for the quadratic assignment problem. *Networks* 56 (3): 188–206.

10 Xia, Y. (2010). An efficient continuation method for quadratic assignment problems. *Computers and Operations Research* 37 (6): 1027–1032.

11 Yang, X., Lu, Q., Li, C., and Liao, X. (2008). Biological computation of the solution to the quadratic assignment problem. *Applied Mathematics and Computation* 200 (1): 369–377.

12 Koopmans, T. and Beckmann, M. (1957). Assignment problems and the location of economic activities. *IEEE Econometrica* 25 (1): 53–76.

13 Munapo, E. (2020). The traveling salesman problem: network properties, convex quadratic formulation, and solution, Chap. 6. In: *Research Advancements in Smart Technology, Optimization, and Renewable Energy*, 88–109. IGI Global ISBN 9781799839705 (hardcover) | ISBN 9781799850397 (paperback) | ISBN 9781799839712 (ebook).

14 Munapo, E. (2016). *Second Proof that N = NP, 5th Engineering Optical Conference*, 19–23. Brazil: Iguassu Falls.

9

Feasibility of Drones as the Next Step in Innovative Solution for Emerging Society

Sadia S. Ali[1], Rajbir Kaur[2], and Haidar Abbas[3]

[1]*Department of Industrial Engineering, College of Engineering, GC, King Abdul-Aziz University, P.O. Box 80204, Jeddah 21589, Saudi Arabia*
[2]*Government Girls College, Harya, Panchkula, Haryana 134001, India*
[3]*Department of Business Administration, Salalah College of Applied Sciences, University of Technology and Applied Sciences, Sultanate of Oman*

9.1 Introduction

Today technology has permeated each and every aspect of our life to the extent that it has revolutionized our daily routines. Right from the way we talk and connect with others, the way we work, study and enjoy our lives, technology has brought drastic changes in all spheres. The ease of conversation, communication, operation, business, service traveling, every aspect of our lives, have undergone possibly the biggest transformation due to the technological revolution. Our smartphones, laptops, iPods, iPads, GPS are not just gadgets, they are the greatest change agent of our civilized lives. The Indian population, which is currently hitting a figure of 1,393,409,038 and is contributing fairly up to 18% of the total world population[1] in the year 2021, signals brighter prospects for drone technology. Among many, this population also includes a critical mass of highly educated people like scientists, engineers, doctors, innovators, etc. They form the vertebra of Research and Development (R&D) in Indian as well as multinational companies[2]. Recently, India has witnessed the adoption of some technology-driven innovations like the Aadhaar Platform, Bajaj Auto's DTS-i (Digital Twin Spark Ignition) technology, blockchain [1, 2], or Radio-frequency Identification (RFID enabled Fastag for directly collecting toll charges from travelers. The DTS-i at Bajaj Auto helps in saving fuel and leads to better engine

---

1 http://www.worldometers.info/world-population/india-population/
2 China and India: Emerging Technological Powers http://issues.org/23-3/dahlman/

Artificial Intelligence in Industry 4.0 and 5G Technology, First Edition.
Edited by Pandian Vasant, Elias Munapo, J. Joshua Thomas, and Gerhard-Wilhelm Weber.
© 2022 John Wiley & Sons, Inc. Published 2022 by John Wiley & Sons, Inc.

performance[3]. Likewise, Vortex Engineering's solar-powered ATMs consume less than 100 Watt and can do without air conditioning to save 1728 units per month[4]. Similarly, market-driven innovation by Tata Ace commercial vehicles that are equipped with lightweight FRP (fiberglass reinforced plastic) wind deflectors, enhances aerodynamics, thereby increasing fuel efficiency[5]. GE India's low-cost ECG machines with one-fourth of its initial prices, achieving greater portability as well as higher flexibility for test data transmission, remote analysis, and storage, emerge as another such instance[6]. There are a plethora of many such innovations in recent years.

India has expanded by 7.3% in the quarter ending 2018, thereby making her world's fastest-growing economy. India will have approximately half of its total population (1.2 billion) under 25 making it the youngest country in the world. This is a good reason for the companies to invest in India[7]. By the end of 2021, a strong middle and emerging class having 900 million people will pave the way for a shift in the commoners' mindset to achieve new value propositions delivered through innovative business models[8]. For such an emergent market, the business models dwelling on innovative technology to keenly offer consumer-centric solutions are setting the trends. According to a report by Price Waterhouse Coopers (PwC), India's current GDP of US$1.9 trillion could reach up to US$10.4 trillion by 2034 by achieving a GDP CAGR of 9% over the next two decades[9]. Among the five emerging markets, India has been adjudged by the "Euromonitor International" with the best middle-class potential due to its large size, income growth prospects, and median per capita income projected to exceed US$10,000 in 2030[10]. Young earning Indians have large disposable income and are ready to pay for more technologically advanced products like cars, mobile phones, personal computers, tablets, etc. Despite numerous complexities related to the infrastructural challenges, regulatory strictures, resource constraints, the tremendous growth potential of the Indian market is constantly convincing the budding entrepreneurs as well as established business barons to develop similar quality-focused and innovation-driven mind-sets to stay competitive in today's globalized environment.

---

3 https://crankit.in/dtsi/
4 http://www.alternative-energy-news.info/solar-powered-atms-by-vortex/
5 http://indianexpress.com/article/auto-travel/cars/tata-superace-mint-launched-at-rs-5-09-lakh/
6 http://www.globalhealth.care/2015/01/ecg-tech-from-india-could-save-us-50.html
7 http://www.cnbc.com/2016/02/08/forget-china-india-reports-higher-gdp.html
8 Future of India: The Winning Leap, *PwC*, 2014. Available at: http://www.pwc.in/en_in/in/assets/pdfs/future-of-india/future-ofindia-the-winning-leap.pdf
9 https://www.pwc.in/assets/pdfs/publications/2015/innovation_driven_growth_in_india_final.pdf
10 http://blog.euromonitor.com/2015/09/top-5-emerging-markets-with-the-best-middle-class-potential.html

9.1.1 Technology and Business

Businesses have remained at the forefront of technology for ages as it is evident from established and successful brands and their philosophy that whatever can speed production will draw in more business. Many successful business growth strategies have technology at their core as it can generate rich new forms of customer insights at a lower cost and faster than conventional methods. Social media technologies, cloud-based technologies, and mobile technologies have given many companies the privilege and opportunities to boost the productivity of their business and access new customers and markets. With the advent of new technologies, companies are learning to manage risk, remove obstacles, and enable capabilities and conditions to create value for their business. No one is immune from the effects of change, let it be any industry or any company or even any geography. The companies, which strategically seek partnerships to gain access to the new markets and new technologies, survive and thrive. However, it's not always easy to find and implement the appropriate technology solutions for financial as well as non-financial restrains reticent. Undoubtedly, the constant technological change pushes organizations to find schemes and initiatives that can help their businesses take advantage and grow. The success of any business is due to the ability of the organization to come up with new ideas to keep operations, products, and services fresh. Firms can apply new ideas to the products or the processes or the combinations of tangible and intangible inputs. In a study done by Accenture, approximately 90% of executives regarded innovation as a critical factor for their success. The majority of business professionals believe that the long-term success of their organization's strategy depends on their ability to develop new ideas.

9.1.2 Technological Revolution of the Twenty-first Century

Right from 3D printing, to online flight booking, to targeted advertising, or education aided by technology; all have crossed borders and continents. Nanotechnology and 3-D simulation are two powerful technologies shaping manufacturing today. Education as well as entertainment sectors are benefitting substantially from advanced technological aids which enable them to transform their traditional outputs into a more engaging, fun, and entertaining process. Technological innovations are disrupting the products and processes of firms irrespective of their scale of operations. The stronger firms take such disruptions as opportunities while their weaker counterparts initially strive to sustain. Consequently, varied experimentations are carried out by the firms, provided that the experiments have become less time-consuming and less expensive. In an environment, where experimentation is very quick and efficient, many traditional practices make less economic sense. Companies have bombarded the customer's market with numerous technological innovations and have performed ultra-well. One such innovation that is disrupting the companies' operations and redefining societal expectations is drone technology.

As this chapter has picked this theme to examine the feasibility of drones as an innovative solution for an emerging society, it initially aims to amass maximum needed information about the technology, the areas and sectors it is benefitting in various parts of the world, the challenges that are associated with its adoption and its future in the Indian market. It conducts a comprehensive review of extant literature which has been knitted around the broader factors that determine the degree of technovation. Employing the Investment methodology for materials (IMM) as used by Whelan [3], the integration of different methodologies and their interdependence has been explored in this study. The Section 9.2 of this chapter offers a detailed account of drone technology, its utilities, associated complexities and prospects in the Indian market. The review of literature is presented in Section 9.3–9.6.

9.2 An Overview of Drone Technology and Its Future Prospects in Indian Market

A delivery drone, also known as a parcelcopter[11], is an unmanned aerial vehicle (UAV) utilized to transport packages, food, or other goods [4]. This technology is offering promising incentives to the industry as well as individual consumers. As a potential replacement to the Indian truck-delivery system, Gabani et al. [5] proposed two drone-delivery models, namely a drone–truck hybrid delivery model and a stand-alone drone-delivery model. Herding sheep, helping farmers manage their crops, guiding lost people, revolutionizing private security, capturing the most beautiful moments during one's wedding, navigating people finding their destinations[12], delivering essentials (medicines, food, or groceries), or perhaps even aerial advertising[13], mining [6, 7]; the uses of drones are endless.

While it may seem that the drones are set to take over our lives, it is not likely to be a cakewalk for the drones' manufacturers for various privacy, security, and stern regulatory reasons. For instance, nine U.S. states have already passed laws restricting drone use, both in the hands of private citizens and law enforcement agencies[14]. In the United States, the Federal Aviation Administration (FAA) controls the National Airspace System (NAS). It has long exempted noncommercial flights of unmanned model airplanes from rules that govern private

---

11 https://en.wikipedia.org/wiki/Delivery_drone
12 http://air-vid.com/wp/20-great-uav-applications-areas-drones/
13 https://www.uavs.org/commercial
14 http://edition.cnn.com/2013/11/03/business/meet-your-friendly-neighborhood-drones/

and commercial aircraft[15]. Drone flights in Canada, Australia, Japan, and many European countries are already regulated. This means unmanned aircraft organizations in these countries know where, what, and when they can fly[16]. The following Sections 9.2.1 and 9.2.2 briefly deal with the major utilities of drone technology, the complexities involved, and the prospects of drone technology in the Indian business environment.

9.2.1 Utilities

This section offers a sufficient account of the major utilities of drone technology in the areas namely delivery, media/photography, agriculture, contingency, and disaster management situations, and civil and military services.

9.2.1.1 Delivery
The wholesale as well as retail distribution industry is likely to emerge more techno-friendly in the next few years. In fact, the first drone delivery happened in America with the approval of the Federal Aviation Administration, Virginia in the second week of July in 2015[17]. Facilitating a driver-free convenient delivery [8, 9] to bringing food, books, medicine, tapes, and pharmacy prescriptions from a commercial online vendor to any location, the idea to deploy drones to take deliveries to the next level seems really exciting and fascinating[18]. Agatz et al. [10] examined various optimization approaches for solving salesman traveling problems through drones. Companies such as Amazon, Wal-Mart, and Alphabet are likely to use it for the last mile delivery of packages [11] within the next few years.

9.2.1.2 Media/Photography
The rising popularity of drones is due to their ability to shoot bird's eye-view photos and videos that have previously been unavailable to photographers and filmmakers[19]. The film makers may grab the aerial shots for high-speed difficult chase or a panoramic view from different angles. Drones are ideal for budget filmmakers or even photo or video enthusiasts. The photography and movie making in specific life events using drone technology are satiating the younger generation's heightened expectations. A camera-equipped drone, the best zoom lens ever invented,

---

15 http://cstwiki.wtb.tue.nl/images/Rubin_(2014).pdf
16 http://www.directionsmag.com/entry/top-five-things-you-need-to-know-about-drones-and-gis/414810
17 http://qz.com/458703/the-first-successful-drone-delivery-in-the-us-has-taken-place/
18 http://www.marketwatch.com/story/drone-delivery-is-already-here-and-it-works-2015-11-30
19 http://thewirecutter.com/reviews/best-drones/

offers a myriad of customization options to businesses. Hobbyists and certain professionals like real estate agents may use these drones to showcase various sites and structures.

9.2.1.3 Agriculture

In the agriculture industry, the large-scale farmers might efficiently utilize the aerial views from UAVs to monitor crop growth, livestock, forestry, and fisheries that need it[20]. On-board sensor equipment can offer valuable information for precision agriculture in various specializations. It can help in spotting schools of fish, navigation (for example, warning of icebergs ahead), conducting environmental research, and providing data for workflow management. Despite India's heavy dependence upon agriculture, it still lags far away from adopting the latest technologies to cultivate optimum productivity [12].

9.2.1.4 Contingency and Disaster Management Scenarios

As a humanitarian response, the "disaster drones" are already being used to temporarily restore communication networks; map and survey; find and destroy landmines, bombs; assess vulnerabilities and damages [13], and deliver essentials like food and medicines[21]. A large number of small and lightweight UAVs had been deployed in the Philippines (after Typhoon Haiyan in 2013), Haiti (following Hurricane Sandy in 2012), Balkans (heavy floods), Oklahoma City, USA (Wild fire), and China (Earthquake)[22]. Various related groups like the Office for the Coordination of Humanitarian Affairs (OCHA) of UN, SkyEye (Philippines), and CartONG (Haiti) are actively training local communities to operate their own UAVs for disaster-preparedness purposes[23]. Joshi and Stein [14] gave an account of emerging drone nations. Ramadass et al. [15] examined the covid-19 prevention practices among masses by observing whether individuals are observing social distancing norms and wearing face masks. In case of violations, the drone alarms the nearest police station as well as the general public, besides carrying the masks and dropping them to the needed people. Likewise, Manigandan et al. [16] highlighted the potential of drone technology to spot the infection, detect the individuals not wearing the masks as well as to sanitize certain areas. Kumar et al. [17] proposed a drone-based system for combating the coronavirus.

---

20 https://www.snowballeffect.co.nz/store/aeronavics-im.pdf
21 http://www.aidforum.org/disaster-relief/top-solutions-that-are-saving-lives-in-humanitarian-response
22 http://money.cnn.com/2013/08/19/technology/innovation/fire-fighting-drones/
23 https://www.virgin.com/virgin-unite/business-innovation/humanitarian-in-the-sky-drones-for-disaster-response

9.2.1.5 Civil and Military Services: Search and Rescue, Surveillance, Weather, and Traffic Monitoring, Firefighting

As a helping hand for law enforcement people to find a missing person, as finding a lost child in woods is much easier with the help of drones, particularly when the terrain is unfriendly or the weather is hostile[24]. Researchers in Michigan are testing drone technology for mapping roads, identifying potholes, and studying traffic patterns [18] [25]. Kumar et al. [17] highlighted the key role played by drone technology in facilitating cooperative communication for critical environments. The usage of drones in military organizations for saving lives and capital is also not a new phenomenon. According to Peter W. Singer, a Brookings Institution drone's expert, the US military had 8000 UAVs in the air and 12 000 on the ground in the year 2012[26]. Over the next decade, the world can expect to see more drones used for things like border control, law enforcement and wildlife research.

9.2.2 Complexities Involved

The adoption of drone technology on a mass level is not devoid of challenges and practical implications. These range from public concerns about safety and privacy (see footnote 15) to changing regulations on commercial use of drones to the costs of deployment [19]. The global leaders should create standard operating procedures[27] with an emphasis on safety risks, rigorous quality, and redundancy requirements, and the use of collision avoidance systems[28]. The most common injuries include blunt trauma to the head, eye injuries, injuries to hands by drone's propeller blades, and skin laceration [20]. There are a host of related legal issues that arouse with the introduction of UAS in the skies. For instance, protection of personal property or a commercial or military aircraft from a trespassing drone[29]. Organized criminal agencies can use drones for unauthorized surveillance of sensitive areas/installations. It all necessitates the countries and international organizations to come up with a stringent regulatory mechanism that could address such potential concerns. The aspects that such regulations must cover may include the usage of drones, airspace ownership, airspace traffic, breach of privacy, and espionage. Silva and Naranjo [21] explored and synthesized the laws and regulations governing drone licensing and operations in Latin American Countries. In the UK, the Civil Aviation Authority (CAA) is charged with regulating civilian drone

---

24 http://edition.cnn.com/2013/11/03/business/meet-your-friendly-neighborhood-drones
25 http://dronelife.com/2014/06/04/drone-tech-embraced-americans-benefits-become-widespread
26 http://www.brookings.edu/research/opinions/2012/12/11-robotics-military-singer
27 http://www.ncpa.org/pdfs/sp-Drones-long-paper.pdf
28 http://www.oliverwyman.com/content/dam/oliver-wyman/global/en/2015/apr/Commercial_Drones.pdf
29 http://remagazine.coop/the-future-of-drones

usage up to a mass of 150 kg[30]. Canada's current regulations categorize commercial drones by weight, with two categories for those that weigh less than 25 kg (55 lb [see footnote 28]).

In addition to safety, security, and regulatory aspects, the cost has to be taken into consideration before this rapidly evolving technology becomes completely in vogue. At times, corporations and agencies may find it more economically expedient to lease drones rather than purchase them. Drone-as-a-service as proposed by Measure (a Federal Aviation Administration approved Company) provides turnkey drone solutions to deliver cost-effective actionable data to enterprise customers[31]. The municipalities or businesses facing only an infrequent and unspecified need may avail such Drone-as-a-service solutions to save costs. With the help of drones, Amazon intended to deliver over a range of 10 mi and a weight of 55 lb each [32].

9.2.3 Drones in Indian Business Scenario

The state defense departments around the world, including in India, are committing huge capital investments in UAVs/UCAVs, as they provide viable intelligence acquisition and combat capabilities at a fraction of the cost of manned combat aircraft. In India, the use of drones is restricted to military, government organizations and police forces largely for surveillance, wildlife surveys, and tournaments coverage purposes. The Indian market was thrilled on May 11, 2014, when Francesco's Pizzeria used a remote-controlled, GPS-enabled four-rotor drone for the first time to deliver a pizza[33]. Though it appeared to be an experimental sortie, it kindled the hopes of Indian customers of a quick and budget-friendly home delivery [22].

Realizing the potential of drone technology for the Indian market, some startups have already begun to explore the likelihood of commercial use of UAVs. Many public and private sector enterprises have experienced the utility of UAVs in recent years. For instance, Aerial photography and cinematography (Drona Aviation, a SINE IIT Bombay); Custom-made drones, drone kits, and accessories (Chennai-based Atoms & Bytes); supply of UAV components to the Defense Research and Development Organization (DRDO) and National Aerospace Labs (Bengaluru-based Edall Systems); city planning and disaster management (a Mumbai-based startup Airpix); disaster management (Netra by IdeaForge IIT Bombay during Uttarakhand floods) (see Footnote 12).

---

30 http://www.information-age.com/it-management/strategy-and-innovation/123459051/flying-high-what-will-it-take-commercial-drone-revolution-lift
31 http://www.measure.aero/
32 http://www.cnet.com/news/amazon-exec-our-drones-will-deliver-in-30-minutes-or-less/
33 12/26. http://yourstory.com/2014/05/pizza-delivery-drones/

In India, the civil aviation ministry's regulations do not approve the drones to fly over 500 m if they are used by private companies[34]. Most such hardware-based ventures are struggling with such regulatory strictures along with the lack of government support. According to Idea Forge, insufficient funding is another significant hurdle that impedes them from competing with established international hardware startups. According to ISRO, despite having the strategy and technology, most of the startups cannot match the requirements and specifications particularly in terms of quality[35]. For UAVs startups to succeed and excel in domestic as well as global markets, the government's support in terms of initial funding, rebates, and regulatory respites is a must.

The researchers have gone through several business reports and quality research papers to find out expert opinions about the scope of the drone industry. Much has not been done as the drone industry is in its nascent stage. A thorough review of extant literature has been carried out and discussed in the Section 9.3 which is followed by an elaboration of various factors that play an important role in the successful adoption of drone technology in India.

9.3 Literature Review

Employing a given set of inputs and process technologies to merely produce and distribute a given set of goods in the marketplace cannot ensure the survival of the business enterprises, in the long run, rather they can constantly innovate and commercialize that will ensure their sustainability and excellence over a long period [23]. These innovations which may be related to the structure, process, policies or product that the firms offer [24], foster growth [25]. Sapuarachchi [26] believed that a change brought about in any aspect of an organization or to its relevant environment is innovation. The product and process innovations [27] determine the firm innovation whereas technological innovations and imitations drive the relative performance of the firms and the evolution of industrial structure. Product innovation has also been considered as a potential source of competitive advantage for many firms. According to Reichstein [28], the firms experiencing a high growth rate are more likely to have been product innovators. Some researchers argue that sociotechnical transitions are required to scale the barriers which eventually help to achieve sustainability through radical innovations ([29, 30]. Bosworth et al. [31] explored and identified various social innovations in European community-led local development initiatives. A Schumpeterian social innovation framework is

---

34 http://yourstory.com/2015/05/startups-seizing-drones-aviation/
35 http://timesofindia.indiatimes.com/tech/tech-news/Young-entrepreneurs-build-bots-drones-to-save-lives/articleshow/50062159.cms

derived which helps in identifying the different processed and outcomes which result in the creation of the social value.

Technological innovations occur when a new or modified product is introduced to the market, or when a new or updated process is used in commercial production. Researchers are also exploring the factors that play an important role in short "Technovations" for raising societal standards [32]. Technovation involves the use of advanced technology or scientific developments to create better products or manufacturing processes through the diffusion process [33] or commercialization [34]. As a process involving inter-related perspectives, it includes a broad range of activities which start from ideation [35] to the production, dispersion, and absorption of the new product, process, or service by the user or adopter throughout the economy. For the production, absorption, and diffusion (commercialization) of any new product or technology, it must be accepted by the market before its classification as an innovation [36].

9.3.1 Absorption and Diffusion of New Technology

The progressive societies hold strong convictions in creating positive social impacts [36]. Kanger et al. [37] believed that positive societal changes can be brought about by the wide diffusion of technology. Some researchers who were optimistic about the social impacts of technology mentioned the significance of electronic mass media for a new decentralized and more democratic world named 'global village'. The bond between social and technical systems has been an interesting research theme for many researchers [38]. Technological innovations can be radical, incremental as well as disruptive [39, 40]. The absorption capacity indicates the acquisition and assimilation of external knowledge in the initial stages and implementation of technology for specific situations. The adoption of a new technology [41], its utilization in infusing the novelty, the commercialization of its distinct output, and evaluation of the benefits through various financial and non-financial parameters indicate the success [42]. The diffusion process [43] ensures the familiarity and adoption of new technovative products in the markets which helps the technovators realize the benefits. Technologies advance and compete through strategic routes which may include the transfer of knowledge, strategic alliances, leadership pulls, or a call for societal change.

9.3.2 Leadership for Innovation

The current business environment characterized by dynamism, rapid technovations and aggressive competition, necessitates the management to be proactive for grabbing the emerging opportunities. Leaders actuate their employees to innovate through their deliberate actions as well as through their more daily routine behaviors [44, 45]. Leading business organizations empower their people to explore the avenues for continual improvements in products as well as in processes so that the improvements with a proven result may be scaled up. In such organizations, leaders create a supportive climate for bottom-up innovations [23], develop a comprehensive analytical framework for continuous innovation, and nurture a culture for selection/facilitation of self-organizing individuals in innovation processes [46]. By using agile management practices and continuous deployment with a culture of trust, delegation, and collaboration, they may innovate and leave the traditional management practitioners behind [47]. Unlike the products having ordinary utilities, drones are created with differentiated techniques meant to deliver unique needs. They are not the toys for some short-time players; rather, they need support from the big layers of the industry. Thus, the national leadership has to play an important role in the promotion and implantation of such technologies especially in a country like India.

9.3.3 Social and Economic Environment

For any organization to become innovative, an organizational culture, which nurtures innovation and is conducive to creativity is a must [24]. Prominent determinants instilling innovation into organizational culture are strategy, structure, support mechanisms, and behavior [48]. The socioeconomic structure of nations is shaped up by technological innovativeness contributing toward their economic growth and social adaptations. Consumer behavior is a direct outcome of their cravings for an alternative pattern of consumption which eventually paves the way for technical innovation and new products. A prevailing approach in the study of social innovation points towards a linkage between technological innovation and social change [38]. The activities that meet the goal of social needs and are developed, promoted, and diffused by social organizations come under the categories of social innovations [49]. The valuable solutions and services provided by these social innovations are more effective, efficient, and sustainable than the existing ones [50]. These social innovations drive technological innovations, industrial growth, and technological development. While the leading economies of the world are pitting against each other and willing to spend enormous amounts to outperform in technological advances, it is safe to presume that society is ready for new technologies. The diversified application of advanced technologies in media, advertisement, defense, finance, and other sectors signals that drone technology will get the suitable push in the right direction.

9.3.4 Customer Perceptions

Technological innovations fill the market with alternatives, highlight the hidden problems, propose novel solutions, and alter the interaction patterns in society. Generally, researchers believe that technology does not come along with zero opposition, for it has pros as well as cons. Many people resist some technological changes for the risks these technologies involve [51]. The threat of unseen, lack of education, capital commitment, and risk aversion tendency, etc. lead the masses to be reticent to accept something new. People have different reasons to either adopt or resist an innovation [52].

Risk perception being the main reason for people's unfavorable attitude towards technology like drone technology [53, 54] identified nine dimensions of public risk perceptions. These include voluntariness of risk-taking, the immediacy of effect, prior knowledge of risk, the severity of consequences, the existence of scientific knowledge about the risks; the controllability of risks using personal skills or diligence; the novelty of risks, and others [53]. With drone technology causing concerns about people's security and privacy, these risks are getting their due attention. With this perspective, Siddappaji and Akhilesh [55] highlighted the role of cyber security, standards, challenges, and requirements in improving the overall security mechanism.

9.3.5 Alliances with Other National and International Organizations

As economies across the world are expanding, a number of large corporations have started seeking new knowledge opportunities worldwide [56]. Diversity and size of competitors have pushed many firms to be innovative via relying on domestic as in-house as external know-how [30]. External sources of knowledge help the business strengthen its ability to develop innovative products. The firms, which are apt in knowledge acquisition from their allies, register improvements in their overall performance, ability to innovate and ultimately acquire the competitive advantage. Thus, they build a new kind of competitive advantage by discovering, accessing, mobilizing, and leveraging knowledge from a number of locations across the globe. At times, resource constraints push the firms for technology sourcing [57] from outside. In such cases, the international R&D firms help the multinational firms to exploit their firm-specific resources, improve their local responsiveness, and ensure sustainable competitive advantages at the global level [58, 59]. Though a firm's ability to use external knowledge depends on its absorptive capacity its competence development gets facilitated by trust and joint learning. It leads to the creation of tacit knowledge in long-term partnerships and has a central role in boundary formation. International technology sourcing is broadly dependent on the economic capacity and inventive performance of countries involved in technology exchange and innovative collaboration.

9.3.6 Other Influencers

Innovation research has long focused on the determinants of organizational success. These determinants largely include structure, information sharing, inventiveness, ability to implement, building cross-functional teams, and organizational learning. Jiang [60] believed that importing foreign technology or capital goods used for production enables firms to innovate and improve their productivities. The organizational structures that are exposed to global competition produce different learning and innovation outcomes [61]. The studies existing on various aspects of product innovation include the process of product innovation adoption [62], the organizational context for product innovation [63], the relations between product innovation process and marketing strategy [64], the relations between the product innovator and the financial performance [65] and the cost of product innovations [66].

All these factors somehow influence an organization's decision to go for new technology. As drone technologies are relatively new in the Indian subcontinent, they demand a positive social environment backed by a strong economic foundation and robust leadership support [67] complemented by strong domestic and international alliances and customer acceptance. These are likely to proffer the essential boost to the growth, absorption, and diffusion of drone technologies in India (Figure 9.1).

Business organizations are constantly scanning the horizons for promising innovations in search of a better interplay between present and future. Most of them are leveraging their capacity to help civil society [68], government, and the private sector overcome sluggish economic growth, financial instability, political upheaval, resource crises, hunger, poverty, and disease. Technology that plays a crucial role in bringing a worthy innovation to the forefront sometimes poses certain challenges as well. Investment decisions get impacted by a number of factors such as the investment avenues (e.g., material, parts, subassemblies, etc.), technological challenges involved, and the future demand of the product. Yoo and Jung [69] examined various factors that determine the US customers' attitude toward the adoption of drone technology. The researchers noted that speed advantage, environmental friendliness, operational complexity, performance risk, and privacy risk interact to collectively influence drone delivery adoption. Many organizations get stuck between a decision to go for new technology and using advanced material for innovation [3].

9.4 Methodology

In order to understand the growth of the drone market, we need an understanding of the design, manufacturing, and operation of a drone. The making of a drone

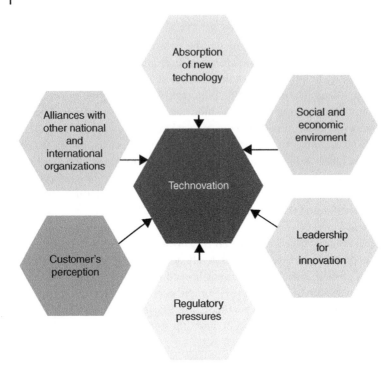

Figure 9.1 Factors for drone's technovation.

requires advanced material which has suitable weight, strength, and corrosion features. The functioning of a drone requires sensing technologies, which can send and receive signals in a suitable electromagnetic wavelength range. Another part of technical requirements is the digital processor which converts the signals into meaningful images or data. Communication technology is necessary to control a drone and this is not possible without gaining an expertise in manufacturing processes. In order to operate, drone standardization and regulatory compliance are mandatory.

Investment methodology for materials (IMM) [3], a useful tool for the material venture, provides insights to managers to assess the feasibility of adopting material innovations by taking into account the technical, commercial, and economical potentials. The concept of IMM has been applied by Maine and Ashby [70] to the case of Meal foams. For innovation to be viable, the organizations are integrating both technology and market-driven methodologies and are willing to explore any possibilities at any level. For the drone market to be viable, the interaction of different methodologies and their interdependence has been explored in this study.

For ease of study, we have taken the concept of IMM that was used by Whelan [3] for exploring the possible outcomes for the success of drones in Indian markets.

The word module has been taken from French or Latin modulus and refers to a 'self-contained, standard unit that can be combined with other different but compatible modules to assemble a wide range of varied end-products[36]. We have used the term module as a unit having factors with related characteristics. In this study, we aim to discover the effect of variables applied for successful implementation of technology on the absorption of Drone Technology in India. For the convenience of the study, we have categorized the dimensions into three different modules. Each variable has its foundation in the previous research studies and is discussed properly in Section 9.5.

9.5   Discussion

Societies develop their standards of living as a result of technological advancement. The pace of technological advancement keeps the product life cycle of technovative products relatively shorter. Drone, though no longer a new technovation, also keeps experience changes and upgradations in terms of its features and value propositions for its users. This chapter examines the feasibility of drone technology as an innovative solution for Indian society. The IMM has been employed to assess the feasibility of adopting material innovations by taking into account the technical, commercial, and economical potentials. For an innovation to be worthy of adoption, the organizations integrate both technology, market-driven methodologies, and their interdependence.

As mentioned in Section 9.5, the first module is the market module consisting of factors related to the dynamics of the market like what and how the trends in the industry are shaping up and impacting the markets. It also includes the other target market suitable for the concerned product where the usage and benefits of the product will drive its demand to yield long-run organizational profits. Besides, it also focuses on the market problems related to the product and thus, needs an amount of benchmarking for taking the product to its required destination. The aim of this module is to facilitate the managers to understand various issues related to the market before they develop the product and frame required marketing strategies. The next module- Technology Module defines the need for the technology required for a product which leads to the transfer of knowledge to bridge the gap. Technology Module is aimed to understand issues related to the technical specification of the products which happens through knowledge transfer. This step starts the chain for product objectives which further leads to process

---

36 http://www.businessdictionary.com/definition/module.html

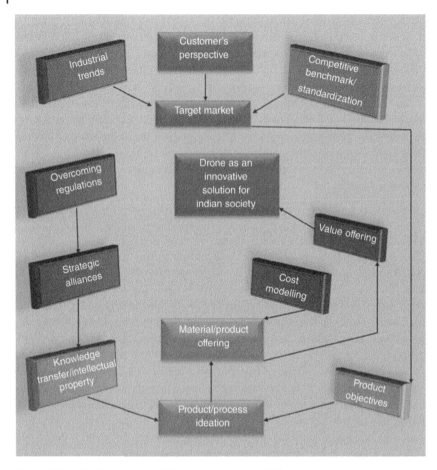

Figure 9.2 Interdependence of three modules related to drone.

or products ideas to match with the objectives. These ideas are materialized to offer the product offerings. Subsequently, one moves to the third module; the Commercial Module. A critical and evaluative approach is required to establish the commercial viability of the material concept here. Once the product is ready to be offered, it has to pass the regulatory hurdles at different levels. Also, in case, the market is not ripe for the launch of products, new markets get explored through strategic alliances. The products' cost is analyzed and finally, the value of the product offering is assessed. For it to be considered as an innovation, the developed products need to meet the tangible success requirements and should be able to deliver and capture the customer value. The relation between the three modules has been shown in Figure 9.2.

Market module

- Industrial trends [8, 56, 71]
- Customer's perspective [8, 72]
- Competitive benchmarking [73]
- Target market [74]

Technology module

- Knowledge transfer [8, 75, 76]
- Product objectives [8, 77, 78]
- Product/Process ideation [8, 79, 80]
- Product/material offering [75, 81, 82]

Commercial module

- Overcoming regulation [8, 68, 83]
- Strategiv alliances [8, 84, 85]
- Cost modelling [8, 86, 87]
- Value offering [4, 8, 68]; Dorling et al. [22]

9.5.1 Market Module

The market module gives us an understanding of the relationship between four constructs; it explains how industrial trends, competitive benchmarking, and understanding of the market or the market problem or customer's perspective help the firms to identify the target markets. It gives the firm an idea about the market requirements.

An organization can become competitive and gain advantage through knowledge and technological skills and experience in the creation of new products [56, 71]. The drivers of this new creation or innovation are the size of the organization, open innovation practices, country of origin, investment in R&D, organizational culture [88]. When industry starts showing such trends, it gives space to technological innovations [89] like drone. As the world is getting connected through swift modes of movements, many organizations believe that a wider drone acceptance could be a breakthrough in giving them a competitive advantage [90]. Among the major clients of drone technology, the main clients are large defense-focused companies and industrial conglomerates[37]. The market module starts with industrial trends. As there is a huge market potential for drones in India, the market opportunity takes precedence over technology. In India, the drone market is most likely to see an escalation in the coming years as trends favor its demand from military

37 http://www.directionsmag.com/entry/top-five-things-you-need-to-know-about-drones-and-gis/414810 accessed on January 28, 2020 at 10:51 pm

units, law and enforcement agencies, surveillance in disastrous times, supervision during major political events, retail, hospital, and hospitality sectors.

Understanding the market means assessing the customers' requirements, matching them with organizational technical capability, and exploring the ways to exploit such emerging opportunities. Given the huge funding requirements and uncertain market response for any innovative product, a pre-assessment of the market is essential. Large market size is a major pull for technical innovations. Based on the market forecast and customers' perspective on the success of an innovation, the viability of those potential market segments gets identified. Identifying customer's perspective in the development of a given technology is very tricky. Individual customers' adoption of new technology passes through five different stages of awareness, interest, evaluation, trial, and adoption [72]. However, an open mindset toward innovation is more important for the success of an innovation [91]. The organization has to identify and understand the customers' problems and develop the offerings to solve these issues [51]. India's young generation is educated and inclined towards technology as the majority of youth is attached to popular social media, which runs on technology. This technology-driven generation from urban areas is willing to accept drones as a new innovative solution for their deliveries at their doorsteps.

Companies can improve their products, processes, strategic planning, forecasting mechanisms, and internal operations by benchmarking them against the industry averages or high performers so as to derive substantial benefits [92]. In order to stay competitive, an organization must exceed the expectation of their customers, particularly in relation to the standards set by their rivals or competitors [93]. The market penetration rate of competent innovative products may be estimated with more precision to further define the market requirements [94]. The firms blessed with financial and technological leverage may take an edge in the nascent Indian drone market.

The likelihood of a venture failure may be curtailed up to 60% by professionally analyzing the target markets [74]. The focus is to identify market segments offering the greatest promise, having a reasonable size, and offering certain insights about the timing and amount of potential revenue flows. The identified target market should be able to capture a significant value so as to justify the investment in innovation [95]. Being relatively new, the drone market may not proffer a clear market segmentation; however, its potential customers in the form of large organizations, startups, and potential professional users make it fascinating [96].

9.5.2 Technology Module

The technology module further elaborates the product ideation based on inputs given by the consumer module, which further gives shape to the product's objectives. This is relatively a tricky phase as the innovation does not stop at product ideation but continues till process ideation. The ultimate purpose of the technology module is to design product concepts or offer and acquire knowledge.

Product innovation involves ideation, evaluation, development, testing, and successful exploitation of new ideas [97]. The ideas or the innovative thought process could apply to both the material attributes and manufacturing processes [79]. On average, out of seven products ideas, only four enter the development stage, two get launched and one finds success. The drone industry is stuck with challenges such as manufacturing hassles, availability of parts, security issues, and varying regulatory systems.

It is through strategic alliances that knowledge transfer happens which eventually sets the stage for the product objectives. New products are responsible for employment, economic growth, technological progress, and high standards of living. An innovative product adds to the growth of the company, its financial performance and plays a crucial role in business planning [97]. Since the drone is meant to offer certain solutions to the present defense, logistics, media, photography, disaster relief industry, the possibility of other offerings cannot be ruled out [77]. The study of the target market is revisited at this point to determine which ideas better fit the drone market requirements.

Knowledge transfer benefits the organizations as most often the firms lack the requisite technical or functional knowledge to develop a new product. It is through knowledge transfer that a firm can achieve its goals, improve its capabilities [98], gain a competitive advantage, and increase the pace of innovation [99]. Ownership of intellectual property is of high importance in extracting value from commercializing a materials innovation. Without patent or trademark protection, it is difficult to maintain a profit margin in the mass production of any product (Teece model). Since the drone technology has not seen full penetration yet, having a patent for the technology protection may serve the profit perspective of this innovation.

The product and process ideas generated along with the knowledge transfer pave the way for the final product offering. Intense international competition, fragmented and demanding markets, diverse and rapidly changing technologies have made product innovation more relevant and meaningful. A firm that manages to offer a product aptly caters to the needs and wants of target customers faster and more efficiently to gain a competitive advantage [75, 81]. Advances in computing and information and communication technology (ICT) in the late twentieeth century have given a rise to a third wave of invention and economic disruption which has heralded social and economic transformations in society. Advanced Robotics, web applications, and intelligent machines have provided society with wonderful creations such as unmanned vehicles, pilotless drones, mobile technology, and many more. Organizations are exploiting rare convergence of network technology, mobility, and human ecosystems to renew the system of wealth creation and to rebuild competitiveness on a completely different scale, at an entirely new pace. According to Myers et al. [78] suggests, such firms grow through mass differentiation or the deconstruction of markets into many niches, or data-driven niches, or customer self-selecting niches. The firms offering drone services are likely to take advantage by choosing their operations for a specific niche market. Also, the organization must look into the cost modeling or cost factor to find out if the cost of the product has exceeded the cost of estimates.

9.5.3 Commercial Module

The commercial module involves a connecting flow process between the market module and the technology module. It determines the commercial viability of the product as defined in the technology module with reference to the market requirements. Drone usage complies with many regulations. With a more prudent approach, nine states in the United States held back on permitting drones into civilian airspace for privacy concerns. Many regulators and policymakers have questioned its operational viability issues on security grounds, noise pollution, vehicle, pilot, and operator licensing, clash avoidance, and drone's possible uses for unlawful purposes [68, 83]. For the benefit of the drone industry, Europe formulated European Aviation Safety Agency (EASA) regulations that take into consideration their risk category. Japan practices de-regulated zones and designates specific radio frequencies for the specific use of drones to enable better transmission of high-resolution videos. In India, the directorate general of civil aviation (DGCA) has also framed broader guidelines for the operation of unmanned aerial vehicles (UAVs) [4, 100]. The drone market is likely to find its applications in a variety of spheres including agriculture [101], wildlife conservation, aerial photography, perimeter security, and even remote monitoring of utilities such as transmission towers, pipelines, highways, railways, etc.

Once the domestic and international regulations are complied with, the Indian organizations may start to think about their developmental strategies. They may forge strategic alliances to avail tangible and intangible benefits [75]. They may also collaborate externally with suppliers and business partners, customers, and internally across business and organizational boundaries [88]. This premeditated association and understanding may lead them to their anticipated growth and operational excellence. Since alliances are different from mergers and acquisitions [59], mutual benefits are derived in the form of reduced cost, access to distinct resources, advance technology, more reliable information, and exploration of the new market. Thus, any organization across the globe may take advantage of such alliances.

Once equipped with better knowledge about the manufacturing process or products features or spare parts, the technical feasibility of material innovation is established to find out the cost of primary and secondary production. Since the development of an innovative product involves high financial costs and uncertain outcomes, the cost and time of the innovation [86] determine a product innovation. Finding a cost estimate for the new product is complicated and requires detailed information about the product specifications and technical specifications [102]. For the drone market especially in a market like this, the potential for technological advances exists. Such approximate estimates can be useful, but a predictive cost model that allows for sensitivity analysis on technical uncertainties is better.

Innovation is meaningful only if it manages to deliver superior customer value such that the customer has a differentiated experience. The organizational commitment of significant investments for an innovative venture is justified only when there is a strong possibility of superior customer value creation through material innovation and potential collaborations. The achievement of a greater customer value and a superior status vis-a-vis the competitors in one or more of several competing dimensions leads an organization towards the required competitive advantage [103]. The material innovations turn attractive once the firms have evaluated technical and economic viability along with profitability. The profits are in the form of tangible economic returns for the firms or intangible benefits for society as a whole. The commercial drones delivering customer value in many parts of the world presents another instance as to how the technology brings societal transformations and address many conspicuous and hidden problems [49, 80]. Over the next couple of years, drone technology is going to receive its due credit.

9.6 Conclusions

There are different reasons for customers to either adopt or resist an innovation [52]. For instance, customer and market requirements or competition among suppliers or evolution of technology [104]. Innovation requires both technical and market capabilities which the drone market is now gradually acquiring. It protects its profit margins through patents, copyrights, trademarks, trade secrets, tacit knowledge, causal ambiguity. Since the market size and shape of the diffusion curve of a product are dependent on its pricing and promotion, the strategies adopted to price and promote the drone will play a crucial role in its success in the Indian market.

This chapter significantly contributes to the body of literature that is relevant to the adoption of drone technology in emerging societies like India. It offers deep insights into the drone as an emerging technology particularly for developing societies. It gently advocates the adoption of drone technology by emerging societies like India where it has plenty of room to get immersed. Besides highlighting the key utilities, it offers across different public and private sector fields, it also highlights the factors that determine the degree and pace of its adoption by countries. Special emphasis has been placed on its adoption and the associated challenges and the prospects in Indian Society. Three modules namely the technology, market, and commercial modules have been discussed in detail vis a vis the feasibility of drone technology in redefining Indian society. Given the growing multimillion youth population in India, it is quite wise to discern it a boom in the years ahead. However, the challenges particularly related to security, privacy, affordability, regulations, etc. will be harsh to deal with.

Being still in the nascent stage of development, it is facing the challenges of cost, technology and knowledge transfer, patents and other regulations. Recent uses of drones in media, disaster management, especially during an earthquake in Uttarakhand have given hope to the techno enthusiasts about their better acceptance for other purposes as well. Recently, the Telangana government has launched the "medicine from sky project" to drone-deliver the COVID-19 vaccines in selected locations[38]. Online food ordering and delivery company Swiggy is thrilled after obtaining the regulatory body's approval for its drone delivery project[39]. The development of drone industry requires rigorous efforts, a robust product innovation strategy, a sound technology strategy and an effective business leadership. Drone technology is ready to carve out its niche in the sphere of transport [105] and various other utility-based services. Thus, the governments must have technical optimism to ensure that a delicate balance is struck between the safety and security of people and the growth of drone technology.

References

1 Holotiuk, F., Pisani, F., and Moormann, J. (2018). Unveiling the Key Challenges to Achieve the Breakthrough of Blockchain: Insights from the Payments Industry. In: *Proceedings of the Hawaii International Conference on System Sciences, Waikoloa*, 3537–3546.

2 Nofer, M., Gomber, P., Hinz, O., and Schiereck, D. (2017). Blockchain. *Business & Information Systems Engineering* 59 (3): 183–187.

---

38 https://www.thehindu.com/news/national/telangana/telangana-launches-medicine-from-the-sky-project-to-drone-deliver-vaccines-medicines-to-remote-areas/article36401406.ece Accessed on October 12m 2021 at 03:22 pm

39 https://economictimes.indiatimes.com/tech/startups/swiggy-anra-get-approval-to-start-drone-trials-for-food-delivery/articleshow/83574991.cms Accessed on October 12m 2021 at 03:27 pm

3 Whelan, G. (2015). Understanding the social value and well-being benefits created by museums: a case for social return on investment methodology. *Arts & Health* 7 (3): 216–230.

4 Khan, M.A., Ectors, W., Bellemans, T., and Janssens and D, Wets G. (2017). Unmanned Aerial Vehicle-Based Traffic Analysis: A Methodological Framework for Automated Multi-Vehicle Trajectory Extraction. *Transportation Research Record Journal: Transportation Research Board.* 32: 1–15. https://doi.org/10.3141/2626-04.

5 Gabani, P.R., Gala, U.B., Narwane, V.S. et al. (2021). A viability study using conceptual models for last mile drone logistics operations in populated urban cities of India. *IET Collaborative Intelligent Manufacturing.* https://doi.org/10.1049/cim2.12006.

6 Bamburry, D. (2015). Drones: Designed for product delivery. *Design Management Review* 26 (1): 40–48.

7 Shahmoradi, J.E., Talebi, E., Roghanchi, P., and Hassanalian, M. (2020). A Comprehensive Review of Applications of Drone Technology in the Mining Industry. *Drones* 4: 34. https://doi.org/10.3390/drones4030034.

8 Ali, S.S., Kaur, R., Gupta, H. et al. (2021). Determinants of an organization's readiness for drone technologies adoption. *IEEE Transactions on Engineering Management*, ahead of print: https://doi.org/10.1109/TEM.2021.3083138.

9 Ali, S.S. and D., Supriya, Jaswal, M., Ahmad, Z. and AlSulami H. (2019). Feasibility of drone integration as last mile delivery, *Emerald Emerging Markets Case Studies*, 9, 3, 1-17. *Emerald Publications.* https://doi.org/10.1108/EEMCS-06-2018-0150.

10 Agatz, N., Bouman, P., and Schmidt, M. (2018). Optimization approaches for the traveling salesman problem with drone. *Transportation Science* 52 (4): 965–981.

11 Aurambout, J.P., Gkoumas, K., and Ciuffo, B. (2019). Last mile delivery by drones: an estimation of viable market potential and access to citizens across European cities. *European Transport Research Review.* 11: 30. https://doi.org/10.1186/s12544-019-0368-2.

12 Mogili, U.R. and Deepak, B.B.V.L. (2018). Review on application of drone systems in precision agriculture. *Procedia computer science* 133: 502–509.

13 Zweglinski (2020). The use of drones in disaster aerial needs reconnaissance and damage assessment – three-dimensional modeling and orthophoto map study. *Sustainability* 12 (15), 6080: doi:10.3390/su12156080.

14 Joshi, S. and Stein, A. (2013). Emerging drone nations. *Survival* 55 (5): 53–78.

15 Ramadass, L., Arunachalam, S., and Sagayasree, Z. (2020). Applying deep learning algorithm to maintain social distance in public place through drone technology. *International Journal of Pervasive Computing and Communications.*

16 Manigandan, S., Ramesh, P.K.T., Chi, N.T.L., and Brindhadevi, K. (2020). Early detection of SARS-CoV-2 without human intervention to combat COVID-19 using drone technology. *Aircraft Engineering and Aerospace Technology.*

17 Kumar, A., Sharma, K., Singh, H. et al. (2021). A drone-based networked system and methods for combating coronavirus disease (COVID-19) pandemic. *Future Generation Computer Systems* 115: 1–19.

18 Yadav, A., Goel, S., Lohani, B., and Singh, S. (2021). A UAV traffic management system for India: requirement and preliminary analysis. *Journal of the Indian Society of Remote Sensing* 49 (3): 515–525.

19 Seharwat, V. (2020). *Drone privacy laws: a comparative of the US, UK, and India.* In Drones and the Law: Emerald Publishing Limited.

20 Bansal, N., Aggarwal, S., & Tiwari, P. (2021). A case report of drone injury and its relevance in India. *Journal of clinical orthopaedics and trauma.* doi:https://doi.org/10.1016/j.jcot.2021.05.027

21 Silva, J.A.C. and Naranjo, D.A.M. (2020). Drone ethics and legal regulation, comparative drone law in Latin American countries. In: *The Politics of Technology in Latin America*, vol. 2, 15–31. Routledge.

22 Dorling, K., Heinrichs, J., Messier, G.G., and Magierowski, S. (2017). Vehicle Routing Problems for Drone Delivery. *IEEE Transactions on Systems, Management, and Cybernetics: Systems* 47 (1): 70–85.

23 Zhao, J., Li, Y. and Liu Y. (2016). Organizational learning, managerial ties, and radical innovation: evidence from an emerging economy. *IEEE Transactions on Engineering Management*, 63 (4) 489-499, doi: https://doi.org/10.1109/TEM.2016.2597290.

24 Kiani, A, Ali, A. Kanwal, S., and Wang, D. (2020). How and when entrepreneurs' passion lead to firms' radical innovation: moderated mediation model, *Technology Analysis & Strategic Management*, 32 (4), 443-456, DOI: https://doi.org/10.1080/09537325.2019.1667972.

25 Tolfree, D. and Walsh, S.T. (2018). An introduction to the field of commercializing emerging materials manufacturing technologies in an IoT world. *Translational Materials Research* 5 (2): 024002.

26 Sapuarachchi, D. B. (2021). Cultural distance and inter-organizational knowledge transfer: a case study of a multinational company. *Journal of Knowledge Management.* doi:https://doi.org/10.1108/JKM-06-2020-0439

27 Islam, N., Marinakis, Y., Majadillas, M.A. et al. (2020). Here there be dragons, a pre-roadmap construct for IoT service infrastructure. *Technological Forecasting and Social Change* 155: 119073.

28 Reichstein, T. (2004). Product Innovation, Interactive Learning and Economic Performance. *Research on Technological Innovation, Management and Policy* 8: 343–361.

29 Slaytona, R. and Spinardib, G. (2016). Radical innovation in scaling up: Boeing's Dreamliner and the challenge of socio-technical transitions. *Technovation* 47: 47–58.

30 Wang, C., Chin, T., & Lin, J. H. (2020). Openness and firm innovation performance: the moderating effect of ambidextrous knowledge search strategy. *Journal of Knowledge Management* doi:https://doi.org/10.1108/JKM-04-2019-0198

31 Bosworth, G., Rizzo, F., Marquardt, D. et al. (2016). Identifying social innovations in European local rural development initiatives. *Innovation: The European Journal of Social Science Research* 29 (4): 442–461.

32 Chong-Feng, Wang, and Y. Chao (2018). *Regional Independent Innovation and Open Innovation Optimization Strategy under Innovation Driving Strategy: Taking the biomedical industry of China as an example* DOI: https://doi.org/10.1109/TEMS-ISIE.2018.8478469. *IEEE International Symposium on Innovation and Entrepreneurship* (TEMS-ISIE).

33 Chiu, C.-Y., Chen, S., and Chen, C.-L. (2017). An integrated perspective of TOE framework and innovation diffusion in broadband mobile applications adoption by enterprises. *International Journal of Management, Economics and Social Sciences* 6 (1): 14–39.

34 Zahra, B., Egide, K., and Poulin, D. (2018). Technology adoption and diffusion: a new application of the UTAUT model. *International Journal of Innovation and Technology Management* 15: https://doi.org/10.1142/S0219877019500044.

35 Cesário, V., Matos, S., Radeta, M., and Nisi, V. (2017). Designing interactive technologies for interpretive exhibitions: enabling teen participation through user-driven innovation. In: *IFIP Conference on Human-Computer Interaction*, 232–241. Cham: Springer.

36 Orji, I.J., Kusi-Sarpong, S., Gupta, H., and Okwu, M. (2019). Evaluating challenges to implementing eco-innovation for freight logistics sustainability in Nigeria. *Transportation Research: Part A* 129: 288–305.

37 Kanger, L., Geels, F.W., Sovacool, B., and Schot, J. (2019). Technological diffusion as a process of societal embedding: Lessons from historical automobile transitions for future electric mobility. *Transportation Research Part D: Transport and Environment* 71: 47–66.

38 Munny, A.A. and Mahtab, Z. (2020). Enablers of social sustainability in the supply chain: An example of footwear industry from an emerging economy. *Sustainable Production and Consumption.* 20: 230–242.

39 Zhu, K., Dong, S.T., Xu, S.X., and Kraemer, K.L. (2006a). Innovation diffusion in global contexts: determinants of post-adoption digital transformation of European companies. *European Journal of Information Systems.* 15 (6): 601–616.

40 Zhu, K., Kraemer, K.L., and Xu, S. (2006b). The process of innovation assimilation by firms in different countries: a technology diffusion perspective on e-business. *Management Science* 52 (10): 1557–1576.

41 Gangwar, H., Date, H., and Raoot, A. (2014). Review on IT adoption: Insights from recent technologies. *Journal of Enterprise Information Management* 27: 488–502. https://doi.org/10.1108/JEIM-08-2012-0047.

42 Andrews D. and Criscuolo, C. (2013). Knowledge-Based Capital, Innovation and Resource Allocation. OECD Economics Department Working Papers, No. 1046, OECD Publishing. doi:https://doi.org/10.1787/5k46bj546kzs-en.

43 Mahajan, N. (2014). Diffusion of Innovation: What Sticks, What Doesn't and Why. *[Online] Available at* http://knowledge.ckgsb.edu.cn/2014/05/06/china-business-strategy/diffusion-ofinnovation-what-sticks-what-doesnt-and-why/.

44 Chen, M.Y.-C., Lin, C.Y.-Y., Lin, H.-E., and Mc.Donough, E.F. (2012). Does transformational leadership facilitate technological innovation? The moderating roles of innovative culture and incentive compensation. *Asia Pacific Journal of Management* 29: 239–264.

45 Metcalf, C.J. (2011). Persistence of technological leadership: Emerging technologies and incremental innovation. *The Journal of Industrial Economics* 59 (2): 199–224.

46 Steiber, A. and Alänge, S. (2013). A corporate system for continuous innovation: the case of Google Inc. *European Journal of Innovation Management* 16 (2): 243–264.

47 Denning, S. (2015). New lessons for leaders about continuous innovation. *Strategy & Leadership* 43 (1): 11–15.

48 Svahn, F., Mathiassen, L. and Lindgren, R. (2017). Embracing Digital Innovation in Incumbent Firms: How Volvo Cars Managed Competing Concerns. *MIS Quarterly.* 41. 239-253. 10.25300/MISQ/2017/41.1.12.

49 Teasdale, S., Roy, M.J., Ziegler, R. et al. (2020). Everyone a changemaker? Exploring the moral underpinnings of social innovation discourse through real utopias. *Journal of Social Entrepreneurship* 1–21.

50 Phills, J., Deiglmeier, K., and Miller, D. (2008). Rediscovering Social Innovation. *Stanford Social Innovation Review* 34-43: 42.

51 Pace, S. (2013). Looking at innovation through CCT glasses: Consumer culture theory and Google glass innovation. *Journal of Innovation Management* 1 (1): 38–54.

52 Widayat, W., Masudin, I., and Satiti, N.R. (2020). E-Money payment: customers' adopting factors and the implication for open innovation. *Journal of Open Innovation: Technology, Market, and Complexity* 6 (3): 57.

53 Fischhoff, B., Slovic, P., Lichtenstein, S. et al. (1978). How safe is safe enough? A psychometric study of attitudes towards technological risks and benefits. *Policy Sciences* 9: 127–152.

54 Mathew, A.O., Jha, A.N., Lingappa, A.K., and Sinha, P. (2021). Attitude towards Drone Food Delivery Services—Role of Innovativeness, Perceived Risk, and Green Image. *Journal of Open Innovation: Technology, Market, and Complexity* 7 (2): 144.

55 Siddappaji, B. and Akhilesh, K.B. (2020). Role of cyber security in drone technology. In: *Smart Technologies*, 169–178. Singapore: Springer.

56 Teece, D. (1998). Capturing Value from Knowledge Assets: The New Economy, Markets for Know-How, and Intangible Assets. *California Management Review* 40 (3): 55–79.

57 Oliveira, T. and Martins, M. R. (2011). Literature Review of Information Technology Adoption Models at Firm Level. 1566-6379, Retrieved from https://www.researchgate.net/publication/258821009_Literature_Review_ of_Information_Technology_Adoption_Models_at_Firm_Level.

58 Boutellier, R., Gassmann, O., and von Zedtwitz, M. (2008). *Managing Global Innovation, Uncovering the Secrets of Future Competitiveness*, 3erd revised edn. Berlin: Springer.

59 Kuemmerle, W. (1999). Foreign direct investment in industrial research in the pharmaceutical & electronic industries – results from a survey of multinational firms. *Research Policy* 28 (2–3): 179–193.

60 Jiang, L., Chen, J., Bao, Y., and Zou, F. (2021). Exploring the patterns of international technology diffusion in AI from the perspective of patent citations. *Scientometrics* 1–17.

61 Guo, H. Shen, R. and Su, Z. (2019). The Impact of Organizational Legitimacy on Product Innovation: A Comparison Between New Ventures and Established Firms, *IEEE Transactions on Engineering Management*, 66 (1), 73-83. doi: https://doi.org/10.1109/TEM.2018.2803061.

62 Rogers, E.M. (2003). *Diffusion of innovations*, 5the. New York: Free Press.

63 Chen, S.L., Chen, J.H., and Lee, Y.H. (2018). A Comparison of Competing Models for Understanding Industrial Organization's Acceptance of Cloud Services. *Sustainability*, 10(3), 673. doi:https://doi.org/10.3390/su10030673.

64 Su, Z. and Yang, H. (2018). Managerial ties and exploratory innovation: An opportunity-motivation-ability perspective, *IEEE Transactions on Engineering Management*, 65 (2), 227-238. doi: https://doi.org/10.1109/TEM.2017.2782730.

65 Capon, N., Farley, J.U., Lehmann, D.R., and Hulbert, J.M. (1992). Profiles of product innovators among large U.S. manufacturers. *Management Science.* 38 (2): 157–169.

66 Mansfield, E. and Rapoport, J. (1975). The costs of industrial product innovations. *Management Science.* 21 (12): 1380–1386.

67 Denti, L. and Hemlin, S. (2012). Leadership and Innovation in Organizations: A Systematic Review of Factors That Mediate or Moderate the Relationship. *International Journal of Innovation Management* 16 (3).

68 Otto, A., Agatz, N., Campbell, J. et al. (2018). Optimization approaches for civil applications of unmanned aerial vehicles (UAVs) or aerial drones: A survey. *Networks* 72 (4): 411–458.

69 Yoo, W. and Jung, J. (2018). Drone delivery: factors affecting the public's attitude and intention to adopt. *Telematics and Informatics.* 35 (6): 1687–1700.

70 Maine, E. and Ashby, M.F. (2002). An investment methodology for materials. *Material Design* 23: 297–306.

71 Tidd, J., Bessant, J., and Pavitt, K. (2005). *Managing Innovation: Integrating technological, market and organizational change*, Thirde. Wiley.

72 Kotler, P., Armstrong, G., Harris, L.C., and Piercy, N. (2013). *Principles of Marketing.* Harlow (England): Pearson Education Limited.

73 Murray, C.C. and Chu, A.G. (2015). The flying sidekick traveling salesman problem: Optimization of drone-assisted parcel delivery. *Transportation Research Part C: Emerging Technologies* 54: 86–109.

74 Hills, G.E. and LaForge, R.W. (1992). *Entrepreneurship Theory and Practice*, (spring), 33–58.

75 Prahalad, C. K. and Hamel, G. (1990). The Core Competence of the Corporation. *Harvard Business Review*, 68(3),79-91 1990. Available at SSRN: http://ssrn.com/abstract=1505251.

76 Zheng, C., Breton, A., Iqbal, W., Sadiq, I., Elsayed, E. and Li, K. (2015). Driving-Behavior Monitoring Using an Unmanned Aircraft System (UAS). In: G. V. Duffy (Ed.), Digital Human Modeling: Applications in Health, Safety, *Ergonomics and Risk Management: Ergonomics and Health: 6th International Conference.* 305-312.

77 Guido, G, Gallelli, V, Rogano, D, and Vitale, A. (2017). Evaluating the accuracy of vehicle tracking data obtained from Unmanned Aerial Vehicles. *International Journal of Transport Science and Technology.* 5(3),136-151. doi:https://doi.org/10.1016/j.ijtst.2016.12.001.

78 Myers, S.C., Nicholas, S., and Majluf, M.N.S. (1984). Corporate Financing and Investment Decisions when Firms Have Information that Investors do not Have. *Journal of Financial Economics* 13 (2): 187–221.

79 Companik, E., Gravier, M., and Farris M.T. (2018). Feasibility of Warehouse Drone Adoption and Implementation. *Journal of Transport Management.* DOI: https://doi.org/10.22237/jotm/1541030640.

80 Dovey, L. (2006). The Competitive Advantage of Interconnected Firms: An Extension of the Resource-Based View. *Academy of Management Review* 31 (3): 638–658.

81 Calantone, R., Vickery, S., and Dröge, C. (1995). Business Performance and Strategic New Product Development Activities: An Empirical Investigation. *Journal of Product Innovation Management* 12: 214–223.

82 Eze, S., Awa, H., Okoye, J. et al. (2013). Determinant factors of information communication technology (ICT) adoption by government-owned universities in Nigeria. *Journal of Enterprise Information Management* 26: 427–443. https://doi.org/10.1108/JEIM-05-2013-0024.

83 Clarke, R. (2014). The regulation of civilian drones' impacts on behavioral privacy. *Computer Law and Security Review* 30 (3): 286–305.

84 Culver, K.B. (2014). From Battlefield to Newsroom: Ethical Implications of Drone Technology in Journalism. *Journal of Mass Media Ethics* 29 (1): 52–64.

85 Kanistras, K., Martins, G., Rutherford, M. J. and Valavanis, K.P. (2013). A survey of unmanned aerial vehicles (UAVs) for traffic monitoring, 2013 *International Conference on Unmanned Aircraft Systems (ICUAS)*, 221-234. DOI: https://doi.org/10.1109/ICUAS.2013.6564694.

86 French S. (2017). Drone delivery economics: Are amazon drones economically worth it? <http://thedronegirl.com/2017/05/07/drone-delivery-economics-amazondrones>. [On-line; accessed 21-January-2020].

87 Iyer, S.R., Sankaran, H., and Walsh, S.T. (2020). Influence of Director Expertise on Capital Structure and Cash Holdings in High-Tech Firms. *Technological Forecasting and Social Change* 158: 120060.

88 Tellis, G.J. (2013). *Unrelenting Innovation: How to build a culture for market dominance.* J-B Warren Bennis Series.

89 Wang, C. and Wang, D. (2017). Research on the complexity of technological innovation supported by Internet. In: *International Conference on Progress in Informatics and Computing (PIC)*, 346–350. Nanjing.

90 Andersen, K.V., Frederiksen, M.H., Knudsen, M.P., and Krabbe, A.D. (2020). The strategic responses of start-ups to regulatory constraints in the nascent drone market. *Research Policy* 49 (10): 104055.

91 Govindarajan, V. and Trimble, C. (2012). Reverse Innovation: Create Far from Home, Win Everywhere, Harvard Business Review Press, 2012. *Management International Review* 54 (2): 277–282.

92 Silvagni, M., Tonoli, A., Zenerino, E., and Chiaberge, M. (2016). Multipurpose UAV for search and rescue operations in mountain avalanche events. *Geomatics Natural Hazards and Risk* 8(1), 18-33: 16.

93 Goldberg, D., Corcoran, M., and Picard, R.G. (2013). *Remotely piloted aircraft systems & journalism: Opportunities and challenges of drones in news gathering.* Reuters Institute for the Study of Journalism: Retrieved from https://ora .ox.ac.uk/objects/uuid:a868f952-814d-4bf3-8cfa-9d58da904ee3.

94 Ham, Y., Han, K.K., Lin, J.J., and Golparvar-Fard, M. (2016). Visual monitoring of civil infrastructure systems via camera-equipped Unmanned Aerial Vehicles (UAVs): A review of related works. *Visualization in Engineering* 4 (1): 1.

95 Holton, A. E., Lawson, S., and Love, C. (2015). Unmanned aerial vehicles: Opportunities, barriers, and the future of drone journalism. *Journalism Practice*, 9, 634–650. doi:https://doi.org/10.1080/17512786.2014.98059.

96 Ferguson, D. A., and Greer, C. F. (2019). Assessing the Diffusion of Drones in Local Television News. *Electronic News.* 13, 23-33. DOI:https://doi.org/10 .1177/1931243119829430.

97 Cooper, R.G. and Edgett, S.J. (2008). Ideation for Product Innovation: What Are the Best Methods? *PDMA Visions Magazine* 1: 12–17.

98 Huizingh, E.K.R.E. (2011). Open innovation: State of the art and future perspectives. *Technovation* 31 (1): 2–9.

99 Revilla, E., Rodriguez-Prado, B. and Cui, Z. (2016) A Knowledge-Based Framework of Innovation Strategy: The Differential Effect of Knowledge Sources, *IEEE Transactions on Engineering Management*, 63, (4), 362–376. doi: https://doi.org/10.1109/TEM.2016.2586300.

100 Salvo, G, Caruso, L. and Scordo A. (2014). Urban Traffic Analysis through an UAV. *Procedia - Social Behavior Science.* 111:1083-1091. doi: https://doi.org/10 .1016/j.sbspro.2014.01.143.

101 Kulkarni, A.A. and Nagarajan, R. (2021). Drone survey facilitated weeds assessment and impact on hydraulic efficiency of canals. *ISH Journal of Hydraulic Engineering* 27 (2): 117–122.

102 Rainey, D.L. (2010). *Sustainable Business Development: Inventing the Future Through Strategy, Innovation, and Leadership; 20 May.* Cambridge University Press.

103 Bakir A. (2016). Innovation management perceptions of principals. *Journal of Education and Training Studies*, 4(7), 1–13. doi:https://doi.org/10.11114/jets.v 4i7.1505.

104 Dachs, B. and Pyka, A. (2010). What drives the internationalization of innovation? *Evidence from European patent data, Economics of Innovation and New Technology* 19 (1): 71–86.

105 Barmpounakis, E.N., Vlahogianni, E.I., and Golias, J.C. (2016). Unmanned Aerial Aircraft Systems for transportation engineering: Current practice and future challenges. *International Journal of Transportation Science and Technology* 5 (3): 111–122.

10

Designing a Distribution Network for a Soda Company: Formulation and Efficient Solution Procedure

Isidro Soria-Arguello[1], Rafael Torres-Esobar[2], and Pandian Vasant[3]

[1]*Departamento de Ingeniería Química, Industrial y de Alimentos, Universidad Iberoamericana, Ciudad de México 01219, Mexico*
[2]*Facultad de Ingeniería, Universidad Anáhuac México, Huixquilucan 52786, Mexico*
[3]*Modeling Evolutionary Algorithms Simulation and Artificial Intelligence, Faculty of Electrical & Electronics Engineering, Ton Duc Thang University, Ho Chi Minh City 700000, Vietnam*

10.1 Introduction

According to international data, Mexico is the biggest consumer of bottled soda with an average consumption of 163 l per person a year, which totals 300 million cases per year in the country, equivalent to the consumption of 163 l per person per year; meaning, 94% of the population currently consumes this type of beverage [1]. That is how the value of the Mexican soft drink market amounts to approximately 15.5 billion dollars. According to complementary information provided in the announcement, there are more than 230 bottling plants in the country that serve more than one million points of sale [2].

The main point of sale of soft drinks in Mexico is the small store, where 75% of these beverages are purchased. Meanwhile, 24% of beverages are sold in restaurants, clubs, discotheques, and hotels, while only the remaining 1% is sold in self-service stores. Nowadays, the tastes of potential consumers have undergone important changes, giving the opportunity to the emergence of new beverages such as flavored waters, sparkling waters, and energy drinks; in addition, the consumption of purified waters, juices, coffee, and bottled tea has been promoted [3].

The beverage bottling group has an important market share, which has generated significant growth for the companies that belong to the before-mentioned bottling group, currently having monthly sales ranging between 8,000,000 and 10,000,000 unit cases depending on the sales season (1 unit case is equal to 5.7 l). These figures represent a daily demand of 140 daily carga of finished product

Artificial Intelligence in Industry 4.0 and 5G Technology, First Edition.
Edited by Pandian Vasant, Elias Munapo, J. Joshua Thomas, and Gerhard-Wilhelm Weber.

from its three bottling plants to its 17 distribution centers, which currently have 513 routes to meet the market demand in its area of coverage [4].

In 2019, a service level of > 99.4% (delivery requested/customer request) was achieved in the distribution centers, but in 2020, the first supply problems emerged due to the production capacity of the plants, which led to the specialization of the production lines of the three manufacturing plants. It ensured the availability of the product in the market and reduced the production costs of each factory.

This led to the lack of control in the primary distribution process, since previously each plant produced and distributed products only for its area of influence; in other words, there were three distribution networks in the bottling group that supplied the requirements of the distribution centers closest to each plant. A fourth plant is considered which corresponds to the purchase of the product as shown in Figure 10.1.

The most important factor that affected the operation of the distribution centers was the change in the pre-sales scheme on the commercial side; previously, there

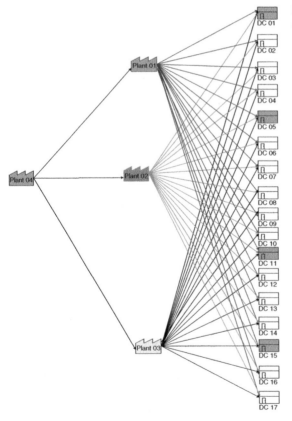

Figure 10.1 Current distribution system.

was a 24-hour scheme, meaning the pre-salesman would pick up the order in a car, and the next day the corresponding route would deliver the product. Now we have a 48-hour scheme, which consists of the same delivery route picking up the order, which is delivered two days after the order is taken.

10.2 New Distribution System

The top management of the bottling group concluded that it would be more convenient to rely on the distribution centers with greater capacity and call them cross-docking warehouses so that they perform the function of consolidating the demand of the other distribution centers but avoiding the product remaining for a long time (with a cross-dock operation) to finally supply the demand of the distribution bases from the cross-docking warehouses.

The main problems faced by the bottling group are as follows:

- Lack of capacity in the minimum storage required for the operation (considering at least one day's sales plus the safety inventory of each product).
- Lack of capacity for receiving freight, due to increased demand.
- Increased time required to assemble orders for route loading, due to increased product diversity.
- Limited personnel and equipment for the current operation.
- Limited attention windows, according to operating times.
- Delays in route departure times due to the lack of capacity to handle freight and routes at the same time.
- Product shortages since it was not possible to receive the entire planned shipment program.
- Limited production capacity and restricted budget.

Operation of the Company: The information listed below corresponds to the company's operation.

- The network has three production plants. A fourth plant is considered which corresponds to the purchase of the product.
- The plants are specialized in their production, which means that each plant does not produce the same products.
- Plant 1 supplies 58% of the products; plant 2, 31%; and plant 3, 9%; the remaining 2% is purchased from other bottling groups (plant 4).
- There are currently 17 distribution centers, five of them are considered cross-docking warehouses due to their size and sales volumes.
- The group's previous distribution was point-to-point, which means that all plants supplied directly to all distribution centers.

- The current demand is about 140 daily transports between origins and destinations, equivalent to a movement of 4487 pallets per day.
- The daily cost of opening and operating the cross-docking warehouses is 1,500,000 Mexican pesos. Since the distribution centers with the largest storage capacity are opened, no additional investment is required to operate the cross-docking warehouses; the bottling group currently has 305 days of operation per year.

The following tables show the data collected corresponding to the 2019 operation in order to calculate the optimal distribution cost of the bottling group.

1. Production capacity of the plants (Table 10.1)
2. Cross-docking warehouses fixed costs and storage capacity (Table 10.2)
3. Distribution costs from plants to cross-docking warehouses (Table 10.3)
4. Distribution costs from cross-docking warehouses to distribution centers (Table 10.4)
5. Monthly demand in pallets for the year 2019 (Table 10.5)
6. Supply of product from each plant (Table 10.6)

Table 10.1 Production capacity per plant.

Plant (P)	Daily production capacity in pallets
P1	4000
P2	2500
P3	1500
P4	5000

Table 10.2 Fixed costs and storage capacity in the cross-docking.

Cross-dock	Daily fixed costs ($)	Daily capacity of pallet storage
M1	300,000	1217
M2	300,000	5311
M3	300,000	2587
M4	300,000	1594
M5	300,000	1540

Table 10.3 Plant distribution costs to cross-docking.

Plant (P)	Distribution costs in Mexican pesos from plants to cross-docking ($/pallet)				
	M1	M2	M3	M4	M5
P1	182	26	29	38	172
P2	182	26	0	38	172
P3	261	172	172	202	0
P4	0	0	0	0	0

Table 10.4 Distribution costs in Mexican pesos ($) per pallet.

Cross-Docking (M)	Distribution costs in Mexican pesos ($) from cross-docking (M) to distribution center (N) per pallet																
	N1	N2	N3	N4	N5	N6	N7	N8	N9	N10	N11	N12	N13	N14	N15	N16	N17
M1	0	262	68	184	269	182	68	252	227	261	182	261	261	93	219	108	229
M2	182	87	262	26	87	0	236	71	46	155	29	185	227	89	38	87	77
M3	182	81	262	26	87	26	236	71	46	155	0	185	227	89	38	75	48
M4	219	173	234	38	50	38	273	34	83	202	38	202	265	126	0	112	85
M5	261	222	212	154	202	172	311	107	206	25	172	19	63	231	202	222	212

Table 10.5 Demand-Supply of each plant for each distribution center.

Plant (Product)	Monthly demand in pallets (d_j) of the distribution centers (N_j) for each product																
	N1	N2	N3	N4	N5	N6	N7	N8	N9	N10	N11	N12	N13	N14	N15	N16	N17
P1	197	201	12	79	11	796	44	97	162	68	378	68	16	64	346	139	137
P2	49	10	78	8	2	55	11	9	6	265	22	476	147	17	18	10	2
P3	29	18	8	13	2	59	7	11	13	45	29	76	13	15	27	11	6
P4	5	5	1	4	1	26	1	5	3	12	8	13	2	4	8	5	2
Total	**280**	**234**	**99**	**104**	**16**	**936**	**63**	**122**	**184**	**390**	**437**	**633**	**178**	**100**	**399**	**165**	**147**

The demand projections corresponding for the years 2020–2025 were developed by a specialized company and supported by the bottling group's marketing and sales areas. Analyzing the problem described, it was concluded that it would be convenient to lean on the distribution centers with the largest capacity (1, 6, 11, 15, and 17) and call them cross-docking warehouses. So that they have the function of

Table 10.6 Supply of product from each plant.

Plant (P)	Total deliveries of product per plant
P1	2592
P2	1408
P3	382
P4	105
Total	**4487**

consolidating the demand of the other distribution centers, but without the product remaining for a long time, as a cross-dock type operation scheme [5–8]. The demand of each of the distribution centers can be supplied from distribution centers with the largest capacity. Under the previous scheme, administration costs are reduced, the performance of the consolidation processes is increased, and inventory levels are reduced.

10.3 The Mathematical Model to Design the Distribution Network

The mathematical model of the situation is described below:

Index sets

"$k \in K$" Sets of sources
"$i \in I$" Sets of cross-docking
"$j \in J$" Sets of distribution center j

Parameters

Q_k = Capacity of source k
K_i = Capacity of cross-docking warehouse i
F_i = Fixed cost of opening cross-docking warehouse in location i
G_{ki} = Transportation cost per pallet of the product from the source k to the cross-docking warehouse i
C_{ij} = Cost of shipping the product from the cross-dock i to the distribution center j
d_{jk} = Demand of the distribution center j for the product k

Decision variables

We have the following continuous variables:

Z_{ki} = Quantity of product shipped from plant k to macro center i
X_{ij} = 1, if the macro center i supplies all the demand of the distribution center j, 0 otherwise.
Y_i = 1, if the macro center i is opened, 0 otherwise.
Y_{ki} = 1, if the plant k sends the product k to the macro center i, 0 otherwise.

Our formula is as follows:

$$\text{Min} \sum_{k=1}^{P} \sum_{i=1}^{M} G_{ki} Z_{ki} + \sum_{i=1}^{M} F_i Y_i + \sum_{i=1}^{M} \sum_{j=1}^{N} C_{ij} d_j X_{ij}$$

Subject to the following constraints:

$$\sum_{i=1}^{M} Z_{ki} \leq Q_k \qquad k = 1, \dots, P \tag{10.1}$$

$$\sum_{j=1}^{N} d_{jk} X_{ij} = Z_{ki} \qquad i = 1, \dots, M, \qquad k = 1, \dots, P \tag{10.2}$$

$$\sum_{i=1}^{M} X_{ij} = 1 \qquad j = 1, \dots, N \tag{10.3}$$

$$\sum_{j=1}^{N} d_j X_{ij} \leq K_i Y_i \qquad i = 1, \dots, M$$

$$p = \min_{j} \{d_j\} \tag{10.4}$$

$$pY_i \leq Z_{ki} \qquad i = 1, \dots, M, \qquad k = 1, \dots, P \tag{10.5}$$

$$Z_{ki} \leq K_i Y_{ki} \qquad i = 1, \dots, M, \qquad k = 1, \dots, P \tag{10.6}$$

$$\sum_{k=1}^{P} Y_{ki} = 4Y_i \qquad i = 1, \dots, M \tag{10.7}$$

where the variables are defined as follows:

$$Y_i \in \{0,1\} \qquad i = 1, \dots, M$$
$$Y_{ki} \in \{0,1\} \qquad k = 1, \dots, P, \qquad i = 1, \dots, M$$
$$X_{ij} \in \{0,1\} \qquad i = 1, \dots, M, \qquad j = 1, \dots, N$$
$$Z_{ki} \geq 0 \qquad k = 1, \dots, P, \qquad i = 1, \dots, M$$

The proposed optimization model presents 91 constraints, 110 binary variables, and 20 continuous variables, which describe the current distribution network and according to its structure corresponds to a mixed-integer programming model. In order to solve it efficiently, the Lagrangian relaxation technique is used.

10.4 Solution Technique

10.4.1 Lagrangian Relaxation

Lagrangian relaxation technique was developed in the early 1970s by Held and Karp to solve the traveling agent problem; nowadays it is one of the most popular decomposition techniques that allow to efficiently deal with problems in which one or more "hard" constraints interfere with the solution of the problem [9].

The original problem (P) is assumed to be in this type of form:

$$(P) \quad \min_{x} \{f(x) \,|\, Ax \le b, Cx \le d, x \in X\}$$

where X contains the sign constraints of x and the constraints that limit x to be an integer. The constraints of the form are assumed to be complicated, in the sense that the original problem (P), without them, may be simpler to solve. The constraints are kept together with x to form the Lagrangian relaxation of (P).

By incorporating the complicated constraints to the objective function by means of a vector of non-negative weights λ we generate a subproblem, that is easier to solve than (P). The vector components λ are called Lagrange multipliers. Because of the above, the calculation process is greatly facilitated and the required computational effort is reduced. It is important to mention that if equality restrictions are relaxed, the vector components λ do not have sign restrictions.

10.4.2 Methods for Finding the Value of Lagrange Multipliers

Historically two iterative methods for finding the optimal value of Lagrange multipliers have been studied most in the literature: column generation and the subgradient method. Both methods alternate between the solution of the Lagrangian relaxation and the dual price adjustment λ. The previously mentioned methods need an initial value of the dual price (λ) to start the iterative solution process [10].

10.4.3 Selecting the Solution Method

In this work, it was decided to select the subgradient method because it is the most suitable for our research. Since it is the easiest to implement and its computational cost is small, in addition, it has been demonstrated that its practical efficiency is very good, especially for large scale optimization problems where a "sawtooth" behavior in the Lagrangian value has been demonstrated during the first iterations, followed by a period of stable improvement that will converge towards a value that will be the optimal Lagrangian coordinate.

The solution methodology based on Lagrangian relaxation and the subgradient method, which allows us to solve the dual Lagrangian problem, was programmed in GAMS [11]. This software was used because the evaluation of the scenarios corresponding to the years 2020, 2021, 2022, and 2023 of the projected operations of the bottling group is relatively easy to carry out since the model built in GAMS allows to easily change the dimensions to its ability to index the variables and equations.

This section describes the pseudocode of the sub-gradient method following the notation used by Martín [12]. The programming of the method was performed in GAMS.

10.4.4 Used Notation

$\lambda = Lagrange\ multipliers$
$\mu^k = $ Step size in the iteration k
$\varepsilon = $ Stopping criteria
$\theta = $ Convergence factor
$\gamma = $ Improvement direction
For convenience, Algorithm 10.1 describes the steps for the subgradient method.

Algorithm 10.1 Pseudocode of the Subgradient Method

Input parameters: An upper bound (L^*) of the optimization problem under study is needed, in addition to an initial value for the Lagrange multipliers $\lambda^0 \geq 0$, the parameter $\varepsilon=0.005$ is used as a stopping criterion.
Initialization: We start with the convergence factor $\theta_0 = 2$, when the optimization does not progress we reduce θ_k systematically in the form geometrically to reach convergence.
{Subgradient iterations}
For $k = 0, 1, \ldots,$ **do**
The direction of improvement at each iteration k is calculated using the following expression:
$\gamma^k = g(x^k)\ \{subgradient\ of\ L(\lambda^k)\}$
Once we have the direction of improvement we must calculate the step size that this direction should take with the formula:
$\mu^k = \frac{\theta_k(L^* - L(\lambda^k))}{\|\gamma^k\|^2}\ \{step\ size\}$
When we have the direction of improvement and the step size we proceed to update the value of the dual variables (Lagrangian multipliers).
$\lambda^{k+1} = max\ \{0,\ \lambda^k + \mu^k\gamma^k\}$

The stopping condition of the iterative process of the subgradient method is carried out:

If $||\lambda^{k+1} - \lambda^k|| < \varepsilon$ **then**

Stop

Ends the if

If the solution does not improve in more tan k iterations we do :

$\theta_{k+1} = \frac{\theta_k}{2}$

If not

We update the value of θ

$\theta_{k+1} = \theta_k$

Ends the if

Finally we update the iteration counter

$k = k + 1$

It ends the for

10.4.5 Proposed Relaxations of the Distribution Model

This section presents the two relaxations proposed to solve the optimization model presented.

10.4.5.1 Relaxation 1

The first relaxation consists of dualizing the restrictions (10.3, 10.7) of the previously mentioned model associated with the fact that the demand of each distribution center will be satisfied by only one macro center and that each macro center will be supplied by the four plants, in order to have the required product mix respectively, the mathematical model of relaxation 1 is shown below:

$$\text{Min} \sum_{k=1}^{P} \sum_{i=1}^{M} G_{ki} Z_{ki} + \sum_{i=1}^{M} F_i Y_i + \sum_{i=1}^{M} \sum_{j=1}^{N} C_{ij} d_j X_{ij} + \sum_{j=1}^{N} \lambda_j \left(1 - \sum_{i=1}^{M} X_{ij} \right)$$

$$+ \sum_{i=1}^{M} \beta_i \left(4Y_i - \sum_{k=1}^{P} Y_{ki} \right)$$

Subject to:

$$\sum_{i=1}^{M} Z_{ki} \leq Q_k \qquad k = 1, \ldots, P \tag{10.8}$$

$$\sum_{j=1}^{N} d_{jk} X_{ij} = Z_{ki} \qquad i = 1, \ldots, M, \qquad k = 1, \ldots, P \tag{10.9}$$

$$\sum_{j=1}^{N} d_j X_{ij} \leq K_i Y_i \qquad i = 1, \ldots, M \tag{10.10}$$

$$pY_i \leq Z_{ki}$$
$$p = \min_{j} \{d_j\} \qquad i = 1, \dots, M, \qquad k = 1, \dots, P \tag{10.11}$$

$$Z_{ki} \leq K_i Y_{ki} \qquad i = 1, \dots, M, \qquad k = 1, \dots, P \tag{10.12}$$

$$X_{ij} \in \{0,1\} \forall_i, \forall_j, \quad Y_{ki} \in \{0,1\} \forall_k \forall_i, \quad Y_i \in \{0,1\} \forall_i, \quad Z_{ki} \geq 0 \ \forall_k, \ \forall_i$$

10.4.5.2 Relaxation 2

The second relaxation consists of dualizing constraints (10.4)–(10.7) of the optimization model presented in section 3. The mathematical model of relaxation 2 is shown below, and it can be clearly observed that the generated subproblem has the structure of a minimum cost flow problem.

$$\text{Min} \sum_{k=1}^{P} \sum_{i=1}^{M} G_{ki} Z_{ki} + \sum_{i=1}^{M} F_i Y_i + \sum_{i=1}^{M} \sum_{j=1}^{N} C_{ij} d_j X_{ij} + \sum_{i=1}^{M} \lambda_i \left(K_i Y_i - \sum_{j=1}^{N} d_j X_{ij} \right)$$

$$+ \sum_{k=1}^{P} \sum_{i=1}^{M} \alpha_{ki} (Z_{ki} - pY_i) + \sum_{k=1}^{P} \sum_{i=1}^{M} \theta_{ki} (K_i Y_{ki} - Z_{ki})$$

$$+ \sum_{i=1}^{M} \beta_i \left(4Y_i - \sum_{k=1}^{P} Y_{ki} \right)$$

Subject to:

$$\sum_{i=1}^{M} Z_{ki} \leq Q_k \qquad k = 1, \dots, P \tag{10.13}$$

$$\sum_{j=1}^{N} d_{jk} X_{ij} = Z_{ki} \qquad i = 1, \dots, M, \qquad k = 1, \dots, P \tag{10.14}$$

$$\sum_{i=1}^{M} X_{ij} = 1 \qquad j = 1, \dots, N \tag{10.15}$$

$$X_{ij} \in \{0,1\} \forall i, \forall j, \qquad Z_{ki} \geq 0 \forall k, \ \forall i$$

10.4.6 Selection of the Best Lagrangian Relaxation

In order to decide which of the two proposed relaxations we will use to solve the optimization model under study, the following analysis is performed. First, we present the optimal results of the current distribution network model, which will serve as a point of comparison to select the most appropriate relaxation for the distribution model of the bottling group. See Table 10.7.

Table 10.7 Distribution costs in the scenario corresponding to the year 2019.

Scenario	Daily network operation cost	Annual network cost (305 operating days)
Actual	$348,082	$111,781,890

Table 10.8 Results obtained with the relaxations proposed for the year 2019.

Relaxation proposed	Constraints relaxed	Number of dualized constraints	The daily cost of the distribution obtained with relaxation	Difference from the optimal solution reported in Table 7.
1	Constraint 3 Constraint 7	22	$190,164	$193,918
2	Constraint 4 Constraint 5 Constraint 6 Constraint 7	50	$343,082	$5,000

To obtain an initial comparison between the two proposed relaxations, we show in Table 10.8 the results obtained with each of them using the algorithm implemented in GAMS.

The program built in GAMS uses as an initial value of the Lagrange multipliers the value of the dual variables of the relaxed problem to make the iterative process of the sub-gradient method more efficient, the step size in each iteration is modified according to the direction of improvement and the adjustment factor with which we initially set is reduced by half when the value of the objective function does not improve in each iteration $\theta_0 = 2$ and is set as a stopping criteria $\|\lambda^{k+1} - \lambda^k\| < \varepsilon$ where $\varepsilon = 0.05$.

It is important to mention that relaxation 2 gives us a better lower bound of the optimal value of the distribution problem under study. Once the previously mentioned is concluded, we must verify whether the solution found is feasible in the original model; otherwise, we must restore the viability through a heuristic algorithm to obtain a good solution for the original distribution model. The heuristic algorithm is detailed in next section.

10.5 Heuristic Algorithm to Restore Feasibility

This part describes the heuristic algorithm used to restore feasibility to the solution found by means of the solution methodology proposed in this work (see Algorithm 10.2). Performing of some tests on the selected Lagrangian relaxation

2 showed that the solution obtained with this relaxation is not feasible for the original problem since it exceeds the storage capacity of the open cross-docking warehouses. Because the plants send to the cross-docking warehouses several products that exceed their capacity, the amount of distribution centers that are supplied by each open macro center is also affected. To solve the problem, the following heuristic algorithm is proposed to restore feasibility to the solution found with the Lagrangian relaxation technique and the subgradient method.

The heuristic procedure used to restore feasibility

Input: Number of cross-docking i open.

Step 1. Rank the cross-docking i that are open ($Y_i = 1$) into two groups: those that have available storage capacity and those that exceed storage capacity; rank the open cross-docking i that violate capacity from highest to lowest.

Step 2. Meet the demand of distribution centers j with the lowest transportation cost without exceeding the capacity of the first cross-docking i in the list ordered in step 1.

Step 3. Rank in decreasing order all open cross-docking i according to their available storage capacity.

Step 4. Reassigned the exceeding products to the cross-docking i with greater storage availability, if this is not feasible, open a new cross-docking i and supply the demand of the remaining distribution centers j without exceeding the storage capacity.

Step 5. Repeat the procedure until the demand of all distribution centers j is satisfied by cross-docking i without violating their storage capacity.

Pseudocode of the proposed heuristic algorithm for obtaining a feasible solution to the distribution model

Used notations

K_i = Available capacity of cross-docking i
CU_i = Capacity used by cross-docking i
C_{ij} = Distribution cost from cross-docking i to distribution center to distribution j
S_i = Excess of product in cross-docking i
DA = Counter used to accumulate the demand, initialized with a value of zero
d_j = Total demand of distribution center j

Algorithm 10.2   Pseudocode of the Proposed Heuristic Algorithm

---

Input parameters: Number of cross-docking warehouses i open, Available Capacity (K_i), Used Capacity (CU_i), Cost of distribution from cross-docking warehouses i to distribution center j (C_{ij}), Cumulative Demand (DA), d_j = total demand of distribution centers j.

Definitions: $i \in A$ cross-docking warehouses violating storage capacity, $i \in B$ cross-docking warehouses with available capacity.

Step 1. For all open cross-docking warehouses (*i such that Yi = 1*),

If $CU_i > K_i$, do $i \in A$ and follow to step 2

Otherwise $i \in B$; follow to step 5

Step 2. Sort $i \in A$ in decreasing order of capacity used (CU_i)

Step 3. Sort C_{ij} from smallet to largest for $i \in A$ and distribution centers j served by the cross-docking warehouses $i \in A$

Paso 4. Reassign the capacity of the cross-docking warehouses $i \in A$

For $i \in A$ and for every j whose demand is met by the cross-docking warehouses $i \in A$

If

$$DA + d_j < K_i$$
$$DA = DA + d_j$$
$$CU_i = DA$$

Otherwise, the demand d_j is not assigned to the cross-docking warehouses $i \in A$ therefore $j \in missing$

Step 5. Sort from highest to lowest all open cross-docking warehouses i by available capacity in the list M

Step 6. Sort from highest to lowest all demands d_j of the distribution centers. $j \in missing$

For every $i \in M$ y $j \in missing$

If $d_j \leq (K_i - CU_i)$

$$CU_i = CU_i + d_j$$

The demand d_j is assigned to the cross-docking warehouses i

Note: The input parameters are obtained from the model programmed in GAMS and the proposed heuristic algorithm was programmed in MATLAB to create an interactive algorithm that restores feasibility regardless of which cross-docking warehouses are open.

10.6 Numerical Analysis

The distribution network used by the bottling group has three production plants, including a fourth plant that purchases the product from other groups, carrying out point-to-point distribution, in other words, all the plants supply all the distribution centers. Plant 1 supplied 58% of the products, plant 2, 31%, plant 3, 9%, and the remaining 2% was covered by the other bottling groups.

The expected demand for 2019 was 4487 pallets per day, which is equivalent to 140 daily transports between origins and destinations, generating an average daily cost of $348,082; considering 305 working days per year, this results in a cost of $111,781,890 Mexican pesos.

With the redesign of the bottling group's supply chain for the year 2019 obtained through the optimization model used in this work, an annual savings of $5,616,880 was achieved, which corresponds to 5% of the distribution cost previously used by the bottling group.

It is important to mention that the new distribution scheme was accepted by the company's management group and was implemented in the daily distribution operation. To adequately respond to changing market conditions, an analysis of four scenarios was performed, representing the behavior of the demand for the next four years (2020, 2021, 2022, 2023).

The data for each study scenario are based on the company's annual strategic rolling 2020–2023; the data and results obtained for each scenario are detailed below. It is important to mention that the plant production capacity data, the fixed costs of opening and operating a macroplant as well as the storage capacity associated with each macroplant remain constant in the four study scenarios. The transportation costs from the plants to the cross-docking warehouses and the distribution costs from each macro center to the distribution centers are shown in Tables 10.3 and 10.4.

10.6.1 Scenario 2020

Table 10.9 shows the behavior of the demand for 2020, prepared by the specialized company and supported by the bottling group's marketing and sales areas. See Table 10.9.

The results obtained with the proposed optimization model for the 2020 scenario are shown in Tables 10.10 and 10.11

Table 10.9 Demand-supply of each plant for each distribution center.

Plant (Product)	Monthly pallet demand (d_j) of the distribution centers (N_j) of 2020 for each product																
	N1	N2	N3	N4	N5	N6	N7	N8	N9	N10	N11	N12	N13	N14	N15	N16	N17
P1	49	200	15	76	11	780	18	95	160	124	366	132	32	63	332	138	138
P2	210	23	80	16	4	115	40	17	17	226	55	440	140	23	51	19	9
P3	31	19	9	14	3	62	8	12	14	48	31	80	14	16	29	12	7
P4	6	6	2	5	2	28	2	6	4	13	9	14	3	5	9	6	3
Total demand	296	248	106	111	20	985	68	130	195	411	461	666	189	107	421	175	157

Table 10.10 Results corresponding to scenario 2020.

Scenario	Cross-docking warehouses that open	Distribution centers operating	Optimal distribution cost per operating day	Dimension obtained with the Lagrangian relaxation	Percentage difference with respect to the optimal cost
2020	1	1,7	$432,479	$431,935.72	0.13%
	2	6,9			
	3	2,4,11,14,16,17			
	4	5,8,15			
	5	3,10,12,13			

Table 10.11 Quantity of products shipped from plant k to cross-docking warehouse i.

Cross-Docking (M) Plant (Product)	1	2	3	4	5
1	67	940	981	438	303
2	250	132	145	72	886
3	39	76	99	44	151
4	8	32	34	17	32

10.6.2 Scenario 2021

Projected demand of the bottling group for scenario 2 corresponding to the 2021 operation. See Table 10.12.

The results obtained with the proposed optimization model for the 2021 scenario are shown in Tables 10.13 and 10.14.

Table 10.12 Demand-Supply of each plant for each distribution center.

Plant (Product)	Monthly pallet demand (d_j) of the distribution centers (N_j) of 2021 for each product																
	N1	N2	N3	N4	N5	N6	N7	N8	N9	N10	N11	N12	N13	N14	N15	N16	N17
P1	51	210	16	80	12	819	19	100	168	131	384	138	34	67	349	145	145
P2	221	24	84	17	4	121	42	18	18	238	58	462	147	24	53	20	9
P3	32	20	9	15	3	66	8	13	15	50	32	84	15	17	30	13	7
P4	6	6	2	5	2	29	2	6	4	14	9	15	3	5	9	6	3
Total demand	310	260	111	117	21	1035	71	137	205	433	483	699	199	113	441	184	164

Table 10.13 Results corresponding to scenario 2021.

Scenario	Cross-docking warehouses that open	Distribution centers operating	Optimal distribution cost per operating day	Dimension obtained with the Lagrangian relaxation	Percentage difference with respect to the optimal cost
2021	1	1,7	$453,814	$453,262.12	0.12%
	2	6,9			
	3	2,4,11,14,16,17			
	4	5,8,15			
	5	3,10,12,13			

Table 10.14 Quantity of products shipped from plant k to cross-docking warehouse i.

Cross-Docking (M) Plant (Product)	1	2	3	4	5
1	70	987	1031	461	319
2	263	139	152	75	931
3	40	81	104	46	158
4	8	33	34	17	34

10.6.3 Scenario 2022

Forecasted demand for scenario 3 corresponding to the bottling group's 2022 operation. See Table 10.15.

Table 10.15 Demand-Supply of each plant for each distribution center.

Plant (Product)	Monthly pallet demand (d_j) of the distribution centers (N_j) of 2022 for each product																
	N1	N2	N3	N4	N5	N6	N7	N8	N9	N10	N11	N12	N13	N14	N15	N16	N17
P1	54	220	17	84	12	859	20	105	176	137	403	145	35	70	366	152	152
P2	232	25	88	18	4	127	44	19	19	249	61	486	154	25	56	21	10
P3	34	21	10	16	3	69	9	13	16	53	34	88	16	18	32	13	7
P4	6	6	2	5	2	31	2	6	4	14	10	16	3	5	10	6	3
Total demand	**326**	**272**	**117**	**123**	**21**	**1086**	**75**	**143**	**215**	**453**	**508**	**735**	**208**	**118**	**464**	**192**	**172**

Table 10.16 Results corresponding to scenario 2022.

Scenario	Cross-docking warehouses that open	Distribution centers operating	Optimal distribution cost per operating day	Dimension obtained with the Lagrangian relaxation	Percentage difference with respect to the optimal cost
2022	1	1,7	$476,227	$475,099.39	0.23%
	2	6,9			
	3	2,4,11,14,16,17			
	4	5,8,15			
	5	3,10,12,13			

Table 10.17 Quantity of products shipped from plant k to the cross-docking warehouses i.

Cross-Docking (M) Plant (Product)	1	2	3	4	5
1	74	1035	1081	483	334
2	276	146	160	79	977
3	43	85	109	48	167
4	8	35	35	18	35

The results obtained with the proposed optimization model for the 2022 scenario are shown in Tables 10.16 and 10.17.

10.6.4 Scenario 2023

Demand projection for scenario corresponding to the bottling group's 2023 operation. See Table 10.18.

The results obtained with the proposed optimization model for the 2023 scenario are shown in Tables 10.19 and 10.20.

After analyzing the results of the scenarios corresponding to the years 2020, 2021, 2022, and 2023, it was observed that the solutions obtained are feasible in the original problem. This is since the subproblem generated by the Lagrangian relaxation for each study scenario assures us that the capacity of the production plants will not be exceeded by the demand of the cross-docking warehouses and that the number of pallets sent by each of the plants to the cross-docking warehouses is equal to the amount received by the distribution centers. In addition, it is guaranteed that the demand of each distribution center will be satisfied by only one macro center.

Table 10.18 Demand-Supply of each plant for each distribution center.

Plant (Product)	Monthly pallet demand (d_j) of the distribution centers (N_j) of 2023 for each product																
	N1	N2	N3	N4	N5	N6	N7	N8	N9	N10	N11	N12	N13	N14	N15	N16	N17
P1	56	231	18	88	13	902	21	110	185	144	423	152	37	73	385	160	160
P2	244	26	93	19	4	133	47	20	20	262	64	510	162	26	59	22	10
P3	36	22	10	16	3	72	9	14	16	55	36	93	16	19	33	14	8
P4	7	7	2	5	2	32	2	7	4	15	10	16	3	5	10	7	3
Total demand	343	286	123	128	22	1139	79	151	225	476	533	771	218	123	487	203	181

Table 10.19 Results corresponding to scenario 2023.

Scenario	Cross-docking warehouses that open	Distribution centers operating	Optimal distribution cost per operating day	Dimension obtained with the Lagrangian relaxation	Percentage difference with respect to the optimal cost
2023	1	1,7	$506,440	$505,295.85	0.22%
	2	6,9			
	3	2,4,11,14,16,17			
	4	5,8,15			
	5	3,10,12,13			

Table 10.20 Quantity of products shipped from plant k to the cross-docking i.

Cross-Docking (M) Plant (Product)	1	2	3	4	5
1	77	1087	1135	508	351
2	291	153	167	83	1027
3	45	88	115	50	174
4	9	36	37	19	36

It is important to mention here that for the four scenarios analyzed, the cross-docking warehouses have sufficient operating capacity to consolidate the demand of the 17 distribution centers; thus, there are no problems with storage capacity.

10.7 Conclusions

By comparing the total cost of the feasible solution against the optimal distribution cost of scenario 5, we obtain a percentage difference of 0.39%, which shows that the heuristic algorithm proposed in this study produces feasible solutions whose associated distribution cost is very close to the optimal distribution cost available. It must be noticed that the values of Lagrangian bounds in this work can be improved with techniques for the refinement of the classical Lagrangian bounds [13] or with a greedy heuristic to the Lagrangian solution [14].

Nowadays, the performance of companies depends to a great extent on the success of the collaboration between the participants in the supply chain. Therefore, the synchronization between the different actors involved in the chain, as well as the relationships established between them, is of vital importance to work in a coordinated manner and to meet the demands of an increasingly competitive market.

In this kind of network design, it is essential to consider a sustainable approach because this company can produce a significant amount of waste with disposable plastic bottles. With this in mind, we can extend the model to a Closed-Loop Supply Chain [15]. Litvinchev et al. [15] developed two mathematical programming models to determine the pricing strategy of the recovered products together with the optimal network that must be designed to be the most profitable closed cycle.

Because of the above and due to the constant increase in transportation costs, efficient channels must be developed in the distribution network, and the coordination of the company's operations must be encouraged to reduce the total cost of distribution and improve the average level of service offered to customers.

References

1 El Universal (2019). "México, primer consumidor de refrescos en el mundo", México. https://www.eluniversal.com.mx/nacion/mexico-primer-consumidor-de-refrescos-en-el-mundo (accessed 31 August 2021).

2 Soria I., (2008). *Rediseño de la cadena de abastecimiento de un grupo embotellador de bebidas,* MS thesis, ITESM Campus Toluca, Toluca, México.

3 Marmolejo, J.A., Soria, I., and Perez, H. A. (2015). "A decomposition strategy for optimal design of a soda company distribution system," *Mathematical Problems in Engineering*, vol. 2015, 8912047.

4 Granados J.A. (2007). *Optimización de la Red de Distribución de un Grupo Embotellador de Bebidas,* Tesis no publicada, ITESM, Campus Toluca, México.

5 Arampantzi, C., Minis, I., and Dikas, G. (2019). A strategic model for exact supply chain network design and its application to a global manufacturer. *International Journal of Production Research* 57 (5): 1371–1397.

6 Azizi, N. (2019). Managing facility disruption in hub-and-spoke networks: formulations and efficient solution methods. *Annals of Operations Research* 272 (1/2): 159–185.

7 Poudel, S., Marufuzzaman, M., Quddus, M.A. et al. (2018). Designing a reliable and congested multi-modal facility location problem for biofuel supply chain network. *Energies* 11 (7): 1682.

8 Yu, V.F., Normasari, N.M.E., and Luong, H.T. (2015). Integrated location-production-distribution planning in a multiproducts supply chain network design model. *Mathematical Problems in Engineering* 2015: 1–13.

9 Fisher, M.L. (2004). The lagrangian relaxation method for solving integer programming problems. *Management Science* 50 (12): 1861–1871.

10 Avriel, M. and Golany, B. (1996). *Mathematical Programming for Industrial Engineers*. New York: Marcel Dekker.

11 GAMS Development Corporation (2021). *The Solver Manuals*. GAMS Development Corporation.

12 Martin, R.K. (1999). *Large Scale Linear and Integer Optimization: A Unified Approach*. Boston: Kluwer Academic Publishers.

13 Litvinchev, I.S. (2007). Refinement of Lagrangian bounds in optimization problems. *Computational Mathematics and Mathematical Physics* 47 (7): 1101–1107.

14 Litvinchev, I., Mata, M., Rangel, S., and Saucedo, J. (2010). Lagrangian heuristic for a class of the generalized assignment problems. *Computers and Mathematics with Applications* 60 (4): 1115–1123.

15 Litvinchev, I., Rios, Y.A., Özdemir, D., and Hernández-Landa, L.G. (2014). Multiperiod and stochastic formulations for a closed loop supply chain with incentives. *Journal of Computer and Systems Sciences International* 53 (2): 201–211.

11

Machine Learning and MCDM Approach to Characterize Student Attrition in Higher Education

Arrieta-M Luisa F[1] and Lopez-I Fernando[2]

[1] Simon Bolivar University, Barranquilla, Atlántico, Colombia
[2] Autonomous University of Nuevo León, Department of Mechanical and Electric Engineering, San Nicolás de los Garza, Nuevo León, México

11.1 Introduction

The definition of student attrition, according to the Colombian Ministry of Education that attrition rates are defined by periods and by cohort. Attrition by periods takes place when a student abandons studies for two consecutive semesters or one year of academic activity [1]; cohort attrition refers to the difference between the number of students starting in a cohort and the number of students graduated from that cohort.

Impact of student attrition: Student attrition has adverse effects not only on educational institutions but also on the life trajectory of those who claudicate in their professional aspiration, in their family, in the educational system as a whole, and in society [2].

Student attrition in the region and the world: 47% of students drop out of in Colombian universities, 57.5% in Latin America, and between 7.4% and 6.1% in the United States [3]. Furthermore, European countries on average have a dropout rate that is even lower than in the United States, while in Africa, only 6% of the population of those that should enter higher education enroll in universities. In China, university attrition is approximately 10% [4].

The problem of mitigating student attrition is difficult to address not only because of the many factors that should be taken into consideration, or the huge amount of data that must be processed for helping in decision making but also because the decisions that should be taken and the artifacts that must be developed (like policies, strategies, among others) are fundamental of subjective nature and highly dependent of the "person(s) in charge" or the decision maker

Artificial Intelligence in Industry 4.0 and 5G Technology, First Edition.
Edited by Pandian Vasant, Elias Munapo, J. Joshua Thomas, and Gerhard-Wilhelm Weber.
© 2022 John Wiley & Sons, Inc. Published 2022 by John Wiley & Sons, Inc.

(DM). As recognized in [5] until now all attempts to characterize student attrition in higher education have failed when only academic factors have been considered.

In this work, the problem of student attrition is addressed by investigating the main factors that affect it, by considering factors proposed by the authors and some other factors also mentioned in other publications, such as academic, social, individual, economic, and institutional. In addition to presenting the problem of student attrition, its impact on society, an approach based on multi-criteria decision making theory (MCDM) [6] and machine learning (ML) [7] is presented to aid in proposing mitigation strategies for student attrition.

These approaches can also be applied to other complex decision problems where the amount of data or information available for decision aiding is large and diverse, which makes it very difficult to be interpreted in terms of preferences by a DM. For example, in portfolio problems like those mentioned in previous studies [8, 9].

11.1.1 Background

For the study of attrition, research has been implemented under an ex post facto holistic approach to identify those variables that play a fundamental role in the permanence of the university student.

Previous work focuses on several factors: vocation [10], academic performance [11, 12], etc. In [10] binary logistic regression analysis has been employed for studying drop out in undergraduate programs they concluded that choosing the degree mainly by vocation can act as a protective factor against abandonment while controlling the rest of the variables. Likewise, through multivariate statistics and discriminant analysis on the pattern of academic performance of engineering students, it is possible to identify that in 79% of the cases, obtaining good grades in mathematics and chemistry is a predicting factor for successfully reaching the fifth year. Authors in [11] concluded that the quality of higher education also affects the phenomenon of attrition due to the influence exerted by university studies on the development of people and the socio-cultural and economic impact that the university has on its environment as well as the importance of the implementation of institutional strategies for attrition management [13].

In addition, student attrition has been predicted from different perspectives employing machine learning through algorithms such as Random Forest, Neural Networks, Support Vector Machines, and Logistic Regression, wherein more than 50% of the attrition cases can correctly be predicted using information related to academic performance and graduation data [14]. To support the proposed work, we have investigated academic grade prediction with uncertainty attributes in performance [18].

11.2 Proposed Approach

Typically MCDM approach has been employed for aiding complex decision problems where the subjectivity of the DM plays a fundamental role [15] and also there are multiple criteria that should be taken into account in the decision process. Nevertheless, this approach requires that the amount of information or data that should be considered by the DM should be reasonably small, so it can be interpreted and evaluated by the DM to make his/her preferences explicit and construct a preference model. In today's scenarios where data-driven decision making is on its way to dominating decision-making processes in the industry, MDCM requires a huge effort to process and simplify data so the DM can represent his/her preferences efficiently in terms of time and effort, here is where ML and data science comes into play, for extracting insight from data reducing the volume of data that the DM must revisit when following an MCDM approach.

In the proposed approach, MCDM is employed for structuring the whole decision process for mitigating student attrition and ML for extracting useful insight from data for helping in decision making.

The proposed approach follows the main stages of a decision process defined by Simon [9]: (i) Intelligence, (ii) Design, (iii) Choice, and (iv) Implementation, but in each stage of the decision making process elements of ML are introduced to cope with a large amount of data. In was follows the stages of the proposed approach are presented, those activities related to the Simon approach are prefixed with an *(S)* and where the new activities related to the proposed approach are prefixed by a *(N)*.

1. *Intelligent phase*: *(S1)* The problem statement is formulated, objectives, goals, and alternatives and their consequences should be defined; decision problem(s) should also be identified by the DM; stakeholders are identified, and also sources of data or information are established.
   (N2) ML strategy should be defined, for deriving insight and synthesizing data volume and complexity in order to get a manageable decision model.
2. *Design phase*: *(N3)* Once the problem is stated, the decision problem type is identified, the ML strategy is defined, and data requirements are pointed out. *(S4)* Suitable alternatives should be designed, and the attributes defining their associated consequences should also be defined, and their values domains, relevant alternatives must be identified and only those should be further taken into consideration for the rest of the process (in a later moment other alternative can be incorporated as relevant or eliminated if proven irrelevant at all for the DM or are not more available). *(N5)* Strategy for ETL (data extraction, transformation and load) must be defined or at least outlined in a workflow design.

3. *Choice phase*: *(S6)* The MCDM method is selected, DM preferences are elicited, the decision model is implemented, or a decision support system (DSS), if required. *(N7)* ETL strategy is implemented; data are extracted, transformed, and loaded into the selected platform; then exploratory data analysis (EDA) is performed and consequently, ML strategy could be adjusted. Then ML strategy is applied for extracting insight from data, reducing data, and preparing it to meet decision model requirements [8]. *(S8)* MCDM decision model is fed with input data; MDCM decision model output is employed to construct the recommendation for the DM. Finally, the recommendation is presented to the DM and adjustments are made to it before implementation if needed.
4. *Implementation phase*: *(S9)* Deployment of the MCDM model and DSS tools is carried out, also any training required to perform the implementation of the final decision, and if needed adjustments to the recommendation. *(N10)* A data pipeline is implemented to guarantee that the decision remains current for a certain period where updates or changes in the system and environment are expected to happen.

11.3 Case Study

University Simon Bolivar (USB) is located in Barranquilla, Colombia and was selected as a case study for applying the methodology. USB is a private university that was founded more than 40 years ago. Currently, more than 10,000 students are registered in at least one of the programs offered by the USB.

Currently, within the Simón Bolívar University, different characterizations of students upon admission are carried out to identify a series of critical moments to guide their educational process. The latter precautions are performed in isolation, which prevents measurement of the effectiveness and positive impact on the actual monitoring of the student, from the different dependencies of the university. Through a management model for student permanence, it is sought to measure the impact of the services offered by the university to the student that aim to be efficient strategies to mitigate dropout and improve the graduation rate (which is currently at 48%). These services consist of a combination of academic support services and social, emotional, and economic efforts.

In this work, only phases 1 to 3 are applied in the case study, as a pilot test, the whole methodology has not officially been implemented yet in the USB, so phase 4 has no point to be included here. In was follows it is described how the methodology phases were applied:

11.3.1 Intelligent Phase

(S1) The problem is being stated as characterization of the main attribute or characteristics that drive student attrition in the USB, for aiding decision making in formulating strategies and policies for mitigating university drop out to help the students to graduate in time and with quality. The alternatives are the records of students' life cycle where their stay at USB. The consequences of each alternative are those related to academic, economic, individual, and institutional states of a student, which are synthesized in the prediction of drop out for the student, so the decision problem is an MCDM evaluation problem where there are two categories: *YES* – meaning the student has a high probability for drop out soon, and *NO* – meaning the student is not foreseen to drop out soon. The DM is the Rector of the USB and academic and supporting personnel are identified as stakeholders. Finally, the Academic information ERP in use in the USB is the main source of information and data for the decision problem.

(N2) As the volume of available information is large and complex, ML is used to perform features reduction, and simplify the amount of data that will be used as input for the decision problem. As there is no evidence of which method performs better, the most popular methods were employed, and their results were compared to extract the most robust features.

11.3.2 Design Phase

(N3) Data requirements were established by considering Academic, Economic, Individual and Institutional factors that were extracted from the ERP system at USB. At the time of conducting current research, the following indicators were available for each registered student at the USB:

- *Academic*: Official Identity Document (OID), an assigned unique identity code inside USB, First year and Semester he/she enrolled in USB (YE,SE), each year and semester he/she registered as a student in USB (YR, SR) with record of faculty (F^*), program (P^*) and courses taken (C^*) (at least 1 course and at most seven courses), as well as grades obtained in each course (G), finally an academic category (AC) was also assigned to each student: *registered*, *graduated* and *dropped out*, if a student did not register in a semester also the main cause of drop out was also registered (MC).
- *Economic*: financial aids is granted to the student in a year and semester (FA^*). Socioeconomic Status, in Colombia, is measured as "Strata" defined in a scale from 1 – 6, where 6 are the richest and 1 the poorest, (SES^*).
- *Individual:* If he/she is considered vulnerable (possess any disability or any other condition recognized which make he/she vulnerable) (V^*).
- *Institutional:* if he/she participates in an institutional companion program (psychological, academical, or financial mentoring) (CP^*).

Those characteristics marked with (*) can have more than one value per student.

(S4) Each unique register for a student in the ERP database was considered as an alternative, but only the ones mentioned in (N3) above were included in the data model.

(N5) Following ETL workflow was implemented:

1. A unique view was defined to store results extracted from the database; SQL was employed (ERP used MS SQL for managing data).
2. Records with incomplete data were discarded.
3. Two calculated fields were added for representing dropout categories: in one field students were labeled for each year (*YR*) and semester (*SR*) as *YES* if they have been dropped out or not *NO* if they haven't dropped out. In the other calculated field students were labeled following the definitions given in Colombia [16, 17]:

   Deserter by discontinuity is a student who does not register in the institution for a period comprising more than two or more contiguous semesters (labeled as *DD*).

   Deserter by cohort is a student who has not graduated between the interval of time of 10 or more semesters since his/her first registration at the institution (labeled as *DC*).
4. In any other case, the student is labeled as no deserter (*ND*).
5. Fields with many values were transformed as discrete with few values, like, for example, grades which in Colombia can take values between 1 and 5 (grades were mapped to *Reproved, Pass, Good, Excellent*).
6. The resulting data were stored as a *csv* file.

11.3.3 Choice Phase

(S6) As the decision problem was recognized as an MCDM evaluation problem ELECTRE TRI was selected as an appropriate method for stating if a student will potentially drop out in near future or not. As a decision-maker was not available no preferences were elicited, but when preferences are available then the preference model should contain: relative importance of attributes (weights), category profiles, at least one for each category (*YES* or *NO*) in terms of minimal requirements for each attribute. This task was not implemented in this work.

(N7) ETL Workflow was implemented, as a result, a *csv* file was created (with data extracted from the database) with more than 500,000 records with student's data collected from 2015 through 2019. For the two new columns calculated if a student has had a grade of NULL in all courses offered in a year and semester, it was considered as a deserter for that period.

EDA was conducted mainly employing the R language. Characterization of attributes by means of descriptive statistics was applied, also correlation analysis was employed, and redundant attributes were discarded. Some attributes were remapped as, for example, *FA* and *CP* which were transformed to a scale of *YES* or *NO*, indicating if a student received financial aid or participated in a companion or not.

As a result of EDA further transformation of data was carried out and the final dataset which was defined as input for ML algorithms was created.

This dataset contained more than 600,000 records and 14 variables, including those representing characterizations of students regarding drop out. In Table 11.1 these variables are listed.

Then different ML classification models were applied for feature reduction. RapidMiner 9.03 was employed with an auto model for classification based on attributes "Is Deserter" and "Deserter YES_NO."

Table 11.1 Dataset variables and calculated variables.

Variables	Explanation
Grade (G)	Assigned to each student when finalizing a course, can take the values: excellent, good, pass, reproved. Or any other scale
Academic state or category (AC)	Take the values active, graduated or dropped out
Year (YR)	Year of date related to current semester and courses in which student is involved
Strategy (CP)	Involvement in University companion program, if applicable
Financial Aid (FA)	External financial Aid granted to student, if applicable
First year of registration (YE)	Year of student's first registration at the University
First semester of registration (SE)	Semester of student's first registration at the University
Faculty (F)	Faculty in which the student is registered
Program (P)	Program in which the student is registered in a year and semester at university
Vulnerable (V)	If the student is socially vulnerable, if applicable
Strata (SES)	Socio economic state or class
Course (C)	Registered course by a student in a year and semester at university
Main cause of drop out (MC)	Given by the student when drop out in a year and semester and registered in ERP
Semester (SR)	Semester, related to current year and registered courses
IS Deserter (calculated)	Deserter by cohort (DC), deserter by discontinuity (DD), Not deserter (ND)
Deserter YES_NO (calculated)	Deserter (YES) or Not deserter (active or graduated) (NO)
Years past since last registration (calculated)	Years past since last semester registration stored in the ERP

11.4 Results

Auto model feature from RapidMiner was applied to the dataset constructed (Figure 11.1).

In Table 11.2 the methods applied for predicting both attributes "*Deserter YES_NO*" and "*Is Deserter*" are listed. In Table 11.2 the accuracy of classification is shown (with 70/30 percent data split for training and testing), where numbers in parentheses represent method ranking by accuracy.

As can be seen in Table 11.2, *Generalized Linear Model* performed best for the prediction of both attributes, followed by *Logistic Regression*, but in general, all

Figure 11.1 RapidMiner auto model feature.

Table 11.2 Accuracy of classification methods.

	DESERTER YES_NO	IS_DESERTER
Method	**Accuracy**	
Naive bayes	90.00% (4)	89.93% (4)
Generalized linear model	90.20% (1)	90.57% (2)
Logistic regression	90.11% (3)	90.52% (3)
Decision tree	90.16% (2)	86.31% (5)
Random forest	89.97% (5)	90.70% (1)
Support vector machine	87.70% (6)	89.33% (6)

methods performed well. Also, the worst algorithms were *Naïve Bayes* and *Support Vector Machines* (although no kernel was employed).

Similar results were obtained by other quality metrics like *AUC*, *Specificity*, *Recall*, *Sensitivity*, *f-Measure*. It is worth noting that classes generated by both columns are highly imbalanced, in such situations metrics like, for example, *AUC*, *Accuracy*, or *Sensitivity*, couldn't give accurate results [9]. Therefore, other metrics were employed like *G-MEAN* and *Balanced Accuracy*, but similar results were obtained.

When examining in detail classification for attribute "Is Deserter" it could be noted that all algorithms performed poorly in identifying class "*DD*," this result can be explained by two facts: there are too few cases for this class and there is no recognizable regular pattern for these students.

As a result of features selection, the most selected features in both situations by all methods were: "Grades," "Years since last enrollment," "Student state," "Economic support," "Year," "Genre." Belonging to *Academic, Economic*, and Individual categories or dimensions. These results agreed in general with responses given to students to a questionnaire related to dropout motives and factors influencing their decision to drop out (as is shown in Figure 11.2).

Finally, RapidMiner auto model was applied only for *Generalized Linear Model* and *Logistic Regression* only for the selected attributes (suggested by the results of features selection in RapidMiner auto model) for classifying both attributes ("*Deserter YES_NO*" and "*Is Deserter*"), results of accuracy can be observed in Table 11.3.

As can be seen, the *Logistic Regression* method performs slightly better than the *Generalized Linear Model* method in classifying the attribute "*Is Deserter*". Where in attribute "*Deserter YES_NO*" the difference between both methods is not significant. Moreover, both methods achieved good results in classifying

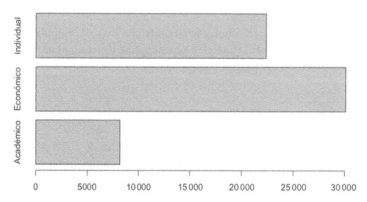

Figure 11.2 The tree main reasons for dropping out.

Table 11.3   Accuracy for the simplified model.

	DESERTER YES_NO	IS_DESERTER
Method	**Accuracy**	
Generalized linear model	**90.16% (1)**	90.50% (2)
Logistic regression	90.09% (2)	**90.60% (1)**

both attributes. Nevertheless, more experimentation is needed for state general conclusions, specifically more data are required for including all indicators of the proposed approach.

11.5   Conclusion

A methodology for decision aiding in the management of student attrition has been presented. The MCDM core of the methodology brings structure to the process while the ML core of the methodology helps in coping with the massive and diverse nature of data, that should be taken into consideration for the development of strategies and policies for effectively addressing student attrition in higher education.

Results of the application of the methodology to the case study have shown that students who eventually drop out are very difficult to classify because they do not exhibit recognizable patterns, at least within the scope of the data collected in USB. Although promissory results were obtained, more experimentation and research should be made for getting robust results in the application of the proposed approach.

RapidMiner auto model has proven to be a valuable tool for starting to study a complex problem like student attrition.

As an interesting research open question remains the task of helping the decision-makers to develop efficient strategies for mitigating the student attrition rates in universities based on the proposed approach.

References

1 Torres Martínez, L. D., Prias Alarcón, V. A., & León González, R. A. (2020). *Analysis and approach to the problem of student attrition within a higher education institution of a technical and technological nature.* Doctoral Dissertation, The Central Technical Institute Technological School, Cooperative University of Colombia, Bogotá, in Spanish.

2 Isaza, L.G., Lubert, C.D., and Montoya, D.M. (2016). Characterization of student attrition at the university of Caldas in the period 2009-2013. In Analysis from the system for the prevention of the dropout of higher – SPADIES. *Latin American Journal of Educational Studies (Colombia)* 12 (1): 132–158.

3 Tight, M. (2020). Student retention and engagement in higher education. *Journal of Further and Higher Education* 44 (5): 689–704.

4 Marioulas, J. (2017). China: a world leader in graduation rates. *International Higher Education* 90: 28–29.

5 Wagner, K., Merceron, A., and Sauer, P. (2020). Accuracy of a cross-program model for dropout prediction in higher education. In: *Companion Proceedings of the 10th International Learning Analytics & Knowledge Conference (LAK 2020)*, 744–749.

6 Alvarez, P.A., Ishizaka, A., and Martínez, L. (2021). Multiple-criteria decision-making sorting methods: a survey. *Expert Systems with Applications* 115368.

7 Ciolacu, M., Tehrani, A. F., Binder, L., & Svasta, P. M. (2018). Education 4.0-Artificial Intelligence assisted higher education: early recognition system with machine learning to support students' success. *IEEE 24th International Symposium for Design and Technology in Electronic Packaging (SIITME)* (23-30). IEEE.

8 Litvinchev, I. and López, F. (2008). An interactive algorithm for portfolio bi-criteria optimization of R&D projects in public organizations. *Journal of Computer and Systems Sciences International* 47 (1): 25–32.

9 Litvinchev, I., López, F., Escalante, H.J., and Mata, M. (2011, 2011). A milp bi-objective model for static portfolio selection of R&D projects with synergies. *Journal of Computer and Systems Sciences International* 50 (6): 942–952.

10 Lottering, R., Hans, R., & Lall, M. (2020). A model for the identification of students at risk of dropout at a university of technology. *International Conference on Artificial Intelligence, Big Data, Computing and Data Communication Systems (icABCD)*, 1-8. IEEE.

11 Dicovskiy, Pedroza (2017). Prediction of dropout and success in students. *Case Study: Agro-Industrial Engineering at UNI Norte*, Nicaragua, 2011-2015.

12 Martín, T.F., Salazar, M.S., Jiménez, M.T.H., and Mora, T.E.M. (2019). A multinomial and predictive analysis of the factors associated with university dropouts. *Educare Electronic Magazine* 23 (1): 3.

13 Domingo, J.R. (2014). University dropout: variables, frames of reference and quality policies. *Journal of University Teaching* 12 (2): 281–306.

14 Kemper, L., Vorhoff, G., and Wigger, B.U. (2020). Predicting student dropout: a machine learning approach. *European Journal of Higher Education* 10 (1): 28–47.

15 Fernandez, E., Lopez, F., Navarro, J. et al. (2009). An integrated mathematical-computer approach for R&D project selection in large public organizations. *International Journal of Mathematics in Operational Research* 1 (3): 372–396.

16 Simon, H.A. (1965). Administrative decision making. *Public Administration Review* 31–37.

17 Forman, G. (2003). An extensive empirical study of feature selection metrics for text classification. *Journal of Machine Learning Research* 3 (2003): 1289–1305.

18 Ali, A.M., Joshua Thomas, J., Nair, G. (2021). Academic and Uncertainty Attributes in Predicting Student Performance. In: Vasant, P., Zelinka, I., Weber, GW. (eds) Intelligent Computing and Optimization. ICO 2020. Advances in Intelligent Systems and Computing, vol 1324. Springer, Cham. https://doi.org/10.1007/978-3-030-68154-8_72.

12

A Concise Review on Recent Optimization and Deep Learning Applications in Blockchain Technology

Timothy Ganesan[1], Irraivan Elamvazuthi[2], Pandian Vasant[3], and J. Joshua Thomas[4]

[1] Member of American Mathematical Society, University of Calgary, Alberta, Canada
[2] Persiaran UTP, Seri Iskandar, Perak
[3] MERLIN Research Centre, TDTU, Ho Chi Minh City, Vietnam
[4] UOW Malaysia, KDU Penang University College, Malaysia

12.1 Background

In recent years, with the rise of cryptocurrency, blockchain technology has become increasingly dominant in the area of global finance and economics. The initial idea of blockchain was presented by a person(s) using the pseudonym Satoshi Nakamoto. Nakamoto provided the description of how the combination of cryptography and open distributed ledger could be applied to develop a reliable digital currency (today known as cryptocurrency) [1]. The central idea of a blockchain is a growing list of records (i.e. blocks) that are connected using cryptographic hashes. Blockchain technology is essentially a type of database that stores records. The data are collected and stored in groups and each group is called a "block." As new data come in, a new block is created and linked to the previous block – creating a chain of blocks (blockchain). The individual blocks contain the cryptographic hash of the previous block, its associated timestamp, and the transaction information. Due to the cryptographic linking of the blocks, the blockchain is inherently resistant to tampering or modification. Thus any individual block could not be modified without changing or altering all its subsequent blocks. This inherent design security endows blockchain technology with the following advantageous features – making it suitable to be applied as under:

(i) an open distributed ledger
(ii) a decentralized system
(iii) a system with complete information transparency

Artificial Intelligence in Industry 4.0 and 5G Technology, First Edition.
Edited by Pandian Vasant, Elias Munapo, J. Joshua Thomas, and Gerhard-Wilhelm Weber.
© 2022 John Wiley & Sons, Inc. Published 2022 by John Wiley & Sons, Inc.

(iv) a tamper-proof and open system

(v) a suitable digital system of finance (cryptocurrency)

The reader is directed to the following literature for more in-depth and comprehensive reading on the fundamentals of blockchain technology: Komalavalli et al. [2], Krishnan et al. [3], Laroiya et al. [4], and Madakam [5].

In addition to its application in financial systems, blockchain technology has widely disseminated into many sectors in the global market e.g. energy [6–8], cybersecurity [9, 10], supply chain [11–13], healthcare [14–16], communication/internet of things (IoT) [17, 18], smart cities [19], smart grids [20], and applied engineering [21].

The area of optimization and deep learning research has progressed alongside blockchain technology. Optimization and deep learning techniques have significantly advanced in recent times – enabling them to take on highly complex academic as well as industrial problems. These complex large-scale problems often contain multiple objectives, nonlinearities, stochasticity/fuzzy uncertainties, nonconvex functions, multiple levels, and many variables [22–31]. A graphical representation of complexities associated with current optimization problems is given in Figure 12.1.

Although blockchain technology in its current state is fully mature and implementable, various improvement and optimization opportunities have been identified across various fields. One such effort in that direction is the optimization of the hashing algorithm. Hashing algorithms are a critical component of modern-day cryptography. As one-way functions, hashing algorithms are employed to ensure and verify the integrity of data – the resulting hash could be thought of as a fingerprint of the original message [32]. Blockchain technology has also been integrated with the domain of IoT to optimize issues related to privacy, security as well as scalability [33]. Another interesting improvement avenue is seen in the optimization of the execution of smart contracts. Smart

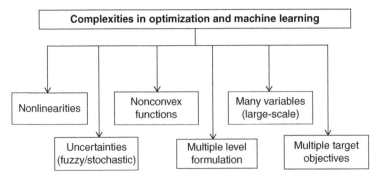

Figure 12.1 Complexities in the field of optimization and machine learning.

contracts are essentially chunks of code that are stored on a blockchain. These codes are then automatically executed when the associated terms and conditions of the smart contracts are satisfied. These smart contracts have very high utility in commercially collaborative environments where they expedite and increase the efficiency of business processes [34] as well as financial systems (e.g. cryptocurrencies). There have also been many blockchain improvement opportunities in the energy sector, particularly in the design and operations of smart grids. Since smart grids require multiple secure transactions on a network – smart contracts running on blockchain technology have emerged as the best solution. Therefore, blockchain technology applied in this respect provides the following advantages [35]:

- Ensuring the execution of contracts between generators and consumers.
- Immutability of transaction history – avoiding transaction disputes.

Thus optimization of the implementation of blockchain technology on smart grids is becoming a necessity. In light of these developments, researchers and industrial practitioners have become motivated to optimize blockchain networks, specifically using artificial intelligence (A.I.) systems [36]. Another interesting area for improving the usage of blockchain technology is in the optimization of supply chains. As supply chains evolve to become more global, managing them has become exponentially complex – in terms of traceability and transactions (validations and time frame) [37]. The effective utility of blockchain technology could play a pivotal role in overcoming some of these management barriers via the usage of smart contracts and other decentralization features. Optimization in these implementations and the advent of a completely digital supply chain could significantly improve conventional supply chain management systems, thereby optimizing business operations. Healthcare systems have also been a target for the application of blockchain technology as the industry continues to be overburdened by high hospital costs, inefficient practices, and constant data breaches [38]. Thus, blockchain technology has been proposed to introduce a secure data management system – which efficiently, reliably, transparently, and securely manages healthcare data. Similar to the application of blockchain technology in other areas, the main idea in healthcare data management systems is to create and maintain a network regulated by members who store and share information instead of by a third party [39]. The potential application of blockchain technology in healthcare involves the optimization of mobile health devices, insurance information storage, sharing and storing medical records, medical monitoring devices, and clinical trial data.

The primary aim of this paper is to provide a brief review of current intersections between the fields of optimization and deep learning with the application of blockchain technology. This paper is organized as follows: Section 12.2 provides current works involving computational optimization frameworks in blockchains while Section 12.3 contains recent optimization efforts in the direction of blockchain technology application in IoT. Section 12.4 discusses the optimization works related to smart grids integrated with blockchain technology. Sections 12.5 and 12.6 provide recent trends with respect to the usage/optimization of blockchain technology in supply chain and healthcare management systems. The paper ends with the outlook Section – which discusses the application of optimization techniques in blockchain-enabled systems.

12.2 Computational Optimization Frameworks

One of the computational bedrock of blockchain technology is the hashing algorithm. The hashing algorithm ensures the integrity of data encryption in blockchains. An interesting move in the direction of optimizing hash algorithm is seen in the work of Fu et al. [40]. In that work, the PRCA (proactive reconfigurable computing architecture) is used to further improve hash algorithms. The heuristic optimization approach focused on enhancing the efficiency of network data transmission as well as communication facilities. This is achieved by combining blockchains with mimic computers. To ensure data integrity and security, Fu et al. [40] strategically employed a lightweight hashing algorithm that does multiple hashing and transforms the hash algorithm structure. The numerical experiments performed in Fu et al. [40] show that the proposed optimization strategies improved the efficiency of data processing as well as the data security of blockchains.

Another step toward algorithmic optimization of blockchains could be observed in the work of Shibata [41] who proposed a decentralized consensus protocol for blockchains. The protocol was developed to address the blockchain energy wastage problem. Using the proposed protocol, the energy wastage is minimized by searching for approximate solutions to problems submitted by any given nodes (i.e. clients). Similar to the proof-of-work methodology, using the proposed protocol a client provides a search program that implements any search algorithm for finding the best approximate solution for adding a new block to the chain. The node that finds the best approximate solution is rewarded by the client.

In Zhang and Hu [42], the hash function specifically employed for bitcoin mining was optimized. The methodology employed a combination of carry-save adder and re-timing method to reduce the number of computational cycles performed by the hash function. The computational results of the simulations performed by the authors showed that the proposed optimization methodology improved the efficiency of hash function computations. A similar optimization effort toward blockchain mining is seen in Adewumi and Liwicki [43]. In that work, the authors propose an inner for-loop population-based approach as an alternative to the brute-force approach for blockchain mining (which is energy-intensive and time-consuming). Comparative analysis performed in Adewumi and Liwicki [43] showed that the proposed approach was slightly faster than the conventional brute-force strategy.

Due to concerns related to the overutilization of energy used in blockchains, various research works have been directed toward tackling this problem. These efforts attack the problem by optimizing the computational overhead, thereby reducing the energy consumption of blockchains. For instance, Chenli et al. [44] proposed a novel blockchain design as an alternative to the energy-intensive proof-of-work (PoW) mechanisms. The proposed design completely recycles the energy and uses it for the computation of deep learning: proof-of-deep-learning (PoDL). The authors performed benchmark simulations and found that the proposed strategy was feasible across many mainstream cryptocurrencies – e.g. Bitcoin, Bitcoin Cash, and Litecoin. Similar research was carried out in the work of Baldominos and Saez [45]. In that research, the authors developed a proof-of-useful-work scheme as an alternative to the PoW mechanism for application over blockchains. The proposed scheme requires that the mining scheme trains deep learning models. Additionally, it also requires that a block is only mined if the performance of the mentioned model exceeds a specified threshold. The proposed system enables model (delivered by miners) verification by the nodes in an easy and efficient way – deciding the time when a block needs to be generated. Another alternative to the PoW mechanism is given in the work of Qu et al. [46]. In that work, the proof-of-federated-learning (PoFL) scheme was proposed where the energy originally used for solving problems in PoW is reinvested for federated learning. The proposed method faced challenges in terms of data privacy. To address these problems the authors introduced a reverse game-based data trading mechanism and a privacy-preserving model verification mechanism. The authors then performed simulations based on real-world data to demonstrate the effectiveness of the proposed methodology (i.e. PoFL).

In this section, it can be surmised that there have been many recent efforts toward computational optimization of blockchains. These developments have been carried out in the spirit of energy optimization, improved computational efficiency as well as enhanced data security/integrity. The following section takes a closer look at recent research works on blockchain optimization with respect to its application in IoT. Table 12.1 provides a summary of the literature discussed in this section:

Table 12.1 Summary of computational optimization framework in blockchain technology.

References	Application	Technique
Fu et al. [40]	Hash algorithm optimization	Proactive reconfigurable computing architecture
Shibata [41]	Energy optimization	Decentralized consensus protocol
Adewumi and Liwicki [43]	Blockchain mining	Inner for-loop population-based approach
Zhang and Hu [42]	Bitcoin mining	Improved hash function
Chenli et al. [44]	Energy optimization	Deep learning
Baldominos and Saez [45]	Blockchain mining	Deep learning
Qu et al. [46]	Blockchain mining	Federated learning, reverse game-based data trading mechanism and privacy-preserving model verification

12.3 Internet of Things (IoT) Systems

In recent times, IoT has become important among the primary platform for the application of blockchain technology – opening various optimization opportunities [47]. An interesting example is seen in the work of Cao et al. [48]. That work focused on the industrial IoT platform and the associated blockchain scalability restrictions. The authors of Cao et al. [48], framed the challenge as a many-objective blockchain-enabled industrial IoT optimization model (with four objective functions). The problem was then solved using an improved Two_Arch2 algorithm – where the computational results proved the solution method effectively optimized the industrial IoT problem.

Another interesting optimization work in IoT could be seen in Liu et al. [49] – where security and efficiency issues of massive industrial IoT data were considered. In that work, the authors proposed a novel deep reinforcement learning (DRL)-based performance optimization framework. The framework was tailored for blockchain-enabled industrial IoT systems. The optimization framework focused on critical aspects of the mentioned IoT systems: decentralization, latency, scalability, and security. In that work, simulations were carried out for benchmark testing of the proposed technique against the conventional method. Simulations results showed that the proposed technique effectively improved the performance of the blockchain-enabled industrial IoT system as well as adapted to its associated dynamics.

In Fu et al. [50], optimization work was focused on maximizing the system energy efficiency of a blockchain-based IoT. In that work, an optimization problem was formulated by optimizing the allocation of cache, computation, and communication resources such that it maximizes the overall system energy efficiency. This was done by employing geometric programming – where the impact of various parameters (involved in the blockchain) on the system performance is analyzed. The authors of Fu et al. [50], then went on to verify the proposed energy efficiency mechanism in the blockchain-based IoT.

A multi-objective optimization methodology for enhancing the security in blockchain-enabled industrial IoT was developed in Cai et al. [51]. In that work, sharing technology that improves the overall throughput and scalability of blockchain networks was considered. The primary aim of the work of Cai et al. [51], was to optimize the performance of blockchain networks and reduce the potential for malicious node aggregation – which results in the optimization of the shard validation validity model. To achieve this, the authors proposed a many-objective optimization algorithm based on dynamic reward and penalty mechanisms. Simulations performed showed that the proposed algorithm significantly improved the sharding model which thereby enhances the security of the blockchain-enabled IoT. In the same spirit of Cai et al. [51], the work of Yun et al. [52] was carried out to improve secure sharded blockchain systems for IoT services. The authors of that work proposed a deep Q network shard-based blockchain (DQNSB) scheme. The scheme dynamically searches for the optimal throughput configuration. To achieve this, deep reinforcement learning agents were trained to find the optimal system parameters in response to the network status. The agents then optimize the system throughput and security level in an adaptive way. The test simulation conducted by Yun et al. [52], showed that the proposed scheme provided higher transactions per second and maintained high-security levels as compared to current deep reinforcement learning-enabled blockchain technology.

A more focused application (cloud storage optimization) in blockchain-enabled IoT is observed in Xu et al. [35]. In that research work, the authors aimed to address problems related to data storage, which often arise when dealing with large-scale networks often seen in blockchain-enabled IoT systems. In Xu et al. [53], the authors constructed a series of objective functions related to storage cost, query probability, and local space occupancy – reducing the problem to block selection. The solution method employed for solving the multi-objective optimization problem was a variant of the non-dominated sorting genetic algorithm (NSGA). Segments of the NSGA technique were modified to account for clustering – e.g. individual selection. The solutions from the Pareto set were then compared using other NSGA algorithms such as NSGA-II and NSGA-III. The comparative analysis showed that the proposed approach NSGA-C (NSGA with modifications for clustering) was superior to NSGA-II and NSGA-III with respect to local space occupancy in the blockchain application.

In Jameel et al. [36], the authors tackle a frequently occurring problem when applying blockchains to IoT networks: the branching problem (i.e. forking event).

Forking introduces excessive overhead in the network which could lead to potential security breaches. The authors employed a deep neural network in the distributed IoT with the target of maximizing the rate of the communication link between the IoT miner and the communication point. The computational results showed that the proposed solution significantly improved the mentioned rate, thus reducing the transmission delays. Similarly, a deep reinforcement learning approach was utilized in Li et al. [54], for a blockchain-enabled IoT with edge computing. A deep learning approach called dueling deep Q-network (DQN) was utilized for optimal selection and decision in caching servers, computing servers, and blockchain systems. The idea was to achieve maximum system rewards by (i) increasing the efficiency of data processing, (ii) enhancing data transfer security, and (iii) lowering network costs. The authors of Li et al. [54] performed extensive simulations and proved that the proposed approach improved the system as compared to existing schemes.

Blockchain-driven IoT applications have also become prevalent in the manufacturing sector. Optimization of such an application in manufacturing could be seen in the work of Leng et al. [55]. In that work, a novel bilevel model was developed (called ManuChain) to remove unbalances/inconsistencies between local execution in individualized manufacturing systems and holistic planning. The lower level problem focuses on blockchain-driven smart contracts and individualized task executions while the upper-level problem deals with the iterative coarse-grained holistic optimization. Table 12.2 gives a summary of the literature discussed in this section.

12.4 Smart Grids Data Systems

The future of the power industry is currently evolving towards the integration of intelligent systems with power grids. These smart grids are more flexible, reliable, secure, and resilient as compared to conventional power grids. Additionally, the advent of blockchain technology has accelerated these developments globally. Recent reviews in the application of blockchain technology in smart grids could be seen in the following works: Alladi et al. [20], Mollah et al. [56], Patil et al. [57], Gai et al. [58], Khan et al. [59], Bera et al. [60], Agung and Handayani [35], Musleh et al. [61], Aderibole et al. [62], Kumari et al. [63], and Hassan et al. [64].

Blockchain-enhanced smart grids have also become a ripe area for the implementation of state-of-the-art optimization techniques. For instance, in Stübs et al. [65], the authors focused on smart grid control optimization (via a virtual power plant framework). The central idea of that work was to determine the optimal solution as a schedule vector over all controllable generators and

Table 12.2 Summary of optimization efforts in blockchain-enabled IoT applications.

References	Application	Technique
Cao et al. [48]	Blockchain scalability restrictions	Improved Two_Arch2 algorithm
Liu et al. [49]	Infrastructure optimization	Deep reinforcement learning (DRL)
Fu et al. [50]	Energy efficiency	Geometric programming
Cai et al. [51]	Network security	Many-objective optimization algorithm based on dynamic reward and penalty mechanism
Yun et al. [52]	Improve secure sharded blockchain systems	Deep Q network shard-based blockchain (DQNSB) scheme
Xu et al. [53]	Data storage optimization	Non-dominated sorting genetic algorithm with clustering (NSGA-C)
Jameel et al. [36]	Branching problem (i.e. forking event)	Deep neural network
Li et al. [54]	Edge computing	Dueling deep Q-network (deep learning)
Leng et al. [55]	Manufacturing system	Bilevel optimization

loads. To achieve this, the authors of Stübs et al. [65], employed blockchain technology and proposed the Mempool model for optimal criteria selection.

Similar work was carried out by the authors of Mureddu et al. [66] – where an optimization framework was introduced for application to smart grids. In that work, a novel decentralized genetic algorithm was presented to optimize the network operation as well as improve the grid's resilience to malfunctions and cyber attacks. The work incorporated a decentralized blockchain which served as a coordinating node among the distributed computing devices. The proposed approach presented in Mureddu et al. [66], was tested on an optimal scheduling problem in a local medium voltage network (with distributed renewable generation and controllable loads).

Optimization efforts on blockchain-enhanced applications in electrical vehicles (EVs) could be seen in the work of Li et al. [67]. In that research work, the concept of hierarchical and zone-based scheduling was proposed for an iterative two-layer model for optimizing the charging/discharging trading of EVs. The main objective of that work was to minimize the overall load variance of the distribution network under (i) vehicle travel demand and (ii) power flow constraints. The problem was formulated as a mixed-integer programming problem and solved using the proposed adaptive inertia weight krill herd (KH) algorithm. The decentralized trading architecture was designed using a consortium blockchain. This architecture ensures the security and privacy of the two-way electricity trading system (between the smart grid and the EVs). The authors of Li et al. [67] then performed the following simulations/analyses to test their proposed systems: power load profiles, security analysis as well as feasibility/efficiency analysis. Another interesting work on EVs could be seen in Wang et al. [68]. In that work, an A.I.-enabled blockchain-based EV integration system where an EV fleet is utilized as a consumer/supplier of electrical energy within a virtual power plant platform. Using a federated learning approach, the proposed scheme was seen to be able to improve power consumption prediction, reduce power oscillations and achieve cost-efficient performance – while providing a secure and transparent service.

In Swain et al. [69] a blockchain-enhanced smart microgrid system was considered. The authors targeted to improve energy efficiency and data security of the system by systematic optimization of the energy model. Similar to the work of Li et al. [67], a decentralized energy trading platform was conceived via the integration of blockchain technology. To optimize energy efficiency, network security, and stability, bio-inspired algorithms were utilized – specifically particle swarm optimization (PSO) and genetic algorithm (GA).

In addition to bio-inspired approaches, deep learning and neural networks have also become critical tools to be implemented in smart grid systems. For instance, in Khalid et al. [70], optimization was carried out on the load demand and storage management of the smart grid – in response to dynamic pricing. An artificial neural network was employed to find the low tariff zone by developing models of load, pricing, and energy storage systems (ESS). The linear programming optimization model is then optimized using the MATLAB-based solver: mixed-integer linear programming. Similarly, the work of Ferrag and Maglaras [71], also utilizes a deep learning approach for improving a blockchain-based smart grid. In that work, the novel deep learning approach called DeepCoin as well as a blockchain-based energy framework for smart grids was proposed. The central theme of that work was to develop a reliable peer-to-peer energy system based on the Byzantine fault tolerance algorithm. The deep-learning neural net acts as an intrusion detection system (IDS) for detecting network attacks and fraudulent transactions in the blockchain-based energy network. The performance of the proposed scheme was analyzed and benchmarked. A brief overview of the research works discussed in this section is given in Table 12.3.

12.5 Supply Chain Management

Blockchain technology has also made its way to supply chain-based applications – digitizing conventional supply chains and in effect optimizing logistics,

Table 12.3 Summary of recent optimization works in blockchain-enhanced smart grids.

References	Application	Technique
Stübs et al. [65]	Smart grid control optimization	Mempool model
Mureddu et al. [66]	Network optimization	Decentralized genetic algorithm
Li et al. [67]	EV charge/discharge trading optimization	Mixed integer programming – adaptive inertia weight krill herd (KH) algorithm
Wang et al. [68]	EV integration system (virtual power plant)	Federated learning
Swain et al. [69]	Energy trading platform optimization	PSO and GA
Khalid et al. [70]	Load demand and storage management optimization	Artificial neural network and mixed integer linear programming
Ferrag and Maglaras [71]	Network resilience and security optimization	Deep learning approach (DeepCoin)

transportation, operations, and transactions. To review some of these recent developments, the reader is directed towards the following literature: Dujak and Sajter [72], Rejeb et al. [73], Choi et al. [74], Merkaš et al. [75], Waller et al. [76], Tan et al. [77], Saurabh and Dey [78], Wamba and Queiroz [79], and Sangeetha et al. [80].

Recent progress in supply chain management has opened up various avenues for optimization efforts. An interesting example is seen in the work of Abidi et al. [81] – where blockchain-based secure information sharing for supply chains was considered. In that work, the optimization for the data sanitation process was carried out to construct a new privacy preservation model. A novel optimization technique of Whale with new crosspoint-based update (WNU) – which is an improvement of the Whale optimization algorithm (WOA) was developed. The constructed model was then analyzed with respect to information preservation rate, hiding failure rate, false rule generation and degree of modification. The authors of Abidi et al. [81] also performed comparative analysis on their proposed blockchain-enhanced model against conventional supply chain management models.

Another interesting application in food supply chains is given in the work of Mao et al. [82]. In that research work, the authors considered the complex system of a food supply chain involving multiple stakeholders – e.g. farmers, retailers, distributors, production factories, and finally consumers. To optimize the supervision and management of food supply chains, the blockchain-based credit evaluation system was proposed. The central idea is for the proposed system to collect credit evaluation texts from traders via smart contracts (held on the blockchain). A deep learning network (long short-term memory – LSTM) was then employed to directly analyze the texts. The optimization technique using LSTM was then compared with other machine learning methods – support vector machine (SVM) and Naive bayesian learning.

Similar work is seen in Khan et al. [83] where an IoT-blockchain enabled optimized provenance system for food supply chain management was proposed. The authors in that work performed analytics and optimization based on the large amount of data generated by the IoT-blockchain data of Industry 4.0 in the food sector. They then employed a recurrent neural network to construct a model in which parameters are then optimized using a GA. The performance of the proposed framework was then evaluated for a different number of users.

In Manupati et al. [84], production allocation problems in multi-echelon supply chains under carbon taxation policy were given focus. The authors of that work accounted for lead time factors under emission rate constraints (as a result of carbon taxation policies) when optimizing production allocation. The optimization problem was constructed as a mixed-integer nonlinear programming problem – which was solved using a distributed ledger-based blockchain approach. Results analysis was then conducted to compare the proposed nonlinear programming solution method against the non-dominated sorting genetic algorithm (NSGA-II).

Route optimization was carried out on a blockchain-secured supply chain network in the work of Rani et al. [85]. The primary objective of that work was to determine the optimal route of a product from a manufacturer to the customer – as well as to ensure traceability of a product back to its origin. A GA-based solution method was utilized for the optimization carried out in the work of Rani et al. [85]. The simulation results were then compared using two consensuses of verification approaches (proof of work and proof of elapsed time). On the other hand, the work of Abbas et al. [86] focused on the pharmaceutical industry – specifically drug supply chains. In that work, the authors proposed a novel blockchain and deep-learning-based drug supply chain management system. The developed system has two central functions in the smart pharmaceutical industry: (i) to continuously track and monitor the drug delivery system and (ii) to generate recommendations on the top-rated medicines to customers. A summary of the literature presented in this section is shown in Table 12.4.

Table 12.4 Summary of recent optimization works in blockchain-integrated supply chains.

References	Application	Technique
Abidi et al. [81]	Data sanitization process	Whale with new crosspoint-based update (WNU)
Mao et al. [82]	Food supply chains	Deep learning network (Long short term memory, LSTM)
Khan et al. [83]	Provenance system for food supply chains	Recurrent neural network and genetic algorithm
Manupati et al. [84]	Production allocation in multi-echelon supply chains (under carbon taxation policy)	Mixed integer nonlinear programming and distributed ledger-based blockchain approach
Rani et al. [85]	Route optimization on a blockchain-secured supply chain network	Genetic algorithm
Abbas et al. [86]	Drug supply chains	Deep-learning

12.6 Healthcare Data Management Systems

In the healthcare industry data privacy, security, and interoperability are the key factors. As seen in the previous sections, blockchain technology has been integrated into many application areas delivering all of the above factors. This makes it the primary candidate for upgrading current healthcare data management systems. The recent overlap of blockchain-based healthcare data management systems with the field of optimization is given in the following literature: Nandi et al. [87], Xu et al. [53], Hasan et al. [88], Onik et al. [89], Agbo et al. [90], and Khan et al. [91].

In Firdaus et al. [92], the authors maximized the security of the blockchain-based medical data management system – specifically, the features/approaches contained in the root exploit malware. The PSO technique was utilized to select the exclusive features that have the android debug bridge (ADB). The PSO technique was used in tandem with machine learning techniques (boost methodologies – e.g. adaboost, realadaboost, logitboost, and multiboost) to improve prediction. The combined methods employed in Firdaus et al. [92] were observed to significantly help improve predictions on the root exploit samples in the system under study. Similar work on medical data security could be seen in the research of Hussein et al. [93]. There the authors tackle the problem of data breaches by introducing a blockchain-based data-sharing system. The proposed approach improves overall data security by using the discrete wavelet transform (DWT). Additionally, the GA was employed to optimize queuing. The authors of Hussein et al. [93] tested their system in terms of execution time and found that the proposed system was (i) scalable, (ii) immune, (iii) robust, and (iv) efficient.

The work of Ni et al. [94] focused on overcoming data security threats such as single point of failure and distributed denial-of-service (DDoS) attacks. The idea was to develop a blockchain-based decentralized data management system (called HealChain). The authors then formulated the consortium blockchain nodes as an optimization problem – optimizing the economic benefit as well as the mining computing power – which was solved using the genetic algorithm. The security analysis/numerical results showed that the proposed system (HealChain) was effective and efficient with respect to data management feasibility and data security.

Another effective move towards integrating healthcare data with blockchain technology is via the IoT framework. This could be seen in the work of Veera-makali et al. [95], Veera-makali et al. [95] and Xu et al. [100] – where deep learning-based secure blockchain-enabled intelligent IoT and healthcare diagnosis model was developed. The mentioned model covers three main areas of healthcare data: hash value encryption, medical diagnosis, and transaction security. The authors employed three techniques in their framework:

- Orthogonal PSO algorithm – for secure sharing of medical images
- Neighborhood indexing sequence algorithm – for hash value encryption
- Deep learning network – for constructing a classification model for disease diagnosis

The authors of Veeramakali et al. [95] conducted detailed experimentation and algorithmic parameter tuning for validation of the proposed method – which was found to be highly sensitive, specific, and accurate.

Recently a healthcare data system that factors in the COVID-19 pandemic was developed in the work of Mohsin et al. [96]. In their proposed framework, the security and confidentiality of the COVID-19 data were taken into considera-tion – where the data confidentiality is carried out by the PSO algorithm as well as the hash function on the blockchain. The evaluation and validation of the proposed framework were presented and discussed in that work.

In addition to pandemic-related data management systems, medical imaging data have also been moving toward a decentralized system. This development can be observed in the work of Kumar et al. [97]; which considers the data infrastruc-ture of computed tomography (CT) images. The authors of that work modeled the data using deep learning (recurrent neural network) – where the weights of the model are shared via smart contracts. Noise reduction on the data was then per-formed using the Bat algorithm (BA) and data augmentation. Rigorous empirical

Table 12.5 Overview of recent optimization works in blockchain-integrated healthcare data management systems.

References	Application	Technique
Firdaus et al. [92]	Data security/privacy	PSO and machine learning (boost)
Hussein et al. [93]	Data security/privacy	Discreet wavelet transform and genetic algorithm
Ni et al. [94]	Data security and optimization (economic and mining computing power)	Genetic algorithm
Veeramakali et al. [95]	Blockchain-enabled IoT system and healthcare diagnosis model	Orthogonal PSO, neighborhood indexing sequence algorithm and deep learning network
Mohsin et al. [96]	COVID-19 data security/privacy	PSO and hash function
Kumar et al. [97]	Medical imaging data	Recurrent neural network, Bat algorithm and data augmentation

studies were performed by Kumar et al. [97] using medical imaging data (lung cancer patients). An overview of the works presented in this section is given in Table 12.5.

12.7 Outlook

Blockchain technology is the fundamental pillar of decentralization of the following industries: finance, energy, healthcare, operations management, computing as well as data infrastructure. Its core features (i.e. efficiency, accuracy, reliability, privacy, and security) make it a primary disruptive technology in many real-world systems. In this review, research works that overlap blockchain technology with the field of optimization (as well as deep learning) are presented and discussed. As a computing paradigm, blockchain technology naturally overlaps with many fields of computing – e.g. cryptography [98], IoT [99], and cloud computing. Thus, an interesting idea for future review works would be to investigate researches that overlap blockchain technology with cybersecurity/encryption, IoT, advanced computing as well as hardware infrastructure. Another interesting area would be to analyze and discuss research works involving the utilization of blockchain technology and its effects on the environment (as compared to current computing/IT infrastructure).

References

1 Nakamoto, S. and Bitcoin, A. (2008). A peer-to-peer electronic cash system. *Bitcoin* https://bitcoin. org/bitcoin.

2 Komalavalli, C., Saxena, D., and Laroiya, C. (2020). Overview of blockchain technology concepts. In: *Handbook of Research on Blockchain Technology*, 349–371. Academic Press.

3 Krishnan, S., Balas, V.E., Golden, J. et al. (2020). *Handbook of Research on Blockchain Technology*. Academic Press.

4 Laroiya, C., Saxena, D., and Komalavalli, C. (2020). Applications of blockchain technology. In: *Handbook of Research on Blockchain Technology*, 213–243. Academic Press.

5 Madakam, S. (2021). Blockchain technology: concepts, components, and cases. In: *Industry Use Cases on Blockchain Technology Applications in IoT and the Financial Sector*, 215–247. IGI Global.

6 Andoni, M., Robu, V., Flynn, D. et al. (2019). Blockchain technology in the energy sector: a systematic review of challenges and opportunities. *Renewable and Sustainable Energy Reviews* 100: 143–174.

7 Lu, H., Huang, K., Azimi, M., and Guo, L. (2019). Blockchain technology in the oil and gas industry: a review of applications, opportunities, challenges, and risks. *IEEE Access* 7: 41426–41444.

8 Wang, Q. and Su, M. (2020). Integrating blockchain technology into the energy sector – from theory of blockchain to research and application of energy blockchain. *Computer Science Review* 37: 100275.

9 Meng, W., Tischhauser, E.W., Wang, Q. et al. (2018). When intrusion detection meets blockchain technology: a review. *IEEE Access* 6: 10179–10188.

10 Mylrea, M. and Gourisetti, S.N.G. (2018). Blockchain for supply chain cybersecurity, optimization and compliance. In: *Resilience Week (RWS)*, 70–76. IEEE.

11 Dutta, P., Choi, T.M., Somani, S., and Butala, R. (2020). Blockchain technology in supply chain operations: applications, challenges and research opportunities. *Transportation Research Part E: Logistics and Transportation Review* 142: 102067.

12 Feng, H., Wang, X., Duan, Y. et al. (2020). Applying blockchain technology to improve agri-food traceability: a review of development methods, benefits and challenges. *Journal of Cleaner Production* 260: 121031.

13 Wang, Y., Han, J.H., and Beynon-Davies, P. (2019). Understanding blockchain technology for future supply chains: a systematic literature review and research agenda. *Supply Chain Management: An International Journal*.

14 Abu-Elezz, I., Hassan, A., Nazeemudeen, A. et al. (2020). The benefits and threats of blockchain technology in healthcare: A scoping review. *International Journal of Medical Informatics* 142: 104246.

15 Radanović, I. and Likić, R. (2018). Opportunities for use of blockchain technology in medicine. *Applied Health Economics and Health Policy* 16 (5): 583–590.

16 Zubaydi, H.D., Chong, Y.W., Ko, K. et al. (2019). A review on the role of blockchain technology in the healthcare domain. *Electronics* 8 (6): 679.

17 Mohanta, B.K., Jena, D., Satapathy, U., and Patnaik, S. (2020). *Survey on IoT Security: Challenges and Solution Using Machine Learning, Artificial Intelligence and Blockchain Technology*, 100227. Internet of Things.

18 Wang, H. (2020). IoT based clinical sensor data management and transfer using blockchain technology. *Journal of ISMAC* 2 (3): 154–159.

19 Bhushan, B., Khamparia, A., Sagayam, K.M. et al. (2020). Blockchain for smart cities: a review of architectures, integration trends and future research directions. *Sustainable Cities and Society* 61: 102360.

20 Alladi, T., Chamola, V., Rodrigues, J.J., and Kozlov, S.A. (2019). Blockchain in smart grids: a review on different use cases. *Sensors* 19 (22): 4862.

21 Alladi, T., Chamola, V., Sahu, N., and Guizani, M. (2020). Applications of blockchain in unmanned aerial vehicles: a review. *Vehicular Communications* 23: 100249.

22 Chuong, T.D. (2020). Optimality conditions for nonsmooth multiobjective bilevel optimization problems. *Annals of Operations Research* 287 (2): 617–642.

23 del Valle, A., Wogrin, S., and Reneses, J. (2020). Multi-objective bi-level optimization model for the investment in gas infrastructures. *Energy Strategy Reviews* 30: 100492.

24 Ganesan, T. (2020). Multi-objective optimization of industrial power generation systems: emerging research and opportunities. *IGI Global* https://doi.org/10.4018/978-1-7998-1710-9.

25 Ganesan, T., Vasant, P., Sanghvi, P. et al. (2020). Random matrix generators for optimizing a fuzzy biofuel supply chain system. *Journal of Advanced Engineering and Computation* 4 (1): 33–50.

26 Ganesan, T. and Vasant, P. (2020). Lévy-enhanced swarm intelligence for optimizing a multiobjective biofuel supply chain. In: *Handbook of Research on Smart Computing for Renewable Energy and Agro-Engineering*, 287–309. IGI Global.

27 Ganesan, T., Vasant, P., Litvinchev, I., and Aris, M.S. (2021a). Extreme value metaheuristics and coupled mapped lattice approaches for gas turbine-absorption chiller optimization. In: *Research Advancements in Smart Technology, Optimization, and Renewable Energy*, 283–312. IGI Global.

28 Ganesan, T., Vasant, P., and Litvinchev, I. (2021b). Chaotic simulator for bilevel optimization of virtual machine placements in cloud computing. *Journal of the Operations Research Society of China* https://doi.org/10.1007/s40305-020-00326-5.

29 Lee, Y.S., Graham, E.J., Galindo, A. et al. (2020). A comparative study of multi-objective optimization methodologies for molecular and process design. *Computers & Chemical Engineering* 136: 106802.

30 Rangaiah, G.P., Feng, Z., and Hoadley, A.F. (2020). Multi-objective optimization applications in chemical process engineering: tutorial and review. *Processes* 8 (5): 508.

31 Sun, X., Shi, Z., Lei, G. et al. (2020). Multi-objective design optimization of an IPMSM based on multilevel strategy. *IEEE Transactions on Industrial Electronics* 68 (1): 139–148.

32 Burnett, M. and Foster, J.C. (2004). *Chapter 4 – Encrypting Private Data, Hacking the Code, Syngress*, 153–204. ISBN: ISBN 9781932266658, https://doi.org/10.1016/B978-193226665-8/50037-0.

33 Kamran, M., Khan, H.U., Nisar, W. et al. (2020). Blockchain and internet of things: a bibliometric study. *Computers & Electrical Engineering* 81: 106525.

34 García-Bañuelos, L., Ponomarev, A., Dumas, M., and Weber, I. (2017). Optimized execution of business processes on blockchain. In: *International Conference on Business Process Management*, 130–146. Cham: Springer.

35 Agung, A.A.G. and Handayani, R. (2020). *Blockchain for smart grid*. Journal of King Saud University-Computer and Information Sciences.

36 Jameel, F., Javaid, U., Sikdar, B. et al. (2020). Optimizing Blockchain networks with artificial intelligence: towards efficient and reliable IoT applications. In: *Convergence of Artificial Intelligence and the Internet of Things*, 299–321. Cham: Springer.

37 Saberi, S., Kouhizadeh, M., Sarkis, J., and Shen, L. (2019). Blockchain technology and its relationships to sustainable supply chain management. *International Journal of Production Research* 57 (7): 2117–2135.

38 Seh, A.H., Zarour, M., Alenezi, M. et al. (2020). Healthcare data breaches: insights and implications. In: *Healthcare*, vol. Vol. 8, 133. Multidisciplinary Digital Publishing Institute.

39 Chen, H.S., Jarrell, J.T., Carpenter, K.A. et al. (2019). Blockchain in healthcare: a patient-centered model. *Biomedical Journal of Scientific & Technical Research* 20 (3): 15017.

40 Fu, J., Qiao, S., Huang, Y. et al. (2020). A study on the optimization of blockchain hashing algorithm based on PRCA. In: *Security and Communication Networks*, 2020.

41 Shibata, N. (2019). Proof-of-search: combining blockchain consensus formation with solving optimization problems. *IEEE Access* 7: 172994–173006.

42 Zhang, X. and Hu, H. (2019). Optimization of hash function implementation for bitcoin mining. In: *3rd International Conference on Mechatronics Engineering and Information Technology (ICMEIT 2019)*, 448–452. Atlantis Press.

43 Adewumi, T.P. and Liwicki, M. (2020). Inner for-loop for speeding up blockchain mining. *Open Computer Science* 10 (1): 42–47.

44 Chenli, C., Li, B., Shi, Y., and Jung, T. (2019). Energy-recycling blockchain with proof-of-deep-learning. In: *IEEE International Conference on Blockchain and Cryptocurrency (ICBC)*, 19–23. IEEE.

45 Baldominos, A. and Saez, Y. (2019). Coin. AI: a proof-of-useful-work scheme for blockchain-based distributed deep learning. *Entropy* 21 (8): 723.

46 Qu, X., Wang, S., Hu, Q., and Cheng, X. (2019). Proof of federated learning: a novel energy-recycling consensus algorithm. *arXiv preprint* arXiv: 1912.11745.

47 Dai, H.N., Zheng, Z., and Zhang, Y. (2019). Blockchain for internet of things: a survey. *IEEE Internet of Things Journal* 6 (5): 8076–8094.

48 Cao, B., Wang, X., Zhang, W. et al. (2020). A many-objective optimization model of industrial internet of things based on private blockchain. *IEEE Network* 34 (5): 78–83.

49 Liu, M., Yu, F.R., Teng, Y. et al. (2019). Performance optimization for blockchain-enabled industrial Internet of Things (IIoT) systems: a deep reinforcement learning approach. *IEEE Transactions on Industrial Informatics* 15 (6): 3559–3570.

50 Fu, S., Zhao, L., Ling, X., and Zhang, H. (2019). Maximizing the system energy efficiency in the blockchain based internet of things. In: *IEEE International Conference on Communications (ICC)*, 1–6. IEEE.

51 Cai, X., Geng, S., Zhang, J. et al. (2021). A sharding scheme based many-objective optimization algorithm for enhancing security in blockchain-enabled industrial internet of things. *IEEE Transactions on Industrial Informatics*.

52 Yun, J., Goh, Y., and Chung, J.M. (2020). DQN-based optimization framework for secure sharded blockchain systems. *IEEE Internet of Things Journal* 8 (2): 708–722.

53 Xu, M., Feng, G., Ren, Y., and Zhang, X. (2020). On cloud storage optimization of blockchain with a clustering-based genetic algorithm. *IEEE Internet of Things Journal* 7 (9): 8547–8558.

54 Li, M., Yu, F.R., Si, P. et al. (2020). Resource optimization for delay-tolerant data in blockchain-enabled iot with edge computing: a deep reinforcement learning approach. *IEEE Internet of Things Journal* 7 (10): 9399–9412.

55 Leng, J., Yan, D., Liu, Q. et al. (2019). ManuChain: combining permissioned blockchain with a holistic optimization model as bi-level intelligence for smart manufacturing. *IEEE Transactions on Systems, Man, and Cybernetics: Systems* 50 (1): 182–192.

56 Mollah, M.B., Zhao, J., Niyato, D. et al. (2020). Blockchain for future smart grid: a comprehensive survey. *IEEE Internet of Things Journal* 8 (1): 18–43.

57 Patil, H., Sharma, S., and Raja, L. (2020). Study of blockchain based smart grid for energy optimization. *Materials Today: Proceedings*.

58 Gai, K., Wu, Y., Zhu, L. et al. (2019). Privacy-preserving energy trading using consortium blockchain in smart grid. *IEEE Transactions on Industrial Informatics* 15 (6): 3548–3558.

59 Khan, F.A., Asif, M., Ahmad, A. et al. (2020a). Blockchain technology, improvement suggestions, security challenges on smart grid and its application in healthcare for sustainable development. *Sustainable Cities and Society* 55: 102018.

60 Bera, B., Saha, S., Das, A.K., and Vasilakos, A.V. (2020). Designing blockchain-based access control protocol in iot-enabled smart-grid system. *IEEE Internet of Things Journal*.

61 Musleh, A.S., Yao, G., and Muyeen, S.M. (2019). Blockchain applications in smart grid – review and frameworks. *IEEE Access* 7: 86746–86757.

62 Aderibole, A., Aljarwan, A., Rehman, M.H.U. et al. (2020). Blockchain technology for smart grids: decentralized NIST conceptual model. *IEEE Access* 8: 43177–43190.

63 Kumari, A., Gupta, R., Tanwar, S. et al. (2020). When blockchain meets smart grid: secure energy trading in demand response management. *IEEE Network* 34 (5): 299–305.

64 Hassan, M.U., Rehmani, M.H., and Chen, J. (2021). optimizing blockchain based smart grid auctions: a green revolution. *IEEE Transactions on Green Communications and Networking* 2102.02583.

65 Stübs, M., Posdorfer, W., and Kalinowski, J. (2020). Business-driven blockchain-mempool model for cooperative optimization in smart grids. In: *Smart Trends in Computing and Communications*, 31–39. Singapore: Springer.

66 Mureddu, M., Ghiani, E., and Pilo, F. (2020). Smart grid optimization with blockchain based decentralized genetic algorithm. In: *IEEE Power & Energy Society General Meeting (PESGM)*, 1–5. IEEE.

67 Li, Y. and Hu, B. (2019). An iterative two-layer optimization charging and discharging trading scheme for electric vehicle using consortium blockchain. *IEEE Transactions on Smart Grid* 11 (3): 2627–2637.

68 Wang, Z., Ogbodo, M., Huang, H. et al. (2020). AEBIS: AI-enabled blockchain-based electric vehicle integration system for power management in smart grid platform. *IEEE Access*.

69 Swain, A., Salkuti, S.R., and Swain, K. (2021). An optimized and decentralized energy provision system for smart cities. *Energies* 14 (5): 1451.

70 Khalid, Z., Abbas, G., Awais, M. et al. (2020). A novel load scheduling mechanism using artificial neural network based customer profiles in smart grid. *Energies* 13 (5): 1062.

71 Ferrag, M.A. and Maglaras, L. (2019). DeepCoin: a novel deep learning and blockchain-based energy exchange framework for smart grids. *IEEE Transactions on Engineering Management* 67 (4): 1285–1297.

72 Dujak, D. and Sajter, D. (2019). Blockchain applications in supply chain. In: *SMART Supply Network*, 21–46. Cham: Springer.

73 Rejeb, A., Keogh, J.G., and Treiblmaier, H. (2019). Leveraging the internet of things and blockchain technology in supply chain management. *Future Internet* 11 (7): 161.

74 Choi, T.M., Wen, X., Sun, X., and Chung, S.H. (2019). The mean-variance approach for global supply chain risk analysis with air logistics in the blockchain technology era. *Transportation Research Part E: Logistics and Transportation Review* 127: 178–191.

75 Merkaš, Z., Perkov, D., and Bonin, V. (2020). The significance of blockchain technology in digital transformation of logistics and transportation. *International Journal of E-Services and Mobile Applications (IJESMA)* 12 (1): 1–20.

76 Waller, M.A., Van Hoek, R., Davletshin, M., and Fugate, B. (2019). Integrating blockchain into supply chain management. In: *A Toolkit for Practical Implementation*. Kogan Page Publishers.

77 Tan, B.Q., Wang, F., Liu, J. et al. (2020). A blockchain-based framework for green logistics in supply chains. *Sustainability* 12 (11): 4656.

78 Saurabh, S. and Dey, K. (2021). Blockchain technology adoption, architecture, and sustainable agri-food supply chains. *Journal of Cleaner Production* 284: 124731.

79 Wamba, S.F. and Queiroz, M.M., 2020. *Blockchain in the Operations and Supply Chain Management: Benefits, Challenges and Future Research Opportunities.*

80 Sangeetha, A.S., Shunmugan, S., and Murugan, G. (2020). Blockchain for IoT enabled supply chain management – a systematic review. In: *Fourth International Conference on I-SMAC (IoT in Social, Mobile, Analytics and Cloud) (I-SMAC)*, 48–52. IEEE.

81 Abidi, M.H., Alkhalefah, H., Umer, U., and Mohammed, M.K. (2021). Blockchain-based secure information sharing for supply chain management: optimization assisted data sanitization process. *International Journal of Intelligent Systems* 36 (1): 260–290.

82 Mao, D., Wang, F., Hao, Z., and Li, H. (2018). Credit evaluation system based on blockchain for multiple stakeholders in the food supply chain. *International Journal of Environmental Research and Public Health* 15 (8): 1627.

83 Khan, P.W., Byun, Y.C., and Park, N. (2020b). IoT-blockchain enabled optimized provenance system for food industry 4.0 using advanced deep learning. *Sensors* 20 (10): 2990.

84 Manupati, V.K., Schoenherr, T., Ramkumar, M. et al. (2020). A blockchain-based approach for a multi-echelon sustainable supply chain. *International Journal of Production Research* 58 (7): 2222–2241.

85 Rani, P., Jain, V., Joshi, M. et al. (2021). A secured supply chain network for route optimization and product traceability using blockchain in internet of things. In: *Data Analytics and Management*, 637–647. Singapore: Springer.

86 Abbas, K., Afaq, M., Ahmed Khan, T., and Song, W.C. (2020). A blockchain and machine learning-based drug supply chain management and recommendation system for smart pharmaceutical industry. *Electronics* 9 (5): 852.

87 Nandi, S., Sarkis, J., Hervani, A.A., and Helms, M.M. (2021). Redesigning supply chains using blockchain-enabled circular economy and COVID-19 experiences. *Sustainable Production and Consumption* 27: 10–22.

88 Hasan, H.R., Salah, K., Jayaraman, R. et al. (2020). Blockchain-based solution for COVID-19 digital medical passports and immunity certificates. *IEEE Access*.

89 Onik, M.M.H., Aich, S., Yang, J. et al. (2019). Blockchain in healthcare: challenges and solutions. In: *Big Data Analytics for Intelligent Healthcare Management*, 197–226. Academic Press.

90 Agbo, C.C., Mahmoud, Q.H., and Eklund, J.M. (2019). Blockchain technology in healthcare: a systematic review. *Healthcare* 7 (2): 56.

91 Khan, F.A., Asif, M., Ahmad, A. et al. (2020c). Blockchain technology, improvement suggestions, security challenges on smart grid and its application in healthcare for sustainable development. *Sustainable Cities and Society* 55: 102018.

92 Firdaus, A., Anuar, N.B., Ab Razak, M.F. et al. (2018). Root exploit detection and features optimization: mobile device and blockchain based medical data management. *Journal of Medical Systems* 42 (6): 1–23.

93 Hussein, A.F., Arun Kumar, N., Ramirez-Gonzalez, G. et al. (2018). A medical records managing and securing blockchain based system supported by a genetic algorithm and discrete wavelet transform. *Cognitive Systems Research* 52: 1–11.

94 Ni, W., Huang, X., Zhang, J., and Yu, R. (2019). HealChain: a decentralized data management system for mobile healthcare using consortium blockchain. In: *Chinese Control Conference (CCC)*, 6333–6338. IEEE.

95 Veeramakali, T., Siva, R., Sivakumar, B. et al. (2021). An intelligent internet of things-based secure healthcare framework using blockchain technology with an optimal deep learning model. *The Journal of Supercomputing* 1–21.

96 Mohsin, A.H., Zaidan, A.A., Zaidan, B.B. et al. (2019). Based blockchain-PSO-AES techniques in finger vein biometrics: a novel verification secure framework for patient authentication. *Computer Standards & Interfaces* 66: 103343.

97 Kumar, R., Wang, W., Kumar, J. et al. (2021). An integration of blockchain and AI for secure data sharing and detection of CT images for the hospitals. *Computerized Medical Imaging and Graphics* 87: 101812.

98 Edastama, P., Lutfiani, N., Aini, Q. et al. (2021). Blockchain encryption on student academic transcripts using a smart contract. *Journal of Educational Science and Technology (EST)* 7 (2): 126–133.

99 Velmurugadass, P., Dhanasekaran, S., Anand, S.S., and Vasudevan, V. (2021). Enhancing Blockchain security in cloud computing with IoT environment using ECIES and cryptography hash algorithm. *Materials Today: Proceedings* 37: 2653–2659.

100 Xu, H., Zhang, L., Onireti, O. et al. (2020). BeepTrace: blockchain-enabled privacy-preserving contact tracing for COVID-19 pandemic and beyond. *IEEE Internet of Things Journal*.

13

Inventory Routing Problem with Fuzzy Demand and Deliveries with Priority

Paulina A. Avila-Torres and Nancy M. Arratia-Martinez

Universidad de las Américas Puebla, Department of Business Administration, Ex-Hacienda Santa Catarina Mártir S/N, San Andrés Cholula, C.P., Puebla 72810, México

13.1 Introduction

Since decades ago, one important challenge is to remain competitive. In order to achieve this, the logistic systems must be significantly improved, and the distribution of goods is the most expensive process. Consequently, making an optimal distribution plan will help to save on transportation costs [1]. Due to this, some authors like Campbell et al. [2] and Moin and Salhi [3] discussed the importance of integration of inventory management and transportation and how these two are interrelated. Yugang et al. [4] mention that the coordination of inventory and transportation decisions in multiple period is the key for the optimization of the inventory routing problem (IRP).

The IRP is an extension of the VRP that involves routing and inventory decisions [5]. In IRP, customers have to be served over a discrete-time horizon by a fleet of capacitated vehicles starting and ending the routes at a depot [6]. Campbell et al. [2] mention the three decisions made in this kind of problem are: (i) When to serve a customer?, (ii) How much to deliver to a customer when it is served?, and (iii) Which delivery route to use?

IRP may be applied in several industries such as gas companies, petrochemicals, clothing, auto parts, etc. [3]. Over the years, many authors have worked in creating reviews of the IRP considering different characteristics. For example, Andersson et al. [7] presented a survey with an emphasis on industrial applications. Coelho et al. [8] presented a survey focused on methodological aspects of the problem, including variants, models, and algorithms. Archetti et al. [6] presented and compared different formulations of IRP and valid inequalities are also presented, considering discrete-time horizons and a fleet of capacitated vehicles. More

Artificial Intelligence in Industry 4.0 and 5G Technology, First Edition.
Edited by Pandian Vasant, Elias Munapo, J. Joshua Thomas, and Gerhard-Wilhelm Weber.

recently, Roldán et al. [9] presented a survey of problems dealing with the stochastic nature of demand and lead times, a factor they identified as critical in real life. Then, Malladi and Sowlati [10] presented a review but considering sustainable aspects in the IRP; they stated that more common objectives are waste management, returnable transport item management, waste prevention and reduction, and emission reduction.

Many authors have incorporated real-life characteristics making the problem more complex. For example, Savelsbergh and Song [11] worked in limited product availabilities at facilities where customers cannot be served using out-and-back tours; they designed tours that cover huge geographic areas and developed a greedy algorithm combined with optimization-based improvement techniques. Yugang et al. [4] studied a multiperiod deterministic IRP with split deliveries and a limited fleet size, they presented an approximate approach that can solve large instances using Lagrangian relaxation, the objective is to minimize the inventory holding cost and transportation cost. Yugang et al. [4] reported that the coordination of inventory and transportation decisions in multiple periods is the key for the optimization of the IRP. The Economic Model Predictive Control (EMPC) strategy proposed by Athanassios and Ganas [12] used a multi-period deterministic model; this strategy captures all attributes of reverse logistics and inventory routing.

Solving instances with a large number of customers with exact methods is very difficult; that is why, some authors have implemented or developed heuristics to solve this kind of problem. For example, Bard and Nananukul [13] compared some heuristics for an IRP. They formulated a mixed-integer program maximizing the net benefits, but optimal solutions were not met with exact methods; so, they developed a two-step procedure that first decides the quantity to deliver and then defines the route. They assume that demand is deterministic, a finite planning horizon and a fleet of same capacity. Yugang et al. [4] studied a multiperiod deterministic IRP with split deliveries and a limited fleet size. They presented an approximate approach that can near optimally solve large instances using Lagrangian relaxation; the objective was to minimize the inventory holding cost and transportation cost. Vansteenwegen and Mateo [14] presented an iterated local search metaheuristic to solve a single-vehicle cyclic IRP.

Other authors have worked with a pickup and delivery inventory routing problem. For example, Iassinovskaia et al. [15] worked on this kind of problems with time windows; for small instances, they present a mixed-integer linear program, and to solve large instances, they proposed a metaheuristic. In this kind of problem, in addition to having transportation and inventory costs, they consider some costs related to the returnable transport items (RTIs). Archetti et al. [16] considered a specific class of this problem where a commodity was available at several origins and demanded at several destinations. They proposed a mathematical model with several valid inequalities. Naccache et al. [17] work in multi-pickup and delivery problem, they also consider time windows, in their problem, they collect and deliver items requested by the customer and solve the problem with an exact method, also they implemented a heuristic.

Authors like Moin et al. [18] considered a many-to-one distribution network, which means they assumed N suppliers where each provides distinct product to an assembly plant. They considered deterministic demand but variable over time, inventory cost at the plant but not at the supplier, unlimited capacitated vehicles. To solve the problem, they used genetic algorithms. Coelho et al. [19] presented an exact method to solve several kinds of IRP as a single vehicle, multi-vehicles, different inventory policies, same or different capacities of trucks, etc.

Some authors as Campbell et al. [2] mentioned that in real life, the IRP is stochastic and dynamic. Also, Roldán et al. [9] mentioned that in practical situations of IRP, demand and lead times are not deterministic and the processes of inventory control and distribution are the keys for the performance in supply chain. Solyali et al. [20] considered the stochastic demand and introduced a robust IRP where a supplier distributes a single product to multiple customers facing dynamic uncertain demands over a finite discrete time horizon. They propose two robust Mixed-Integer Programming (MIP) formulations and a new MIP formulation for the deterministic problem with backlogging demand allowed. They developed a branch-and-cut algorithm using the proposed formulation. Bertazzi et al. [21] also considered stochastic demand and order-up-to level policy and proposed a hybrid rollout algorithm. Moin and Halim [22] considered an IRP with stochastic and dynamic demand modeled as a Markov decision process; a hybrid rollout algorithm where a MILP is also used, and an ant bee colony algorithm is implemented.

Other authors have also dealt with the stochastic nature of the IRP. For example, Niakan and Rahimi [23] presented a model for healthcare issues considering fuzzy demand and transportation, representing the uncertainty using triangular fuzzy numbers. Qamsari et al. [24] proposed a model for the IRP with fuzzy time windows considering customer satisfaction; they divided the customers into categories and gave them a priority for visiting them. They proposed a column generation technique in order to obtain an optimal solution. They represented the fuzzy time windows with trapezoidal fuzzy numbers. Also, Micheli and Mantella [25] incorporated the stochastic demand assuming that is normally distributed. Authors like [26] used the fuzzy methodology in urban transport planning, considering demand and waiting time as triangular fuzzy numbers; they implemented two fuzzy methods (k-preference and second index of Yager) and compared the results obtained. As we can see, most of the authors considered stochastic demand and this becomes the problem even more complex.

The objective of this research is to formulate a mathematical model that creates the weekly plan according to the inventory level and the uncertain demand of customers using triangular fuzzy numbers and giving priority to some customers. The model will be applied in a case study. Some of the findings of this chapter are as follows: the solution time increases when the number of customers with priority is high, and the higher the security degree, the lower the total cost. The remaining chapter is organized as follows: the problem description is presented in Section 13.2, the mathematical formulation in Section 13.3, the results of computational experiments are discussed in Section 13.4, and finally, the conclusions and future work are presented in Section 13.5.

13.2 Problem Description

This chapter is inspired by a producer and distributor company of gases in the northern region of Mexico. These kinds of companies deliver products such as oxygen, nitrogen, and argon in different ways, for example, big trucks, small trucks, and trucks with cylinders (Figure 13.1).

Some decisions related to this particular business are as follows:

- *Production*: How much, where, and what time to produce in order to save energy cost.
- *Assignment*: Decide from which plant each customer is going to be served.
- *Assignment*: Decide how to manage the fleet (own trucks and subcontracted).

The current decision process has six main steps (see Figure 13.2). First, the company has a responsible team to plan all routes for different plants and products, each member of the team has been assigned a plant and a product, this is a daily plan. Second, as part of their job, every day they have to verify the level of inventory and the consumption of each customer assigned to them. Third, they decide which customers are to be included in the route for the next day, based on the inventory level and consumption, which can vary momentarily. Fourth, once they have chosen the customers to visit, they create manually the order in which to visit the customers. Fifth, when the route they are creating is for hospitals, then those customers have to be first in the route. Sixth, when they create the route, they also must consider the window time of product reception with the customer.

There are two main characteristics in this research, one is uncertain demand and the other is to give priority to the customers who required oxygen. In the next example (Figure 13.3) we explain how a route changes when some customers must be served first. In Figure 13.3 there are four customers (1–4) and one depot (0). On the left side of Figure 13.3, there are no customers with priority, so the routes are $0 \rightarrow 3 \rightarrow 4 \rightarrow 1 \rightarrow 2 \rightarrow 0$ (blue line) and $0 \rightarrow 3 \rightarrow 4 \rightarrow 2 \rightarrow 0$ (red line). Then, when

Figure 13.1 The north region of Mexico for distribution.

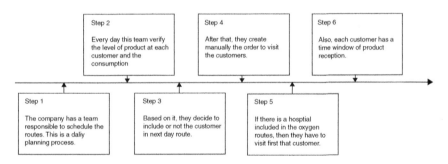

Step 2	Step 4	Step 6
Every day this team verify the level of product at each customer and the consumption	After that, they create manually the order to visit the customers.	Also, each customer has a time window of product reception.

Step 1	Step 3	Step 5
The company has a team responsible to schedule the routes. This is a daliy planning process.	Based on it, they decide to include or not the customer in next day route.	If there is a hosptial included in the oxygen routes, then they have to visit first that customer.

Figure 13.2 Current decision process.

we assigned priority (right side of figure) to customers 1 and 2, the routes change. We first create an exclusive route for customer 2 ($0 \to 2 \to 0$, blue line), and then simultaneously create a second route to visit the other customers ($0 \to 3 \to 4 \to 0$, blue line), in the second period the model creates a route to visit customers with a low inventory level ($0 \to 1 \to 4 \to 0$, red line).

In order to deal with demand uncertainty, we use fuzzy linear programming. In the following, we assume triangular fuzzy numbers, which, in general, will be denoted by $(\overline{a}, a, \underline{a})$ where $\overline{a}$ and $\underline{a}$ denote the lower and upper limits, respectively, of the support of a fuzzy number $\widetilde{a} \in N(R)$ [27].

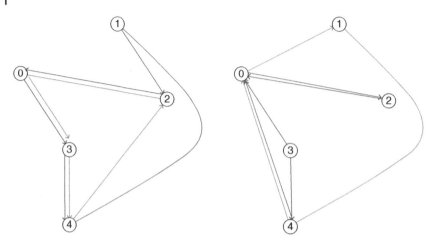

Figure 13.3 Example of the route both with and without priority.

Here, we implement the approach of Adamo [28]. This method evaluates the fuzzy number based on the rightmost point of the $\alpha-$ cut given for a given α [29]. Let us denote $\tilde{u}(v, u) \in L^{H_v \times [0,1]}$ the fuzzy relation so determined. We impose the corresponding utility function to be an increasing function in the following particular sense [28]:

For any $v1 \in H_v$, $v2 \in H_v$,

$$v_1 \geq v_2 \leftrightarrow u_1^\alpha \geq u_2^\alpha \forall \, \alpha \, L$$

where,

$$u_1^\alpha = \text{Max} \left\{ u \in [0,1] | \, \tilde{u}(v_1, u) \geq \alpha \right\}$$

$$u_2^\alpha = \text{Max} \left\{ u \in [0,1] | \, \tilde{u}(v_2, u) \geq \alpha \right\}$$

Given this property, any crisp utility function u^α derived from $\tilde{u}(v, u)$ and defined by $u^\alpha(v) = \text{Max} \left\{ u \in [0,1] | \, \tilde{u}(u, v) \geq \alpha \right\}$ is an increasing function.

Brunelli and Mezei [29] mentioned that the approach of Adamo is the only one that satisfies all properties for ordering fuzzy numbers. They summarized the Adamo approach as follows:

$$\text{AD}_\alpha(A) = a_\alpha^+$$

where $\alpha \in [0, 1]$ and $a_\alpha^+ = A_{i\alpha}$. By definition, all the α-cuts of fuzzy numbers are intervals; therefore, one can conveniently define them by means of their endpoints [29].

13.3   Mathematical Formulation

The mathematical model presented in this section is based on Archetti et al. [30] and Archetti et al. [6] formulations.

We consider a directed network where the product is shipped from a depot to a set of customers (N') over a time horizon T. Also, we consider that in $t = 0$, distribution and consumption of product do not exist.

The objective function (Eq. (13.1)) minimizes the inventory cost at depot (h_0). I_{0t} represents the inventory level at the depot for each period t, the objective function also minimizes the inventory cost at customers. (h_i), and I_{it} indicates the inventory level at customer i for each period t and also minimizes the distribution cost c_{ij} from node i to node j, where the binary decision variable y_{ij}^{kt} is equal to 1 if customer j is visited from customer i in the route k at the time t.

$$\min \sum_{t \in T} h_0 \cdot I_{0t} + \sum_{i \in N'} \sum_{t \in T} h_i \cdot I_{it} + \sum_{k \in K} \sum_{i \in N'} \sum_{j \in N'} \sum_{t \in T} c_{ij} \cdot y_{ij}^{kt} \qquad (13.1)$$

Constraint (13.2) is about the inventory level at the depot in the current period I_{0t} that is equal to the inventory level at the depot in the previous period I_{0t-1} plus the product available at the depot at the current period r_{0t} plus the total product delivered to all customers q_{it}^k at the current period t in the route k.

$$I_{0t} = I_{0t-1} + \widetilde{r_{0t}} - \sum_{k \in K} \sum_{i \in N'} q_{it}^k \quad \forall t \in T : t > 0 \qquad (13.2)$$

Inventory level at customers (I_{it}) is represented with constraint (13.3); the inventory level at period t is equal to the inventory level the customer had at the previous period I_{it-1} minus the consumption of the product at the current period r_{it} plus the total product delivered to the customer at the current period with any route q_{it}^k. This inventory level can be represented as in the following constraint (13.3).

$$I_{it} = I_{it-1} - \widetilde{r}_{it} + \sum_{k \in K} q_{it}^k \quad \forall i \in N' \ \forall t \in T : t > 0 \qquad (13.3)$$

In order to guarantee the correct operation of the tank, the inventory level at the customer must be greater or equal to a minimum inventory level, $I_i^{\min}$ as expressed in constraint (13.4).

$$I_{it} \geq I_i^{\min} \quad \forall i \in N \ \forall t \in T \qquad (13.4)$$

The constraints (13.5–13.7) are related to the order-up-to-level policy. Constraints (13.5–13.6) indicate that the quantity delivered to the customer q_{it}^k must be greater than or equal to the difference between the maximum capacity of the tank U_i and the inventory level at the previous period I_{it-1} but only if the customer was visited $\left(z_{it}^k = 1\right)$ in period t on route k. According to constraint (13.7), the quantity delivered to the customer q_{it}^k has to be less or equal than tank capacity U_i if the customer was visited $\left(z_{it}^k = 1\right)$.

$$\sum_{k \in K} q_{it}^k \geq U_{i \sum_{k \in K} z_{it}^k} - I_{it-1} \quad \forall i \in N' \; \forall t \in T : t > 0 \tag{13.5}$$

$$\sum_{k \in K} q_{it}^k \leq U_i - I_{it-1} \forall i \in N' \; \forall t \in T : t > 0 \tag{13.6}$$

$$q_{it}^k \leq U_i \cdot z_{it}^k \quad \forall k \in K \forall i \in N' \; \forall t \in T : t > 0 \tag{13.7}$$

Constraint (13.8) guarantees that the truck capacity (Q) will not be exceeded.

$$\sum_{i \in N'} q_{it}^k \leq Q \cdot z_{0t}^k \quad \forall k \in K \, T : t > 0 \tag{13.8}$$

Each customer will be visited just by one truck as expressed in constraint (13.9).

$$\sum_{k \in K} z_{it}^k \leq 1 \quad \forall i \in N' \; \forall t \in T : t > 0 \tag{13.9}$$

The flow constraint (13.10) indicates that if the customer was visited, then for every arc entering to a customer, there must be one arc leaving the same customer.

$$\sum_{j \in N : j \neq i} y_{ij}^{kt} + \sum_{j \in N : j \neq i} y_{ji}^{kt} = 2 \cdot z_{it}^k \quad \forall i \in N, \forall k \in K, \forall t \in T \tag{13.10}$$

The sub-tour elimination constraint is represented in constraint (13.11) where $E(S)$ is the set of all possible sub tours.

$$\sum_{(i,j) \in E(S):} y_{ij}^{kt} \leq \sum_{i \in E(S)} z_{it}^a - z_{kt}^a \quad \forall s \subseteq N', \forall k \in S, \forall a \in K, \forall t \in T \tag{13.11}$$

Constraint (13.12) allocates routes at the beginning to customers with priority deliveries (P_j).

$$\sum_{j \in N : i \neq j} y_{ji}^{kt} \cdot P_j \geq P_j \cdot z_{it}^k \quad \forall i \in N', \forall k \in K, \forall t \in T : t > 0 \tag{13.12}$$

The non-negativity and integrality of variables are represented with constraints (13.13–13.15)

$$z_{it}^k \in \{0,1\} \quad \forall i \in N, \; \forall k \in K, \; \forall t \in T \tag{13.13}$$

$$q_{it}^k \geq 0 \quad \forall i \in N', \forall k \in K, \forall t \in T \tag{13.14}$$

$$y_{0j}^{kt} \in \{0,1,2\} \quad \forall j \in N', \; \forall k \in K, \; \forall t \in T \tag{13.15}$$

Also, the valid inequalities proposed by Archetti et al. [30] were implemented in the current framework.

13.4 Computational Experiments

First, a brief computational experiment was carried out with the main objective to study the effect on solution time when we add the priority rules that are part of this problem and the simple way, that means without considering uncertainty on demand.

Three groups of five instances with 5, 10, and 15 customers, respectively, were generated based on Archetti et al. [30] instances. All groups were created with different levels (0%, 25%, 50%, and 75% of customers) of priority.

The lineal mathematical model was implemented with ILOG CPLEX Optimization Studio version 12.8. The problem has a total of $|N|^2 * (1+|K|*|T|) + 2(|N-1|*|T|*|K|$ variables. The time of solution is shown in Figure 13.4. In Figure 13.4a the solution time for the instances with 5 customers is presented, as you can see the solution time is very small, with less than a second. In this case,

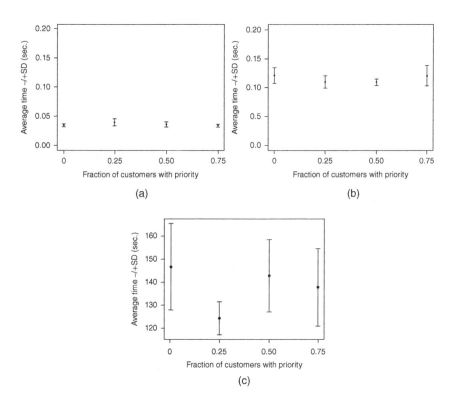

Figure 13.4 Solving time for instances with 5, 10, and 15 customers and different levels of priority (0, 0.25, 0.5, 0.75). (a) Solving time for instances with 5 customers. (b) Solving time for instances with 10 customers. (c) Solving time for instances with 15 customers.

the solution time in instances with 0.25 fraction of the customers with priority takes more time, but still remains insignificant.

In Figure 13.4b,c, the solution time of instances with 5 and 10 customers, respectively, are presented, the average time of solution for instances with 0.25 and 0.50 as a fraction of priority are smaller than those without any priority.

13.4.1 Numerical Example

In order to have a better understanding of the IRP problem and the application of the mathematical model elaborated in the previous section, we provide a semi-real numerical example. First, we solve the case under certainty and then show the results when uncertainty exists.

To apply the mathematical model we first recollect the data about the consumers. Then, we get the driving distance between each pair of customers through google.com services. After the driving distance was reached, we established the inventory levels, the product consumption rate, and the complete data. Finally, the mathematical model was solved through the branch and bound algorithm in ILOG CPLEX Optimization Studio 12.8.

So, let us define a set of 14 consumers from two sectors, namely industry and health. The customer locations were recollected from real locations (open data) of the metropolitan region of Nuevo León, México. The locations are represented in Figure 13.5. The blue nodes are consumers in the industry sector and the red nodes are consumers in the health sector. We selected randomly a location to be depot location. According to the implementation in the software, this case has 1,883 variables and 806,782 constraints.

In Table 13.1, the complete data of the 14 customers are presented. In the first column, we assigned an ID number. In the second column, the sector in

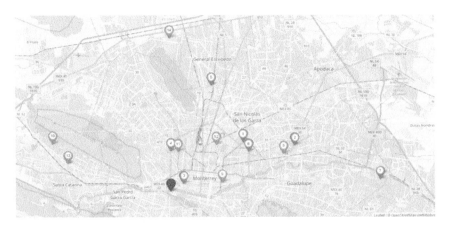

Figure 13.5 Locations of the consumers.

Table 13.1 Data of the customers.

ID of consumer	Type	Coordinates	Rate of product consumption (tons)	Initial inventory level (tons)	Maximum inventory level (tons)	Inventory cost (tons)
—	Depot	25.668253, −100.351380	—	55	—	0.30
1	Health	25.714369, −100.274426	1.6	3.6	9.02	0.32
2	Health	25.710962, −100.220026	1.1	2.6	4.3	0.33
3	Health	25.675227, −100.337898	2.6	7.9	8.5	0.23
4	Health	25.705879, −100.351176	2.3	3.9	9.7	0.18
5	Health	25.766933, −100.307896	0.6	4	7.8	0.29
6	Industria	25.705371, −100.268336	0.5	1.6	6.15	0.42
7	Industria	25.679350, −100.129464	1.9	2.9	6.61	0.42
8	Industria	25.703018, −100.231821	1.2	4.3	5	0.24
9	Industria	25.676414, −100.296572	0.6	3.6	10.54	0.43
10	Industria	25.712387, −100.477145	1.6	4.3	11.84	0.18
11	Industria	25.705246, −100.343517	1.7	4.3	15.86	0.22
12	Industria	25.711917, −100.302886	0.8	2.5	6.22	0.24
13	Industria	25.694062, −100.460915	2.4	6.5	6.82	0.31
14	Industria	25.810689, −100.352579	2.1	3.8	7.83	0.22

which the consumer belongs is stated. In the third column, the real coordinates (latitude, longitude) are given. The fourth, fifth, and sixth columns include the average rate of product consumption daily (here assumed as constant), the initial inventory level, and the maximum inventory level, respectively. Finally, the last column includes the inventory cost by the unit of measure. The data in Table 13.1 in columns 4–7 shows are randomly generated values. We considered the regular capacities of trucks to state the letter values.

The minimum inventory for all consumers was considered to be zero. In the depot, the initial inventory level was 55 tons, the rate of production was 76 tons, and the inventory cost per ton is considered as $ 0.30 by a ton.

In Table 13.2, the driving distance between two consumers is presented in km. For this case, we recollected the driving distance in one direction and assume a symmetric matrix of distances. In addition, the distances were multiplied by a cost factor of 1.06$ by kilometer approximately to get the transport cost.

The delivery of the product was programmed over three periods and the maximum capacity in the truck was stated as 30 tons (just one vehicle).

Table 13.2 Driving distance between locations of customers in kilometers.

ID	Depot	1	2	3	4	5	6	7	8	9	10	11	12	13	14
Depot		14.1	14.7	3.4	2.3	13.8	8.8	14.5	15.2	10.1	24.6	5.1	9.6	13	21.5
1	14.1		8.8	7.5	12.4	10.1	1.7	20.6	5.9	5.7	36.1	4.6	3.4	24.4	17.8
2	14.7	8.8		13	16	16	4.9	17	4.6	6.8	27	6.3	8.8	27	24
3	3.4	7.5	13		4.5	16	12	28	15	6.8	19	5	7.5	15	24
4	2.3	12.4	16	4.5		12	14	33	18	12	21	6.3	7.3	18	19
5	13.8	10.1	16	16	12		13	32	18	15	31	13	11	28	8.2
6	8.8	1.7	4.9	12	14	13		19	5.5	6.7	28	5.9	5.1	24	21
7	14.5	20.6	17	28	33	32	19		13	8.1	27	4.6	8	7.8	13
8	15.2	5.9	4.6	15	18	18	5.5	13		10.5	24.6	7.5	8.4	28	22.9
9	10.1	5.7	6.8	6.8	12	15	6.7	8.1	10.5		23.4	5.7	4.2	19.9	21.7
10	24.6	36.1	27	19	21	31	28	27	24.6	23.4		17.7	25.8	3.7	28.2
11	5.1	4.6	6.3	5	6.3	13	5.9	4.6	7.5	5.7	17.7		3.3	15.8	15.1
12	9.6	3.4	8.8	7.5	7.3	11	5.1	8	8.4	4.2	25.8	3.3		22.5	22.5
13	13	24.4	27	15	18	28	24	7.8	28	19.9	3.7	15.8	22.5		29.5
14	21.5	17.8	24	24	19	8.2	21	13	22.9	21.7	28.2	15.1	22.5	29.5	

13.4.1.1   The Inventory Routing Problem Under Certainty

The solution for this case was obtained in 231.7 seconds. Two routes were created for the periods first and second. In the third period, there weren't programmed deliveries.

- In the first period, the visit order of consumers was: Depot → C0 → C4 → C3 → C1 → C2 → C11 → Depot. The total product delivered with this route was 25.08 tons (83.6% of vehicle capacity).
- The second route was: Depot → C3 → C10 → C13 → C7 → C14 → C11 → Depot. And the total of products delivered is 27.9 (93% of vehicle capacity).

In the generated routes, the consumers C1, C2, C3, and C4 are first visited (health sector) and then the route can visit the costumers of the industry sector. The consumers C5, C6, C8, C9, and C12 are not visited along the periods, but for example, when six periods of planning are contemplated, the C5 and C9 are not programmed in route by their low rate of consumption. The total cost obtained was 330.34.

13.4.1.2   The Inventory Routing Problem Under Uncertainty in the Consumption Rate of Product

After assessing the uncertainty over the consumption rate of the product, a representation of the values through triangular fuzzy numbers was applied. To establish the left and right values for the triangular fuzzy number, the ± percentage of variation of 5%, 10%, and 15% over the original consumption rate was applied. The security degrees considered for this case for the given 4 values were 0, 0.50, 0.75, and 1, respectively.

To solve each case considering uncertainty in the consumption rate of the product, the constraints that involve these values were transformed in order to have a model with crisp values. The method applied was as defined by Adamo [28] as the righter value when a $\alpha-$ cut is defined under a degree of security.

In Table 13.3, the solution times and total cost are presented. The solution time in most cases was less than 5 minutes.

In the case when up to a percentage of decrement and increment occurs over the rate of consumption of a product, the total cost is impacted. When the percentage of variation is 5%, and the security degree is 0, the total cost is 339.9. Then with a 10% and a 15% of variation were 358.6 and 365.8, respectively. When the security degree increases, then the total cost reduces, then the total cost is reduced. Naturally, the highest cost is obtained when there is more variation and the security degree is the lowest (0) given that, a lower degree of security shows a lower level of knowledge. Also, the case when the highest cost was obtained (365.8) also implied the highest solution time (685.2).

Table 13.3 Solution times and total costs.

Variation ($\pm$%)	Security degree (α)	Solution time (seconds)	Total cost
5	0	202.5	339.9
	0.50	201.3	333.3
	0.75	165.1	332.1
10	0	620.6	358.6
	0.50	166.7	339.9
	0.75	195.0	333.3
15	0	685.2	365.8
	0.50	567.3	353.9
	0.75	244.8	334.6
5, 10, 15	1	231.7	330.34

13.5 Conclusions and Future Work

A mathematical model inspired by previous IRP formulations was developed incorporating priority in the deliveries for a certain group of customers. In the first experiment, we observed that the computational time of the solution increases very fast as well as the number of customers increases. However, one of the main objectives of this study was to show how the incorporation of uncertainty in demand besides the priority of customers in the health sector can be treated by the representation of fuzzy numbers and applying a defuzzification method.

In general, when the method of defuzzification is applied considering a α-cut, with α as the security degree, the results show that the more α is reduced, the more the total cost (of inventory and transportation) increases. Naturally, the solution will be impacted by the grade of knowledge or the security degree that is established.

As future research, we are looking to develop new formulations that allow us to understand the behavior of the problem with the priority in deliveries as the new characteristic proposed in this chapter. Also, to make the problem closer to reality, it is important to incorporate other characteristics like vehicle fleet of different capacities, integrate IRP with the production planning problem, and consider aspects related to transport companies.

It is well known that IRP is a complex problem even with tens of customers and it will be required to develop non-exact methods to solve bigger real instances in the future.

References

1 Herer, Y.T. and Levy, R. (1997). *The metered inventory routing problem, an integrative heuristic algorithm. International Journal of Production Economics* 51 (1): 69–81.

2 Campbell, A., Clarke, L., Kleywegt, A., and Savelsbergh, M. (1997). *The Inventory Routing Problem in Fleet Management and Logistics* (ed. C.T. Gabriel and L. Gilbert), 95–113. Spring US.

3 Moin, N.H. and Salhi, S. (2007). *Inventory routing problems: a logistical overview. Journal of the Operational Research Society* 58 (9): 1185–1194.

4 Yugang, Y., Haoxun, C., and Feng, C. (2008). *A new model and hybrid approach for large scale inventory routing problems. European Journal of Operational Research* 189 (3): 1022–1040.

5 Cordeau, J., Laport, G., Savelsbergh, M., and Vigo, D. (2007). *Vehicle Routing in Handbooks in Operations Research and Management, edited by Barnhart Cynthia, Laporte Gilbert, 367-428.* Elsevier.

6 Archetti, C., Bianchessi, N., Irnich, S., and Speranza, M.G. (2014). *Formulations for an inventory routing problem. International Transactions in Operational Research* 21 (3): 353–374.

7 Andersson, H., Hoff, A., Christiansen, M., and Hasle, G. (2010). *Industrial aspects and literature survey: Combined inventory management and routing. Computers & Operations Research* 37 (9): 1515–1536.

8 Coelho, L.C., Cordeau, J., and Laporte, G. (2014). *Thirty Years of Inventory Routing. Transportation Science* 48 (1): 1–19.

9 Roldán, R.F., Basagoiti, R., and Coelho, L.C. (2017). *A survey on the inventory-routing problem with stochastic lead times and demands. Journal of Applied Logic* 24 (1): 15–24.

10 Malladi, K.T. and Sowlati, T.S. (2018). *Sustainability aspects in inventory routing problem: a review of new trends in the literature. Journal of Cleaner Production* 197 (1): 804–814.

11 Savelsbergh, M. and Jin-Hwa, S. (2007). *Inventory routing with continuous moves. Computers & Operations Research* 34 (6): 1744–1763.

12 Athanassios, N. and Ioannis, G. (2017). *Economic model predictive inventory routing and control. Central European Journal of Operations Research* 25 (3): 587–609.

13 Bard, J.F. and Nananukul, N. (2009). *Heuristics for a multiperiod inventory routing problem with production decisions. Computers & Industrial Engineering* 57 (3): 713–723.

14 Vansteenwegen, P. and Mateo, M. (2014). *An iterated local search algorithm for the single-vehicle cyclic inventory routing problem. European Journal of Operational Research* 237 (3): 802–813.

15 Iassinovskaia, G., Limbourg, S., and Riane, F. (2017). *The inventory-routing problem of returnable transport items with time windows and simultaneous* pickup and delivery in closed-loop supply chains. *International Journal of Production Economics* 183 (1): 570–582.

16 Archetti, C., Christiansen, M., and Speranza, M.G. (2018). *Inventory routing with pickups and deliveries. European Journal of Operational Research* 268 (1): 314–324.

17 Naccache, S., Cordeau, J.F., and Coelho, L.C. (2018). *The multi-pickup and delivery problem with time windows. European Journal of Operational Research* 269 (1): 353–362.

18 Moin, N.H., Salhi, S., and Aziz, N.A.B. (2011). *An efficient hybrid genetic algorithm for the multi-product multi-period inventory routing problem. International Journal of Production Economics* 133 (1): 334–343.

19 Coelho, L.C., Cordeau, J., and Laporte, G. (2012). *Consistency in multi-vehicle inventory-routing. Transportation Research Part C: Emerging Technologies* 24 (1): 270–287.

20 Solyali, O., Cordeau, J.F., and Laporte, G. (2014). *Robust inventory routing under demand uncertainty. Transportation Science* 46 (3): 327–340.

21 Bertazzi, L., Bosco, A., Guerriero, F., and Lagana, D. (2013). *A stochastic inventory routing problem with stock-out. Transportation Research Part C: Emerging Technologies* 27 (1): 89–107.

22 Moin, N., Hasnah, H., and Huda, Z.A. (2018). *Solving inventory routing problem with stochastic demand. AIP Conference Proceedings* 1974 (1).

23 Niakan, F. and Rahimi, M. (2015). *A multi-objective healthcare inventory routing problem; a fuzzy possibilistic approach. Transportation Research Part E: Logistics and Transportation Review* 80 (1): 74–94.

24 Qamsari, A., Saeed, N., Motlagh, S. et al. (2020). *A column generation approach for an inventory routing problem with fuzzy time windows. Operational Research.*

25 Micheli Guido, J.L. and Mantella, F. (2018). *Modelling an environmentally-extended inventory routing problem with demand uncertainty and a heterogeneous fleet under carbon control policies. International Journal of Production Economics* 204: 316–327.

26 Avila-Torres, P., Caballero, R., Litvinchev, I. et al. (2018). *The urban transport planning with uncertainty in demand and travel time: a comparison of two defuzzification methods. Journal of Ambient Intelligence and Humanized Computing* 9 (3): 843–856.

27 Campos, L. and Verdegay, J.L. (1989). Linear programming problems and ranking of fuzzy numbers. *Fuzzy Sets and Systems* 32 (1): 1–11.

28 Adamo, J.M. (1980). *Fuzzy decision trees. Fuzzy Sets and Systems* 4: 207–219.

29 Brunelli, M. and Mezei, J. (2013). *How different are ranking methods for fuzzy number? A numerical study. Internal Journal of Approximate Reasoning* 54: 627–639.

30 Archetti, C., Bertazzi, L., Laporte, G., and Speranza, M.G. (2007). *A branch-and-cut algorithm for a vendor-managed inventory-routing problem. Transportation Science* 41 (3): 382–391.

14

Comparison of Defuzzification Methods for Project Selection

Nancy M. Arratia-Martinez[1], Paulina A. Avila-Torres[1], and Lopez-I Fernando[2]

[1]*Universidad de las Américas Puebla, Ex-Hacienda Santa Catarina Mártir, Department of Business Administration, San Andrés Cholula, Puebla, C.P. 72810, México*
[2]*Autonomous University of Nuevo León, Department of Mechanical and Electronic Engineering, San Nicolás de los Garza, Nuevo León, México*

14.1   Introduction

Resources of an organization (people, time, money, equipment, etc.) are never endless. As such, a constant and continuous challenge for decision makers is to decide which projects should be given priority in terms of receiving critical resources in a way that the productivity and profitability of an organization are best guaranteed [1].

Projects selection is typically a hard task to solve problems present in most organizations that drive investments and resources assignment based on project's portfolios on several of their business areas and activities. One main reason why a project's portfolio selection is hard to solve is that there are generally more projects to support than resources to distribute among them. Projects selection is also an important problem that managers have to face in large public organizations (like government institutions, universities, foundations, etc.) that are involved in funding or supporting research and development in some way [2].

One of the main tasks of management at different levels in many organizations, in public or private sectors, is to decide on how to invest and manage resources to support potential projects in different areas associated with specific benefits and (social) impacts. To accomplish this task effectively, it is necessary to implement rational, efficient, and effective decision-making processes, for ensuring the achievement of the main objectives of the organization [3]. Typically portfolio selection is a periodical activity, involving a group of projects with several impacts

Artificial Intelligence in Industry 4.0 and 5G Technology, First Edition.
Edited by Pandian Vasant, Elias Munapo, J. Joshua Thomas, and Gerhard-Wilhelm Weber.

or benefits competing for financial support, resources, etc. [4]. More often subjectivity of the decision-makers is very important and must be taken into account; otherwise, the solution could be of poor quality or not taken into consideration by the decision-maker (DM) [2]. In addition, other important tasks such as the criteria definition and project evaluation should be carried out before the project selection.

In general, the project portfolio selection problem can be defined as follows: it is necessary to choose a set of projects proposals optimizing certain impact measures. These are proposed by the decision-maker under resource limitations within a planning horizon ([5, 6]. Those proposals that meet the minimum requirements (defined by the decision maker) will be eligible for project funding or support. However, not all the proposals meeting the minimum requirements get into the portfolio because of budget or resource limitations [5]. It is also desirable that the selected portfolio fulfills certain constraints determined by public regulation, organization policies, and achieves a reasonably good tradeoff between certain portfolio impact measures (e.g. social, political, economic, etc.). Therefore, portfolio selection is a very challenging task for decision-makers [7].

The problem of portfolio selection of social projects is characterized as follows: the resources are not sufficient to support all the proposals, there may be interactions between projects affecting resource assignment, it is desirable to select a portfolio of projects that maximizes several impact measures, some of which may be subjective (not quantitative) [7]. The main characteristics of portfolio selection are projects tasks, different resource allocation policies, interdependence between tasks and/or projects, portfolio balancing rules, uncertainty in the overall budget, and uncertainty in the number of resources requested by tasks [5].

Other characteristics of portfolio selection problems found in the published scientific literature are: (i) projects funds or resources are assigned to a set of general activities that are common to all projects (e.g. equipment inversion, scholarships, etc.), (ii) each project is associated to a certain area, region, department, category, etc., (iii) all projects start at the same date and are expected to be finished at the same date as well, and (iv) it is desirable to select a portfolio that maximizes some impact measures [7].

Real-world problems are very complex and the complexity involves a degree of uncertainty, as uncertainty increases, so does the complexity of the problem [8]. In order to better address the inexactness and vagueness of data required for modeling real-world problems, it has been suggested to use fuzzy numbers as parameters in fuzzy linear programming problems [9]. Other recent research addressing project portfolio selection under uncertainty is [10], where the authors combine multicriteria analysis, Monte Carlo simulation in order to deal with the uncertainty in project evaluation. They use the Iterative Trichotomic Approach (ITA) by classifying the projects into three sets.

Academic research in this area usually represents uncertain parameters by fuzzy numbers leading to fuzzy optimization problems [11]. In practice, research and development project portfolio selection problems are complicated due to the fact that the quality as well as the estimated numerical data of cash flows vary at every stage of the research and development process [12]. In addition, the projects are selected before they are implemented, the information available may be characterized by imprecision and uncertainty. In particular, the budgets or the resources required by each project and the expected benefits may vary considerably, since their value is simply estimated before the projects are running. In this sense, uncertainty regarding certain parameters in the model has to be taken into account in such a way that the solutions obtained are reliable even within contexts characterized by change. In recent years, there has been an increase in studies on scheduling and selecting project portfolios that use fuzzy techniques to deal with uncertainty [13].

Fuzzy programming considers the uncertain parameters as fuzzy numbers and the constraints are treated like fuzzy sets [14], also a violation of the constraint is allowed and the satisfaction degree is defined by a membership function [15]. A fuzzy model simulates the uncertainty based on previous behavior [16]. Stochastic programming models the uncertainty through random variables and scenarios and is based on estimations or historic data. Both fuzzy and stochastic programming deal with the optimization under uncertainty, the main difference is how the uncertainty is modeled.

The use of fuzzy programming instead of stochastic programming is recommended because, according to the research in [17–20] fuzzy programming is more appropriate when there is vagueness and imprecision in data or information missing and there is insufficient statistical information. Also, the decision makers feel more comfortable in specifying vague data than crisp data, all these are good reasons to represent the uncertain parameters with fuzzy numbers [8].

Some recent studies [21, 22] apply fuzzy logic for decision-making. In [21], the authors propose a fuzzy logic model based on the Technique for order of preference by similarity to ideal solution (TOPSIS) to incorporate sustainability under uncertainty, the objective is to capture information about the vague environment and give the decision-maker numerical recommendations based on sustainability indicators. In [22], the authors propose a model for project selection and scheduling considering resource management, cash flow, among others. They introduce a new solution approach based on triangular interval-valued fuzzy random variables to incorporate fuzziness and randomness.

The remaining chapter is organized as follows: in Section 14.2 the description of the problem addressed here is presented; in Section 14.3 the mathematical formulation is described; in Section 14.5 a brief discussion of the methods applied is exposed; then the results are presented in Section 14.6; and finally, in Section 14.7 the conclusions are stated.

14.2   Problem Description

Selecting a project portfolio is a complex process involving many factors and considerations from the time it is proposed to the time the project portfolio is finally selected. Due to this reason, it is important to develop good mathematical models so the organization can achieve its objective [13].

In this research, we address the problem of portfolio selection with two objectives: the impact and the number of projects selected. Also, we consider the following constraints:

- not to exceed the total available budget;
- to ensure the amount allocated to the projects belonging to each area is between the minimum and maximum of the defined budget;
- to ensure the resources allocated to the projects are between the minimum and the maximum required since the partial allocation is allowed;
- to ensure the resources allocated to the tasks of projects are between the minimum and the maximum required, where the total assigned to all the tasks of a project is equal to the resources assigned to the project;
- constraints of dependence of tasks and projects.

In addition, we incorporate uncertainty in the total budget and the budget assigned to balance the funding in different areas. For this, some constraints are reformulated in order to include the representation of uncertainty values.

As mentioned before, uncertainty in some data in the portfolio selection problem is an important issue that has to be taken in consideration. In this problem, a triangular fuzzy number represents the uncertainty in the total budget, therefore we selected two methods to compare fuzzy numbers and to see the impact in the solution obtained. Those methods are the definition of inequalities-based k-preference [18] and integral value [23]. To solve this model, we implemented the SAUGMECON algorithm with these methods.

14.3   Mathematical Model

In this section, we describe a proposed bi-objective model inspired by the project selection model previously proposed by Litvinchev et al. [7] but extended to incorporate the decision of selection and resource allocation at the task level.

14.3.1 Sets and Parameters

J — Set of projects.

K — Set of areas of knowledge.

J_k — Set of projects belonging to the k area.

I_j — Set of tasks of the project $j \in J$.

P — Total budget.

$P_k^{\min}$ — Minimum amount of resources available for the area k.

$P_k^{\max}$ — Maximum amount of resources available for the area k.

$M_j^{\min}$ — Minimum amount of resources to be allocated to the project j and still ensuring his success.

$M_j^{\max}$ — Maximum amount of resources to be allocated to the project j and still ensuring his success.

$R_{j,i}^{\min}$ — Minimum amount of resources to be allocated to the task i belonging to the project j.

$R_{j,i}^{\max}$ — Maximum amount of resources to be allocated to the task i belonging to the project j.

$\omega_{j,b}$ — Value of the benefit type b of the project j (when there is just one benefit category the index is omitted).

$\rho_{j,i,b}$ — Relative value of the importance of the task i belonging to the project j over the benefit b.

14.3.2 Decision Variables

The established decision variables allow to identify which project and tasks should be selected, as well as the allocation of resources to project tasks.

$x_{j,i}$ — Resources assigned to the task i of the project j.

y_j — If a project j is selected and is sufficiently funded (0-1).

$z_{j,i}$ — If a task (j, i) is selected and sufficiently funded (0-1).

14.3.3 Objective Functions

The objective functions to be optimized can be formulated as measures of the portfolio quality over a specific benefit category. And the determination of the benefit categories depends on the real case situation of the organization's needs and preferences. Let B be the set of benefit categories and

$$F_b(x, y, z) \tag{14.1}$$

a general representation of an objective function associated with the benefit $b \in B$.

In this study, two objective functions are defined and employed: the total social impact and the number of selected projects.

- The total impact measure:

$$F_{\text{impact}} = \sum_{j \in J} \omega_j \left(\sum_{i \in I_j} (\alpha_{j,i} z_{j,i} + \beta_{j,i} x_{j,i}) \rho_{j,i} \right) \tag{14.2}$$

where ω_j is the total impact of project j and the parameter $\rho_{j,i}$ is the index of the relative importance of the task i of the project j. The parameters $\alpha_{j,i}$ and $\beta_{j,i}$ are defined as the same way that [7] (but extended to tasks formulation):

$$\alpha_{j,i} = \gamma - \frac{R_{j,i}^{\min}(1-\gamma)}{R_{j,i}^{\max} - R_{j,i}^{\min}} \quad \beta_{j,i} = \frac{(1-\gamma)}{R_{j,i}^{\max} - R_{j,i}^{\min}}$$

The expression $\sum_{i \in I_j}(\alpha_{j,i} z_{j,i} + \beta_{j,i} x_{j,i})$ allows calculating the grade in which a project is financed according to the range to allocate resources to its tasks $\left(\left[R_{j,i}^{\min}, R_{j,i}^{\max} \right] \right)$. This value ranges from a minimum level parameter (γ) to one, $\gamma \leq \sum_{i \in I_j}(\alpha_{j,i} z_{j,i} + \beta_{j,i} x_{j,i}) \leq 1$. The value of one is obtained when a project is selected and the maximum resources are allocated to all its tasks.

- The number of selected projects:

$$F_{\text{projects}} = \sum_{j \in J} y_i \tag{14.3}$$

14.4 Constraints

$$\sum_{j \in J} \sum_{i \in I_j} x_{j,i} \leq P \tag{14.4}$$

$$P_k^{\min} \leq \sum_{j \in J_k} \sum_{i \in I_j} x_{j,i} \leq P_k^{\max}, k \in K \tag{14.5}$$

$$M_j^{\min} y_j \leq \sum_{i \in I_j} x_{j,i} \leq M_j^{\max} y_j, j \in J \tag{14.6}$$

$$R_{j,i}^{\min} z_{j,i} \leq x_{j,i} \leq R_{j,i}^{\max} z_{j,i}, j \in J, i \in I_j \tag{14.7}$$

$$y_j \leq \sum_{i \in I_j} z_{j,i} \tag{14.8}$$

$$\sum_{i \in I_j} z_{j,i} \leq |I_j| y_j, j \in J \tag{14.9}$$

In constraints (14.4)–(14.8) some resource control constraints are presented. In (14.4), the total resources assigned are limited by the total budget. In (14.5) the resources assigned to projects belonging to each knowledge area are limited by a minimum and maximum amount of resources. The constraints (14.6) allows assigning enough resources to each selected project. And the constraints (14.7) allows the assignment of enough resources to each task belonging to the selected projects. Some logical relations between tasks and projects are defined in (14.8) and (14.9), that is, a project cannot be selected if no one of its tasks is selected, and a project is selected if at least one of its tasks are funded.

And finally, the nature of the decision variables,

$$z_{j,i} \in \{0,1\}, y_j \in \{0,1\}, x_{j,i} \geq 0 \tag{14.10}$$

When the total budget is considered uncertain, the affected constraints are reformulated, resulting as in the inequality (14.11).

$$\sum_{j \in J} \sum_{i \in I_j} x_{j,i} \leq \widetilde{P} \tag{14.11}$$

where, as we mentioned before, the triangular fuzzy number $\widetilde{P}$ is established through the values $\widetilde{P} = (\underline{P}, P, \overline{P})$; and the extreme values of the support of $\widetilde{P}$ are $\underline{P} = P - \varepsilon_1$ and $\overline{P} = P + \varepsilon_2$. In this case, P is an estimation of the total budget and $\underline{P}$ and $\overline{P}$ the extreme possible values.

To limit the resources to support projects belonging to each area, the constraints in (14.12) incorporate the fuzzy representation.

$$\widetilde{P_k^{\min}} \leq \sum_{j \in J_k} \sum_{i \in I_j} x_{j,i} \leq \widetilde{P_k^{\max}}, k \in K \tag{14.12}$$

The corresponding values of $\widetilde{P_k^{\min}}$ and $\widetilde{P_k^{\max}}$ are established in the same way as of the total budget $\widetilde{P}$.

14.5 Methods of Defuzzification and Solution Algorithm

In this section, the two methods selected to rank fuzzy numbers are described in detail. The first one is the K preference method taken from [18] and the method of the Integral value taken from [23].

14.5.1 *k*-Preference Method

Models suggested by fuzzy linear programming problems with fuzzy coefficients would be the more adequate to describe the situation of imprecise or inexact information, for a general structure a fuzzy linear program can be defined as follows [18]:

$$\max y = cx$$

s.t.

$$\tilde{a}_i x \leq \tilde{b}_i, i \in M = \{1, 2, \ldots, m\}$$

$$x \geq 0$$

where for each $i \in M$, $\tilde{a}_i = (\tilde{a}_{i1}, \tilde{a}_{i2}, \ldots, \tilde{a}_{in})$ is a vector of fuzzy numbers and $\tilde{b}_i$ is another fuzzy number on the right-hand side of the constraint i. On the other hand, the objective function is assumed to be well known by the decision-maker, that is, $c \in \mathbb{R}^n$ will be the vector of costs.

In the following, we assume, for the sake of commodity and without any loss of generality, triangular fuzzy numbers, which, in general, will be denoted by $\tilde{a} = (\underline{a}, a, \overline{a})$ where $\underline{a}$ and $\overline{a}$ denote the lower and upper limits of the support of a fuzzy number $\tilde{a} \in N(R)$, being $N(R)$, the set of real fuzzy numbers.

In the case of two triangular numbers, $\tilde{a} = (\underline{a}, a, \overline{a})$ and $\tilde{b} = (\underline{b}, b, \overline{b})$, and assuming fixed k in [0, 1] as the level of security or certainty, it is possible to obtain F_k index defined as under:

$$F_k(\tilde{a}) = \max\{x | \mu_{\tilde{a}} \geq k\} \text{ and } F_k(\tilde{b}) = \max\{x | \mu_{\tilde{b}} \geq k\}$$

We say that $\tilde{a} \leq \tilde{b}$ with a degree $k \in [0, 1]$ if $F_k(\tilde{a}) \leq F_k(\tilde{b})$.

Then, for the triangular fuzzy numbers the latter can be reformulated by the following inequalities [18]:

$$ka + (1 - k)\overline{a} \leq kb + (1 - k)\overline{b}$$

$$ka + (1 - k)\underline{a} \leq kb + (1 - k)\underline{b}$$

To apply this inequality, we reformulate the constraints (14.11) and (14.12). For the constraint (14.11) the resultant inequalities are constraints (14.13) and (14.14).

$$k \sum_{j \in J} \sum_{i \in I_j} x_{j,i} + (1 - k) \sum_{j \in J} \sum_{i \in I_j} x_{j,i} \leq kP + (1 - k)\overline{P} \tag{14.13}$$

$$k \sum_{j \in J} \sum_{i \in I_j} x_{j,i} + (1 - k) \sum_{j \in J} \sum_{i \in I_j} x_{j,i} \leq kP + (1 - k)\underline{P} \tag{14.14}$$

Since for this equation, the left-hand side is a crisp value, the latter constraints are simplified with the Eq. (14.15).

$$\sum_{j \in J} \sum_{i \in I_j} x_{j,i} \leq kP + (1 - k)\underline{P} \tag{14.15}$$

The same procedure should be applied for the group of constraints (14.12).

14.5.2 Integral Value

The method proposed by Liou and Wang [23] based on the total integral value of a fuzzy number A with the index of optimism α is defined as

$$I_T^\alpha(A) = \alpha I_R(A) + (1 - \alpha) I_L(A)$$

Where $I_R(A)$ and $I_L(A)$ are the right and left integral values of A, respectively, and $\alpha \ \varepsilon \ [0, 1]$.

The index of optimism (α) represents the degree of optimism of a decision-maker. A larger α indicates a higher degree of optimism.

From the previous equation and given $\alpha \ \varepsilon \ [0, 1]$, the total value of the triangular fuzzy number $A = (a, b, c)$ is

$$I_T^\alpha(A) = \frac{1}{2}\alpha(b + c) + \frac{1}{2}(1 - \alpha)(a + b)$$

$$I_T^\alpha(A) = \frac{1}{2}[\alpha c + b + (1 - \alpha)a].$$

To apply this inequality, we have to reformulate the constraints (14.11) and (14.12) as in the previous method. For the constraint (14.11) the resultant inequalities are constraints (14.16).

$$\sum_{j \in J}\sum_{i \in I_j} x_{j,i} \leq \frac{1}{2}[\alpha \underline{P} + P + (1 - \alpha)\overline{P}] \tag{14.16}$$

The same procedure should be applied for constraints (14.12).

14.5.3 SAUGMECON Algorithm

The SAUGMECON is an algorithm proposed by Zhang and Reimann [24] to generate all non-dominated solutions of integer programming problems with multiple objectives. This method was inspired by the augmented ε-constraint (AUGMECON), a method proposed by Mavrotas [25] and later improved by Mavrotas and Florios [26].

The AUGMECON, since it is an algorithm based on the epsilon-constraint method, transforms inequality constraints of constrained objectives into equality constraints by adding slack variables. While the objective function is augmented with the weighted sum of these slack variables. Then the AUGMECON2 was presented as an improved version of the previous algorithm resulting in the reduction of the number of subproblems solved to non-dominated solutions [24].

The SAUGMECON algorithm improves two mechanisms proposed by Mavrotas [25]:

- *An acceleration algorithm with early exit*: This mechanism exits from loops early when the problem becomes infeasible. In order to implement this algorithm, the constrained objectives must start with less restricted values, gradually moving to the more restricted values [24].

- *An acceleration algorithm with bouncing steps*: This is an innovation to the algorithm. It is easy to see that in the situation of feasibility the optimal solution will keep unchanged when a loop-control variable moves from its current value to the relatively worst value of its associated objective [24].

This method combines features of the traditional ε-constraint method and those of the AUGMECON model. Similar to the traditional ε-constraint method, the SAUGMECON uses inequalities for the objectives added to the constraint space. On the other hand, similar to the AUGMECON model, this method incorporates the sum of weighted constrained objectives into the objective function [24].

14.6 Results

In this section, we present the main results obtained when three sets of ten instances were solved using the mathematical formulation in Section 14.3, maximizing the objective functions in (14.2)–(14.3) with constraints (14.6)–(14.12). The model with the methods for the representation of uncertain data and the SAUGMECON algorithm was implemented in IBM ILOG CPLEX Optimization Studio V12.8.

The three sets of instances were named as P#T#A# where the symbol # represents the number of projects (for P), tasks (for T), and the number of areas (for A), respectively. Then, the sets of ten instances used are: (i) P32T16A8, (ii) P64T16A8 and, (iii) P128T16A4.

The methods were applied to all three sets, and the optimal solution for instances with 32 and 64 projects was obtained. For the set of 128 projects, a tolerance of 5% was established.

The representation through a triangular fuzzy number for the budget is an estimation of the real value (previously denoted as P) with ±10% to state the minimum and maximum values in the support of the triangular fuzzy number. Also, the $\widetilde{P}_k^{\min}$ and $\widetilde{P}_k^{\max}$ values are established with an estimation of the budget limits for each area with ±10% to state the extreme values in the triangular fuzzy number support.

14.6.1 Results of *k*-Preference Method

The sets of instances with 32, 64, and 128 projects were solved with the *k*-preference method considering five levels of security over the level of security in the available budget, 0%, 25%, 50%, 75%, and 100%.

Table 14.1 Average number of points in the pareto-front and average runtime for instance sets P32T16A8, P64T16A8, and P128T16A4.

Instance set	Security level (%)	Number of portfolios		Runtime	
		Average	Standard deviation	Average	Standard deviation
P32T16A8	0	6.2	1.08	42.70	23.75
	25	5.9	1.38	33.94	14.03
	50	6.1	1.58	38.49	28.74
	75	6.1	1.30	27.70	10.29
	100	6.3	1.35	38.12	21.53
P64T16A8	0	12.5	1.36	577.31	528.47
	25	12.6	1.68	404.66	380.97
	50	12.7	1.55	1050.27	1477.64
	75	13	1.79	443.04	295.10
	100	13.2	1.47	418.77	326.44
P128T16A4	0	23.4	1.50	102.56	12.81
	25	23.7	1.68	101.79	15.42
	50	23.6	1.50	107.42	17.56
	75	23.9	1.97	106.00	25.89
	100	24.1	1.76	107.43	20.67

The information of the average of the number of points in the Pareto-front and average runtime of the SAUGMECON method is presented in Table 14.1. In general, when we solved the instances with 32 candidate projects the Pareto-fronts had six points (portfolios) on average, and the solution time on average did not exceed a minute. For 64-projects instances, we got between 12 and 13 portfolios in the Pareto-front and the runtime increased. Finally, for instances with 128 candidate projects for the Pareto-front search, a tolerance gap of 5% was applied. On average, the Pareto-fronts obtained are composed of 23 and 24 points-portfolios.

As a reference solution to the analysis, we show the Pareto-fronts for one instance in Figure 14.1, for simplicity we selected the instance P32T16A8-1.

For each one of the five levels of security one Pareto-front is obtained (Figure 14.1). With the implementation of the k-preference method, we can see that if we define a higher level of security over the budget, more resources were used increasing the measure of impact and the number of projects in portfolios. The Pareto-front with more supported projects was obtained with the 100% of

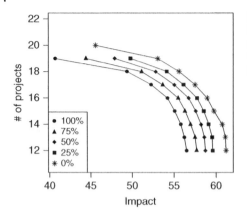

Figure 14.1 Pareto-fronts for instance P32T16A8-1 with the method of *k*-preference and security levels of 0%, 25%, 50%, 75%, and 100% in budget.

security. The Pareto-front for 75% as a degree of security is comparable with the <75% security level Pareto-fronts at least for the objective of a number of projects, but the objective of impact measure decreases.

In order to analyze in detail, the portfolios with a determined level of security in the budget. Table 14.2 shows the projects selected in each one of the eight portfolios for instance P32T16A8-1 with 100% of security.

Portfolio one corresponds to the portfolio with the maximum impact value and portfolio eight corresponds to the portfolio with the highest number of projects. The quantity in cells indicates the budget assigned to the corresponding project. Because the objective functions are in conflict, if the number of projects increases then the impact measure of the portfolio decreases. This decrease in the impact value is the result of reducing the number of resources assigned to the projects. Table 14.2 shows how projects 7, 11, 13, 16, 21, 22, 24, 28–31 are prevalent in all project portfolios, consuming a significant amount of the total budget. In contrast, projects 8, 14, 18 are included only in one or two portfolios.

It is important to note that the projects not included in Table 14.2 are projects that are not selected by any Pareto-front, such as projects 2, 5, 6, 9, 12, 17, 19, 23, 25–27, and 32.

The portfolios in Pareto-front for instance P32T16A8-1 with 75% as the degree of security are presented in Table 14.3. The Table shows that the projects included in portfolios with 75% of security remain selected in portfolios with 100% of security except for the projects 8 and 18 that were left out of the portfolios in Pareto-front. Also, Tables 14.2 and 14.3 show that the resources allocated for one particular project can change for a different level of security, which means that no portfolio is exactly the same in both configurations (75% and 100% security level).

Table 14.2 Portfolios of the pareto-front for the instance P32T16A8-1 with 100% level of security in the total budget value.

	Portfolios in Pareto-front							
Projects	1	2	3	4	5	6	7	8
1	0	0	0	1142.6	1142.6	1142.6	851.92	814.26
3	0	584.88	681.56	681.56	681.56	681.56	681.56	496
4	0	0	0	0	1039.5	968.08	941.49	941.49
7	2280.74	2280.74	2280.74	2280.74	1520.68	1520.68	1520.68	1416.76
8	0	0	0	0	0	0	0	2260.76
10	2564.5	2564.5	2564.5	2564.5	2564.5	2564.5	2564.5	0
11	1549.52	1549.52	1165.72	1165.72	1165.72	1165.72	1165.72	903.29
13	1460.8	1460.8	1460.8	1460.8	1460.8	1460.8	1072.6	913.971
14	0	0	0	0	0	0	2232.12	2232.12
15	0	0	987.72	952.06	987.34	952.06	925.57	844.27
16	2090.84	1927.74	1992.92	1971.28	1992.92	1992.92	1924.52	1924.52
18	0	0	0	0	0	0	0	1933.73
20	0	0	0	0	0	1932.64	1932.64	1925.17
21	1414.98	1414.98	1414.98	1414.98	1414.98	1414.98	1414.98	1374.08
22	2516.82	2151.38	2151.38	1417.24	1417.24	1031.8	683	592.3
24	3203.32	3203.32	3203.32	2852.16	2515.8	1802.34	719.38	580.52
28	1857.72	1857.72	1857.72	1857.72	1857.72	1857.72	1857.72	1731.46
29	2118.78	2118.78	1738.5	1738.5	1738.5	1011.46	1011.46	934.21
30	2607.86	2607.86	2607.86	2607.86	2607.86	2607.86	2607.86	2361.51
31	1877.78	1877.78	1492.28	1492.28	1492.28	1492.28	1492.28	1419.579

14.6.2 Results of Integral Value Method

The integral value method was implemented to solve instances of 32, 64, and 128 projects with 16 tasks and 8 areas. The method was configured with optimism levels of 60%, 80%, and 100%. The average number of portfolios for the instances of 32 projects was 6.49 while the average running time was 41.35 seconds (see Table 14.4). For instances of 64 projects, we got 13 projects on average and the running time is 239.88 seconds on average, increasing in a significant way. These two groups of instances were solved optimally, but the instances of 128 projects used a gap of 5% and the average of projects was 24.43, while the running time was 117.34 seconds. It is important to point out that the gap of 5% in instances of 128 projects helps reduce the execution time.

Table 14.3 Portfolios of the pareto-front for instance P32T16A8-1 with 75% of security level in the total budget value.

Projects	Portfolios in pareto-front						
	1	2	3	4	5	6	7
1	0	0	0	1142.6	1142.6	992.68	851.92
3	0	681.56	681.56	681.56	681.56	681.56	496
4	0	0	0	0	1039.5	941.49	941.49
7	2280.74	2280.74	2280.74	1520.68	1520.68	1520.68	1520.68
10	2564.5	2564.5	2564.5	2564.5	2564.5	2564.5	2528.46
11	1549.52	1549.52	1165.72	1165.72	1165.72	1165.72	1165.72
13	1423.74	1460.8	1460.8	1460.8	1460.8	1460.8	988.42
14	0	0	0	0	0	0	2232.12
15	0	0	1066.02	1050.48	1036.22	879.09	863
16	1924.52	1956.84	2008.76	1992.92	1992.92	1924.52	1924.52
20	0	0	0	0	0	1932.64	1925.17
21	1414.98	1414.98	1414.98	1414.98	1414.98	1414.98	1414.98
22	2516.82	2151.38	1417.24	1417.24	1417.24	1031.8	683
24	3203.32	3203.32	3203.32	2852.16	2515.8	1480.22	719.38
28	1857.72	1857.72	1857.72	1857.72	1857.72	1857.72	1731.46
29	1738.5	1738.5	1738.5	1738.5	1049.62	1011.46	1011.46
30	2607.86	2607.86	2607.86	2607.86	2607.86	2607.86	2469.94
31	1877.78	1492.28	1492.28	1492.28	1492.28	1492.28	1492.28

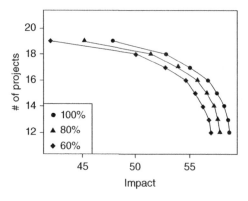

Figure 14.2 Pareto-fronts for instance P32T16A8-1 with integral value method and optimism levels of 60%, 80%, and 100%.

Table 14.4 Average number of points in the pareto-front and average runtime for instances P32T16A8.

Instance set	Optimism level (%)	Number of portfolios		Runtime	
		Average	Standard deviation	Average	Standard deviation
P32T16A8	60	6.37	1.68	35.44	15.30
	80	6.37	1.76	39.88	23.77
	100	6.75	1.58	48.73	33.38
P64T16A8	60	12.9	1.52	300.58	201.82
	80	13	1.63	358.94	223.52
	100	13.1	1.85	378.02	294.32
P128T16A4	60	23.9	2.02	104.13	28.04
	80	24.5	1.95	122.64	38.78
	100	24.9	1.72	125.25	48.06

We obtained a pareto-front for each level of optimism (Figure 14.2). The portfolios with the levels of optimism (60%, 80%, and 100%) show the same minimum and the maximum number of projects in the range of 12–19. For the impact measure, as the level of optimism increases also does the impact measure. The latter occurs given that with a high level of optimism, more resources are distributed. In achieving this, both methods (k-preference method and the integral value method) remain similar.

In Table 14.5, can be seen the founds assigned to each project in the different portfolios of the instance 1 with 32 projects and 100% of optimism level. The eighth portfolio eight have financed the higher number of projects, in all portfolios the total amount assigned is the same $26,800. Then, in Table 14.6 with the same instance but with a level of optimism is 60% we see that the total budget assigned is $25,856. In the solution with the 100% of optimism 19 different projects are financed in at least one portfolio, while with 60% of optimism 20 projects are financed in at least one portfolio, differing in project 18, which is only included in the 60% configuration for the portfolio with the highest number of projects. Projects that remain funded in most portfolios can be considered valuable in terms of the impact they provide.

Overall, the runtime in small instances (32-projects) results in short times for both methods, though they address different environments of decision making. For medium instances (64 projects) there was a significant difference in the averages. For large instances (128 projects), the obtained runtimes were very similar

Table 14.5 Portfolios of the pareto-front for instance P32T16A8-1 with 100% of optimism level in the total budget value.

	Portfolios in pareto-front							
Project	1	2	3	4	5	6	7	8
1	0	0	0	1142.6	1142.6	1142.6	992.68	814.26
3	0	681.56	681.56	681.56	681.56	681.56	681.56	496
4	0	0	0	0	1070.22	1039.5	941.49	941.49
7	2620.26	2620.26	2280.74	2280.74	2280.74	1520.68	1520.68	1520.68
8	0	0	0	0	0	0	0	2260.76
10	2564.5	2564.5	2564.5	2564.5	2564.5	2564.5	2564.5	2459.62
11	1941.7	1549.52	1549.52	1549.52	1165.72	1165.72	1165.72	1016.44
13	1460.8	1460.8	1460.8	1460.8	1460.8	1460.8	1072.6	988.42
14	0	0	0	0	0	0	2235.23	2232.12
15	0	0	1066.02	1066.02	1066.02	1016.16	952.06	844.26
16	1989.5	2027.7	2045.04	2014.94	2067.1	1992.92	1924.52	1924.52
20	0	0	0	0	0	1932.64	1932.64	1925.17
21	1793.38	1793.38	1414.98	1414.98	1414.98	1414.98	1414.98	1374.08
22	2516.82	2516.82	2151.38	1804.66	1417.24	1417.24	1031.8	683
24	3203.32	3203.32	3203.32	3203.32	2852.16	2199.54	1480.22	719.38
28	1857.72	1857.72	1857.72	1857.72	1857.72	1857.72	1857.72	1731.46
29	2118.78	2118.78	2118.78	1738.5	1738.5	1373.3	1011.46	1011.46
30	2607.86	2607.86	2607.86	2607.86	2607.86	2607.86	2607.86	2444.59
31	2205.36	1877.78	1877.78	1492.28	1492.28	1492.28	1492.28	1492.28

for both methods, this can be caused by the 5% tolerance applied. Note that if one wishes to explore different levels of security or optimism indices, the time consumption will increase because each level is a different configuration that results in a full pareto-front.

The k-preference method reflects the level of security as a grade of knowledge over the parameter. In the worst case, a 0% of security level can be stated given as a result a more imprecise value of the represented parameter. While the integral value method allows establishing an attitude of optimism-pessimism of the decision-maker. A larger value of optimism means, that, there is an increased confidence that more resources will be released; and in solved instances it was observed a greater amount of resources distributed.

Table 14.6 Portfolios of the pareto-front for instance P32T16A8-1 with 60% of optimism level in the total budget value.

	Portfolios in pareto-front							
Project	1	2	3	4	5	6	7	8
1	0	0	0	1142.6	1142.6	1142.6	851.92	814.26
3	0	681.56	681.56	681.56	681.56	681.56	584.88	496
4	0	0	0	0	994.42	941.49	941.49	941.49
7	2280.74	2280.74	2280.74	2280.74	1520.68	1520.68	1520.68	1520.68
8	0	0	0	0	0	0	0	2260.76
10	2564.5	2564.5	2564.5	2564.5	2564.5	2564.5	2564.5	0
11	1549.52	1549.52	1549.52	1165.72	1165.72	1165.72	1165.72	912.32
13	1460.8	1460.8	1460.8	1460.8	1460.8	1460.8	1072.6	912.52
14	0	0	0	0	0	0	2232.12	2232.12
15	0	0	928.32	1066.02	952.06	952.06	891.45	844.27
16	2024.78	2087.06	1924.52	2062.26	1992.92	1959.25	1924.52	1924.52
18	0	0	0	0	0	0	0	1933.73
20	0	0	0	0	0	1932.64	1932.64	1925.17
21	1793.38	1414.98	1414.98	1414.98	1414.98	1414.98	1414.98	1374.08
22	2516.82	2151.38	2151.38	1804.66	1417.24	1031.8	683	592.3
24	3203.32	3203.32	3203.32	2515.8	2852.16	2118.6	1106.18	660.64
28	1857.72	1857.72	1857.72	1857.72	1857.72	1857.72	1857.72	1731.46
29	2118.78	2118.78	1738.5	1738.5	1738.5	1011.46	1011.46	934.21
30	2607.86	2607.86	2607.86	2607.86	2607.86	2607.86	2607.86	2361.51
31	1877.78	1877.78	1492.28	1492.28	1492.28	1492.28	1492.28	1483.96

14.7 Conclusions

In this research, we addressed the problem of portfolio selection with uncertainty using triangular fuzzy numbers. We used SAUGMECON as a solution method and implemented two methods to rank fuzzy numbers: k-preference that works with levels of security and the integral value method that uses a level of optimism. In both methods, the greater the confidence or optimism, the more resources were allocated. The two methods can be applied in different situations of decision making and guarantee the obtention of solutions to the project selection problem in reasonable times.

Despite the differences between the methods, their behavior of both remain similar, and the projects included in portfolios also remain similar by their contribution to the impact measure.

As part of the research, other methods were explored for possible application but in cases when symmetric triangular fuzzy numbers are defined, some methods are simplified to the center value in the support of the triangular fuzzy number. This imply to solve the problem only with a central value ignoring the amplitude of the estimation of the triangular fuzzy number resulting in a disadvantage.

Research remains open to measuring the real impact of considering fuzzy methods in real-world problems, since the application of different methods to classify fuzzy numbers may provide different solutions. In cases of inversion or with high impact of solutions, these methods should be analyzed before the implementation in the real world in order to understand their operation. Other future directions of research are the consideration of fuzzy constraints and the fuzzy coefficients in the objective function to model the imprecise information of resources requested or needed by the project.

References

1 Jafarzadeh, H., Akbari, P., and Abedin, B. (2018). A methodology for project portfolio selection under criteria prioritisation, uncertainty and projects interdependency–combination of fuzzy QFD and DEA. *Expert Systems with Applications* 110: 237–249.

2 Litvinchev, I. and López, F. (2008). An interactive algorithm for portfolio bi-criteria optimization of R&D projects in public organizations. *Journal of Computer and Systems Sciences International* 47 (1): 25–32.

3 Litvinchev, I., Arratia, N., and López, F. (2013). Large scale portfolio selection with synergies. *Journal of Computer and Systems Sciences International* 52 (6): 980–985.

4 Litvinchev, I., Lopez-Irarragorri, F., Arratia-Martínez, N.M., and Marmolejo, J.A. (2014). Selecting large portfolios of social projects in public organizations. *Mathematical Problems in Engineering* 2014.

5 Arratia-Martinez, N.M., Caballero-Fernandez, R., Litvinchev, I., and Lopez-Irarragorri, F. (2018). Research and development project portfolio selection under uncertainty. *Journal of Ambient Intelligence and Humanized Computing* 9 (3): 857–866.

6 Carazo, A.F., Núñez, T.G., Casas, F.M.G., and Fernández, R.C. (2016). Evaluation and classification of the techniques used by organizations in the last decades to select projects. *Rev Met Cuant Para Econ Empresa* 67–115.

7 Litvinchev, I., López, F., Escalante, H.J., and Mata, M. (2011). A milp bi-objective model for static portfolio selection of R&D projects with synergies. *Journal of Computer and Systems Sciences International* 50 (6): 942–952.

8 Avila-Torres, P., Caballero, R., Litvinchev, I. et al. (2018). The urban transport planning with uncertainty in demand and travel time: a comparison of two defuzzification methods. *Journal of Ambient Intelligence and Humanized Computing* 9 (3): 843–856.

9 Pérez-Cañedo, B., Verdegay, J.L., Concepción-Morales, E.R., and Rosete, A. (2020). Lexicographic methods for fuzzy linear programming. *Mathematics* 8 (9): 1540.

10 Mavrotas, G. and Makryvelios, E. (2021). Combining multiple criteria analysis, mathematical programming and Monte Carlo simulation to tackle uncertainty in research and development project portfolio selection: a case study from Greece. *European Journal of Operational Research* 291 (2): 794–806.

11 Perez, F. and Gomez, T. (2016). Multiobjective project portfolio selection with fuzzy constraints. *Annals of Operations Research* 245 (1): 7–29.

12 Carlsson, C., Fuller, R., Heikkilä, M., and Majlender, P. (2007). A fuzzy approach to R&D project portfolio selection. *International Journal of Approximate Reasoning* 44 (2): 93–105.

13 Pérez, F., Gómez, T., Caballero, R., and Liern, V. (2018). Project portfolio selection and planning with fuzzy constraints. *Technological Forecasting and Social Change* 131: 117–129.

14 Nasseri, S.H. and Behmanesh, E. (2013). Linear programming with triangular fuzzy numbers – a case study in a finance and credit institute. *Fuzzy Information and Engineering* 5 (3): 295–315.

15 Sahinidis, N.V. (2004). Optimization under uncertainty: state-of-the-art and opportunities. *Computers & Chemical Engineering* 28 (6–7): 971–983.

16 Riedewald, F. (2011). Comparison of deterministic, stochastic and fuzzy logic uncertainty modelling for capacity extension projects of DI/WFI pharmaceutical plant utilities with variable/dynamic demand. Doctoral dissertation, University College Cork.

17 Brito, J., Martinez, F.J., Moreno, J.A., and Verdegay, J.L. (2012). Fuzzy optimization for distribution of frozen food with imprecise times. *Fuzzy Optimization and Decision Making* 11 (3): 337–349.

18 Campos, L. and Verdegay, J.L. (1989). Linear programming problems and ranking of fuzzy numbers. *Fuzzy Sets and Systems* 32 (1): 1–11.

19 Kumar, M., Sarkar, P., and Madhu, E. (2013). Development of fuzzy logic based mode choice model considering various public transport policy options. *International Journal for Traffic and Transport Engineering* 3 (4): 408–425.

20 Teodorović, D. and Lučić, P. (2005). Schedule synchronization in public transit using the fuzzy ant system. *Transportation Planning and Technology* 28 (1): 47–76.

21 Ma, J., Harstvedt, J.D., Jaradat, R., and Smith, B. (2020). Sustainability driven multi-criteria project portfolio selection under uncertain decision-making environment. *Computers & Industrial Engineering* 140: 106236.

22 Zolfaghari, S. and Mousavi, S.M. (2021). A novel mathematical programming model for multi-mode project portfolio selection and scheduling with flexible resources and due dates under interval-valued fuzzy random uncertainty. *Expert Systems with Applications* 182: 115207.

23 Liou, T.S. and Wang, M.J.J. (1992). Ranking fuzzy numbers with integral value. *Fuzzy Sets and Systems* 50 (3): 247–255.

24 Zhang, W. and Reimann, M. (2014). A simple augmented ε-constraint method for multi-objective mathematical integer programming problems. *European Journal of Operational Research* 234 (1): 15–24.

25 Mavrotas, G. (2009). Effective implementation of the ε-constraint method in multi-objective mathematical programming problems. *Applied Mathematics and Computation* 213 (2): 455–465.

26 Mavrotas, G. and Florios, K. (2013). An improved version of the augmented ε-constraint method (AUGMECON2) for finding the exact pareto set in multi-objective integer programming problems. *Applied Mathematics and Computation* 219 (18): 9652–9669.

15

Re-Identification-Based Models for Multiple Object Tracking

Alexey D. Grigorev[1], Alexander N. Gneushev[1,2], and Igor S. Litvinchev[2]

[1]*Moscow Institute of Physics and Technology, Department of Control and Applied Mathematics,
Dolgoprudny, Moscow Region, Russia*
[2]*Federal Research Center "Computer Science and Control" of Russian Academy of Sciences, Dorodnicyn
Computing Center, Moscow, Russia*

15.1   Introduction

The problem of multiple object tracking (MOT) is very relevant in the field of computer vision due to video surveillance systems that have developed a lot over the past decade. Tracking tasks arise both in security systems, automation control, self-driving, and robotics. Online processing of incoming data provides an opportunity to reduce the computational load on the system and process incoming data locally, regardless of the availability of a remote server.

MOT is associated with objects' detection in a sequence of video frames that are an integral part of systems for identifying and verifying objects, assessing their spatial position. Special detection models are used for coarse localization of objects in the image; however, these detectors require significant computational resources [1–3]. Because many objects are presented in the image, tracking models help clarify their coordinates by determining a separate track of each object's movement by measuring its position in each frame and estimating movement in future moments in time. The predicted object position can significantly reduce image areas for subsequent image analysis and increase the computational efficiency of the whole system. A tracking system alleviates the accumulation of multiple images that contain the same object at different frames; therefore, a tracking procedure can significantly increase the accuracy of object identification based on the approaches of recognition probability accumulating and feature aggregation.

The tracking problem is solved by the reconstruction of objects' tracks in a video based on the object detection at each frame under the constraint that the number of objects is not known in advance. Of particular interest is the tracking problem

Artificial Intelligence in Industry 4.0 and 5G Technology, First Edition.
Edited by Pandian Vasant, Elias Munapo, J. Joshua Thomas, and Gerhard-Wilhelm Weber.
© 2022 John Wiley & Sons, Inc. Published 2022 by John Wiley & Sons, Inc.

which includes the online mode limitation imposed by practical applications. In this mode, at each moment, information that is obtained only from the previous frames in time is available.

In the majority of scenarios, the problem of MOT in the online mode can be represented in the form of several subtasks: initialization of objects' tracks, object detection on the frame and comparing object position with an estimated track, prediction of the track for the next frame, and termination of the track. These subtasks are solved sequentially for each frame of the video sequence.

Several online object tracking techniques are used in a variety of applications with additional constraints. Such methods should be computationally efficient and should operate in real-time (i.e. the frame rate of the algorithm should be not less than the frame rate of the video). These limitations often lead to the simplification of the methods used in each subproblem. In particular, a linear model of the object movement is used for the subtask of track predicting for the next frame [4]. The subproblem of detection and localization of an object is solved using various detection systems that are trained for the desired class of objects. Nowadays, the most effective neural network detectors are based on popular architectures such as single shot detector (SSD) [1], You Only Look Once (YOLO) [2], and RetinaNet [3]. These models provide compelling quality and relatively acceptable speed for most practical applications in which specialized matrix processors can be used. However, the instability of the detectors, false detections, and omissions in conditions of a high object density in the frame, and the impossibility of quick identification of each newly found object underline the need to compare the found object with a unique track for its re-identification, position refinement, and motion prediction.

The subproblem of comparing a detected object with its track is reduced to a linear assignment problem with a certain cost function which is effectively solved by the methods described previously [5, 6]. There are various approaches to choosing a cost function. The method [4] assumes a linear motion model of an object the position of which undergoes significant changes in successive frames. The cost function in the assignment problem is defined as the ratio of the intersection area to the union area of the prediction and detection bounding boxes of the object, i.e. IoU (intersection over union) or the Jaccard similarity coefficient. This approach provides sufficient stability for simple tracking scenarios, though in some cases it yields track breaks related to objects that were missed by the detector in the video for several consequent frames. This problem is associated with the complex movement of such objects which do not fit the linear model.

Re-identification of objects temporarily disappearing from the frame is proposed to solve the problem of track breaks [7]. A characteristic description (descriptor) of the object image is constructed to make it possible to compare image areas containing an object, repeat object recognition, and bind them to the existing track. A convolutional neural network (CNN) is utilized to obtain the object descriptor; the outputs of its last fully connected layer form the feature space describing images that contain objects that are relevant to track. The cosine similarity measure of feature vectors is introduced in this linear space to match the image areas that are predicted to contain an object as track extensions image areas related to new object detections. The application of the neural network leads to a decrease in incorrect object re-identifications but at the cost of an increase in the computational complexity of the tracking algorithm.

The procedure of descriptors pre-selection based on their "quality" indicator for the re-identification task is proposed to increase the computational efficiency and to reduce the probability of track breaks and incorrect associations [8]. The indicator of "quality" is defined by the measure of its usefulness for the recognition problem, which is proposed to be estimated based on the confidence of the detector or specially developed algorithms evaluating the biometric indicators of image quality.

15.2 Multiple Object Tracking Problem

Let $\{\mathbf{I}_t\}_{t=1}^T$ be the image sequence of the video stream with a fixed discretization step in the time domain where T is the number of frames in the video sequence, $\mathbf{I}_t \in \mathbf{I} \subset \mathbf{R}^{C \times H \times W}$, and C, H, W denote the number of color channels, the height, and the width of the image, respectively. The set of objects $O = \{j \mid j \in \overline{1, N}\}$ where N denotes the number of objects is presented in the video sequence $\{\mathbf{I}_t\}_{t=1}^T$.

Let $\mathbf{y}_{j,1}, \ldots, \mathbf{y}_{j,T}$ be the observations of the object j at each moment of time t, while $\mathbf{x}_{j,1}, \ldots, \mathbf{x}_{j,T}$ are the true states of this object at all of the time moments t that are needed to be estimated. Let $\mathbf{z}_{1,t}, \ldots, \mathbf{z}_{M_t,t}$ be observations of unknown objects at the moment of time t where M_t denotes the number of detected objects of unknown identity in this frame. The problem of MOT in the online mode relates to finding the joint estimates of the true states $X_t = \{\mathbf{x}_{j,t}, \mid j = 1, \ldots, N\}$ for all N objects based on all observations of these objects in previous frames $Y_t = \{\mathbf{y}_{j,1}, \ldots, \mathbf{y}_{j,t-1}, \mid j = 1, \ldots, N\}$ and the observations $Z_t = \{\mathbf{z}_{1,t}, \ldots, \mathbf{z}_{M_t,t}\}$ of unknown objects in this frame t.

The states of the object j are assumed to have the property of the Markov process, that the true state $\mathbf{x}_{j,t}$ depends only on the true state $\mathbf{x}_{j,t-1}$ at the previous moment, and the observations $\mathbf{y}_{j,t}$ are determined exclusively by the true states $\mathbf{x}_{j,t}$ of this object at the same moment of time. Given the assumption that true state $\mathbf{x}_{j,t}$ obeys the linear relationship, this process for each object j is described by the model of a linear dynamic system (LDS) and all distributions are defined by the Gaussian model:

$$p(\mathbf{x}_{j,t} \mid \mathbf{x}_{j,t-1}) = N(\mathbf{A}_j \mathbf{x}_{j,t-1}, \boldsymbol{\Gamma}_j),$$
$$p(\mathbf{y}_{j,t} \mid \mathbf{x}_{j,t}) = N(\mathbf{B} \mathbf{x}_{j,t}, \boldsymbol{\Sigma}_j),$$
$$p(\mathbf{x}_{j,1}) = N(\boldsymbol{\mu}_{j,0}, \boldsymbol{\Gamma}_{j,0}),\tag{15.1}$$

where $N(\boldsymbol{\mu}, \boldsymbol{\Gamma})$ is the normal distribution with the expectation $\boldsymbol{\mu}$ and the covariance matrix $\boldsymbol{\Gamma}$; $p(\mathbf{x}_{j,t} \mid \mathbf{x}_{j,t-1})$ describes the evolution of the object states and defines its linear motion model with the matrix $\mathbf{A}_j$ and the random noise with the covariance matrix $\boldsymbol{\Gamma}_j$; $p(\mathbf{y}_{j,t} \mid \mathbf{x}_{j,t})$ describes the linear relationship between the true state of the object and its observations with the transformation matrix B and the addition of the normal random noise with the covariance matrix $\boldsymbol{\Sigma}_j$; $p(\mathbf{x}_{j,1})$ denotes the prior normal distribution of the true states at the initial moment of time with the mean value $\boldsymbol{\mu}_{j,0}$ and the noise with the covariance matrix $\boldsymbol{\Gamma}_{j,0}$.

The problem of object states estimation is solved by maximization of the joint a posteriori probability density of their true states X_t given the known past observations of these objects Y_t and the observations of unknown objects Z_t detected in this frame:

$$\max_{X_t} p(X_t \mid Y_t, Z_t)\tag{15.2}$$

The true states are defined by the vector $\mathbf{x} = (u, v, s, a, \dot{u}, \dot{v}, \dot{s})^{\mathrm{T}}$, where u and v denote horizontal and vertical coordinates of the center point of the object, s and a represent the area and the aspect ratio of the rectangle bounding the object, $\dot{u}, \dot{v}, \dot{s}$ are the change rates of the corresponding parameters.

Therefore, the MOT problems in the online mode correspond to the estimation of the true states X_t for all objects N by solving the problem (15.2) with the introduced assumptions.

15.3 Decomposition of Tracking into Filtering and Assignment Tasks

The observed objects are assumed to not interact with each other and move independently. Let us introduce an additional assumption related to the independence of the true states of different objects from each other and the independence of the true state of the object from the past observations of other objects. Thus, the joint posteriori probability density of the true states X_t can be rewritten in the form factorized by objects:

$$p(X_t \mid Y_t, Z_t) = \prod_{j=1}^{N} p(\mathbf{x}_{j,t} \mid Y_t, Z_t) = \prod_{j=1}^{N} p(\mathbf{x}_{j,t} \mid \mathbf{y}_{j,1}, \dots, \mathbf{y}_{j,t-1}, Z_t)\tag{15.3}$$

Let us introduce a realistic assumption that the object state $\mathbf{x}_{j,t}$ is statistically related only to the observations of one object detected in the frame which is unknown in advance, and the states of other objects $\mathbf{x}_{i,t}$, $i \neq j$ do not depend on the observations of this object. Then taking this assumption into account, equation (15.3) can be rewritten as follows:

$$p(X_t \mid Y_t, Z_t) = \prod_{j=1}^{N} \left\{ \sum_{i=1}^{M_t} a_{i,j} p(\mathbf{x}_{j,t} \mid \mathbf{y}_{j,1}, \dots, \mathbf{y}_{j,t-1}, \mathbf{z}_{i,t}) \right\} =$$

$$= \prod_{j=1}^{N} p \left(\mathbf{x}_{j,t} \mid \mathbf{y}_{j,1}, \dots, \mathbf{y}_{j,t-1}, \sum_{i=1}^{M_t} a_{i,j} \mathbf{z}_{i,t} \right) =$$

$$= \prod_{j=1}^{N} p(\mathbf{x}_{j,t} \mid \mathbf{y}_{j,1}, \dots, \mathbf{y}_{j,t-1}, \mathbf{y}_{j,t}) \qquad (15.4)$$

under the constraints:

$$a_{i,j} \in \{0,1\}, i \in \overline{1, M_t}, j \in \overline{1, N},$$

$$\sum_{i=1}^{M_t} a_{i,j} = 1, j \in \overline{1, N},$$

$$\sum_{j=1}^{N} a_{i,j} = 1, i \in \overline{1, M_t}, \qquad (15.5)$$

where

$$\mathbf{y}_{j,t} = \sum_{i=1}^{M_t} a_{i,j} \mathbf{z}_{i,t} \qquad (15.6)$$

denotes the observation that is assigned to the object j. Constraint (15.5) determines the correspondence of one observation $\mathbf{z}_{i,t}$ to one state $\mathbf{x}_{j,t}$, i.e., the coefficient $a_{i,j} = 1$ if $\mathbf{x}_{j,t}$ depends on the observation $\mathbf{z}_{i,t}$ else $a_{i,j} = 0$. Thus, the coefficients $a_{i,j}$ determine the choice of the state posteriori probability density of the object that the observation $\mathbf{z}_{i,t}$ is associated with.

To estimate the true states X_t, the optimal choice of the observation $\mathbf{y}_{j,t}$ among all observations of unknown identities $\mathbf{z}_{i,t}$ for each object j could be organized by maximization of the likelihood function of the posterior distribution (15.4). Let there be some state estimate $\mathbf{x}_{j,t} = \widetilde{\mathbf{x}}_{j,t}$ for the current frame that can be obtained based on the previous state $\mathbf{x}_{j,t-1}$ via the layer-wise deep stacking (LDS) model (15.1). Therefore, in accordance with (15.4), the likelihood function $L_{\widetilde{\mathbf{x}}_{j,t}}$ for object j could be defined as follows:

$$L_{\widetilde{\mathbf{x}}_{j,t}}(\mathbf{y}_{j,1}, \dots, \mathbf{y}_{j,t-1}, \mathbf{z}_{i,t}) = p(\widetilde{\mathbf{x}}_{j,t} \mid \mathbf{y}_{j,1}, \dots, \mathbf{y}_{j,t-1}, \mathbf{z}_{i,t}). \qquad (15.7)$$

Considering (15.6) the expression for maximization of the likelihood function $L_{\tilde{\mathbf{x}}_{j,t}}$ of object j could be rewritten into the following form:

$$
\max_{\mathbf{z}_{i,t}\in Z_t} L_{\tilde{\mathbf{x}}_{j,t}}(\mathbf{y}_{j,1},\dots,\mathbf{y}_{j,t-1},\mathbf{z}_{i,t}) = \max_{a_{i,j}} L_{\tilde{\mathbf{x}}_{j,t}}\left(\mathbf{y}_{j,1},\dots,\mathbf{y}_{j,t-1},\sum_{i=1}^{M_t}a_{i,j}\mathbf{z}_{i,t}\right) =
$$

$$
= \max_{a_{i,j}}\sum_{i=1}^{M_t}a_{i,j}L_{\tilde{\mathbf{x}}_{j,t}}(\mathbf{y}_{j,1},\dots,\mathbf{y}_{j,t-1},\mathbf{z}_{i,t}) \tag{15.8}
$$

under constraint (15.5).

The likelihood function of the joint posterior distribution (15.4) needs to be optimized in order to find joint assignments of observations Z_t to all states X_t. Taking into account the expression (15.8), equation (15.4) can be rewritten to obtain estimations $X_t = \tilde{X}_t$ for the current frame:

$$
\max_{a_{i,j}} p(\tilde{X}_t \mid Y_t, Z_t) = \max_{a_{i,j}} \prod_{j=1}^{N}\left\{\sum_{i=1}^{M_t}a_{i,j}L_{\tilde{\mathbf{x}}_{j,t}}(\mathbf{y}_{j,1},\dots,\mathbf{y}_{j,t-1},\mathbf{z}_{i,t})\right\} \tag{15.9}
$$

under the constraint (15.5). Let us transform the problem (15.9) into the following form:

$$
\max_{a_{i,j}} \log p(\tilde{X}_t \mid Y_t, Z_t) = \max_{a_{i,j}} \sum_{j=1}^{N}\log\left(\sum_{i=1}^{M_t}a_{i,j}L_{\tilde{\mathbf{x}}_{j,t}}(\mathbf{y}_{j,1},\dots,\mathbf{y}_{j,t-1},\mathbf{z}_{i,t})\right) =
$$

$$
= \max_{a_{i,j}} \sum_{j=1}^{N}\sum_{i=1}^{M_t}a_{i,j}\log L_{\tilde{\mathbf{x}}_{j,t}}(\mathbf{y}_{j,1},\dots,\mathbf{y}_{j,t-1},\mathbf{z}_{i,t})
$$

$$
= \min_{a_{i,j}} \sum_{j=1}^{N}\sum_{i=1}^{M_t}a_{i,j}C_{i,j} \tag{15.10}
$$

Under the constraint (15.5), where $C_{i,j} = -\log L_{\tilde{\mathbf{x}}_{j,t}}(\mathbf{y}_{j,1},\dots,\mathbf{y}_{j,t-1},\mathbf{z}_{i,t})$ denotes the cost matrix that corresponds to the assignment of i-th observation $\mathbf{z}_{i,t}$ to the object j, $a_{i,j}$ is the matrix of assignment of i-th observation to object j.

Considering that the solution of the problem (15.10) is represented in the form (15.6) and taking into account the expression (15.4), the tracking problem (15.2) can be rewritten as follows:

$$
\max_{X_t} \log p(X_t \mid Y_t, Z_t) = \max_{\mathbf{x}_{1,t},\dots,\mathbf{x}_{N,t}} \log \prod_{j=1}^{N}\{p(\mathbf{x}_{j,t} \mid \mathbf{y}_{j,1},\dots,\mathbf{y}_{j,t-1},\mathbf{y}_{j,t})\} =
$$

$$
= \sum_{j=1}^{N}\max_{\mathbf{x}_{j,t}} \log p(\mathbf{x}_{j,t} \mid \mathbf{y}_{j,1},\dots,\mathbf{y}_{j,t-1},\mathbf{y}_{i,t}).
$$

Furthermore, based on the solution of assignment problem (15.10), the joint estimation problem (15.2) of objects' states X_t leads to subtasks of individual estimation of the true state for each object j:

$$x_{j,t}^* = \arg\max_{x_{j,t}} \log p(x_{j,t} \mid y_{j,1}, \ldots, y_{j,t-1}, y_{j,t}), \tag{15.11}$$

where $x_{j,t}^*$ denotes the estimation of the true state for the object j in the current frame t.

Thus, the original problem (15.2) of joint estimation of objects' true states is reduced to three sequential problems: the problem of obtaining an initial estimate $x_{j,t} = \widetilde{x}_{j,t}$ for the current frame from the previous one based on the LDS model (15.1), the problem (15.10) of assignment of the set of observations Z_t on a given frame, and a set of objects with the true states $\widetilde{X}_t$ and the problem (15.11) of refining the estimate of the object true $x_{j,t}^*$ based on the observation (15.6) assigned to the object j in the previous step (15.10).

The solution of the problems related to obtaining the preliminary estimate $\widetilde{x}_{j,t}$ of the state for the current frame from the previous one (predict) and its update $x_{j,t}^*$ via maximization of the logarithm of the posterior distribution (15.11) for each object j under the introduced assumptions (15.1) is specified by the linear Kalman filter [9, 10] and consists of two steps. At the first step, the distribution of the true states $x_{j,t}$ is predicted based on the information from previous frames:

$$p(x_{j,t} \mid y_{j,1}, \ldots, y_{j,t-1}) = N(x_{j,t} \mid \widetilde{\mu}_{j,t}, \widetilde{V}_{j,t}),$$
$$\widetilde{\mu}_{j,t} = A\mu_{j,t-1},$$
$$\widetilde{V}_{j,t} = \Gamma + AV_{j,t-1}A^\mathsf{T}. \tag{15.12}$$

At the second step, the preliminary estimate is updated in accordance with the new information:

$$p(x_{j,t} \mid y_{j,1}, \ldots, y_{j,t}) = N(x_{j,t} \mid \mu_{j,t}, V_{j,t}) = N(y_{j,t} \mid Bx_{j,t}, \Sigma)N(x_{j,t} \mid \widetilde{\mu}_{j,t}, \widetilde{V}_{j,t}), \tag{15.13}$$

$$\mu_{j,t} = \widetilde{\mu}_{j,t} + K_{j,t}(y_{j,t} - B\widetilde{\mu}_{j,t}),$$
$$V_{j,t} = (I - K_{j,t}B)\widetilde{V}_{j,t},$$
$$K_{j,t} = \widetilde{V}_{j,t}B^\mathsf{T}(B\widetilde{V}_{j,t}B^\mathsf{T} + \Sigma)^{-1}, \tag{15.14}$$

where $y_{j,t}$ is defined by the (15.6) and the solution of the assignment (15.10).

15.4 Cost Matrix Adjustment in Assignment Problem Based on Re-Identification with Pre-Filtering of Descriptors by Quality

As objects are detected and localized for the given frame, the image areas that are predicted to contain objects of unknown identities are to be compared with the previously found objects related to the corresponding tracks by solving the assignment problem (15.10). Such a procedure is needed to update (15.14) the preliminary estimate of each object tracking based on the assigned observation. The updated track predictions are stored for all tracks, and the next frame is then analyzed.

The assignment problem can be effectively solved using the existing algorithms [5, 6]. In accordance with (15.10), the elements of the cost matrix $\mathbf{C}_{i,j}$ are defined via the logarithm of the posterior distribution likelihood function (15.13). The prediction $\widetilde{\mathbf{x}}_{j,t} = \widetilde{\boldsymbol{\mu}}_{j,t}$ obtained with the Kalman filter (15.12) from the previous frame is utilized as the preliminary estimate $\widetilde{\mathbf{x}}_{j,t}$ for the likelihood function related to the current frame. Thus, in accordance with (15.10) and (15.13), the cost matrix $\mathbf{C}_{i,j}$ is defined by the following expression:

$$
\begin{aligned}
\mathbf{C}_{i,j} &= -\log N(\mathbf{z}_{i,t} \mid \mathbf{B}\widetilde{\boldsymbol{\mu}}_{j,t}, \boldsymbol{\Sigma}) = \\
&= \frac{1}{2}(\mathbf{z}_{i,t} - \mathbf{B}\widetilde{\boldsymbol{\mu}}_{j,t})^{\mathsf{T}} \boldsymbol{\Sigma}^{-1}(\mathbf{z}_{i,t} - \mathbf{B}\widetilde{\boldsymbol{\mu}}_{j,t}) - \frac{m}{2}\log 2\pi - \frac{1}{2}\log |\boldsymbol{\Sigma}|
\end{aligned}
\tag{15.15}
$$

where m denotes the dimension of the vector $\mathbf{x}$. Thus, the elements of the assignment cost matrix are determined by the distance measure, the proximity of the observation of the unknown object, and the estimated true state of the object corresponding to the track.

Given complex observation conditions, such as intersections of different tracks in the case of their dense accumulation, or track breaks due to missing detections of this object in intermediate frames, the criterion (15.15) is not adequately reliable. To increase the stability of assignments, deep learning applied to the object tracking (DeepSORT) method [7] defines the elements of the cost matrix via comparisons of image descriptors that contain the characteristic features of the target objects. The mapping of image space into the descriptor space is defined to obtain the desired image descriptors. The descriptor space is a linear space of a fixed dimension, on which the scalar product operation is defined:

$$
\mathbf{d}_i = \mathbf{h}(\mathbf{I}_i \mid \boldsymbol{\theta})
$$

where $\mathbf{d}_i$ denotes the descriptor of the image $\mathbf{I}_i$ related to the object i and $\boldsymbol{\theta}$ represents the parameters of the mapping function. The CNN with L_2-normalized outputs can be utilized as the mapping $\mathbf{h}(\mathbf{I}_i \mid \boldsymbol{\theta})$ where $\boldsymbol{\theta}$ is the trainable parameters. Such neural model $\mathbf{h}(\mathbf{I}_i \mid \boldsymbol{\theta})$ can be trained for the task of desired objects classification with the special loss function, for instance, Cosine Softmax Cross-Entropy [11]. This choice of the model and the training method provides an opportunity for image descriptors comparison based on the cosine similarity measure:

$$c(\mathbf{d}_i, \mathbf{d}_j) = \frac{\langle \mathbf{d}_i, \mathbf{d}_j \rangle}{\|\mathbf{d}_i\| \|\mathbf{d}_j\|}$$

The re-identification approach can be applied to minimize the error in the assignment problem in difficult observation conditions. Therefore, the criterion defining the assignment of the unknown object to the existing tracks is determined based on the comparison of the descriptor of its image with the descriptors of images related to previous observations in the corresponding tracks. Object descriptors from frame to frame are subjected to nonlinear distortions, the model of which is unknown. Therefore, the entire set of descriptors obtained from all previous observations presented in the track should be used in the re-identification procedure. Thus, the measure of the distance between the descriptor of the unknown object i and the track j is determined based on the choice of the most similar descriptor from history in accordance with the following expression:

$$\ell_{i,j} = [1 - \max \{ c(\mathbf{d}_{i,t}, \mathbf{d}_{j,1}), \ldots, c \left(\mathbf{d}_{i,t}, \mathbf{d}_{j,t-1} \right) \}] \tag{15.16}$$

where $\ell_{i,j}$ denotes the matrix, composed of the corresponding distances the object i and the track j $\mathbf{d}_{i,t} = \mathbf{h}(\mathbf{I}_{i,t} \mid \boldsymbol{\theta})$, $\mathbf{I}_{i,t}$ is the image containing object i in the frame t.

The cost matrix for the assignment problem can be adjusted by adding the distance measure between the object descriptor i to the track j to the expression (15.15):

$$\widetilde{\mathbf{C}}_{i,j} = \mathbf{C}_{i,j} + r\ell_{i,j} \tag{15.17}$$

where r is the coefficient. Thus, the modified cost matrix takes into account not only the proximity of the new observations to the object state predictions obtained by the Kalman filter but also the proximity of the object descriptor to the previously accumulated descriptors related to the corresponding track.

To obtain a modified cost matrix (15.17), as in the DeepSORT approach [7], comparisons (15.16) with all descriptors from the corresponding track are needed to be made. Therefore, to overcome this disadvantage, there is a point to limit the number of descriptors from the track that are utilized for comparison. Such an adjustment reduces the complexity of cost matrix $\widetilde{\mathbf{C}}_{i,j}$ computation and limits the dependence of the re-identification algorithm complexity on the length of the track.

Since not all descriptors accumulated in the track are equally informative, they should be selected via the special weighting procedure [8]. To organize the weighting process, let us define the "quality" as the function $q_{j,\tau} = q(\mathbf{I}_{j,\tau})$ which estimates the informativeness of the image $\mathbf{I}_{j,\tau}$, $\tau = 1, \ldots, t-1$ containing the object j in terms of the object recognition task. Let $q_{j,(k)}$ be the k-th order statistic of the set $\{q_{j,1}, \ldots, q_{j,t-1}\}$ that is computed by sorting the "quality" indicators in descending order for the track j. Then the weight $w_{j,\tau}$ of each observation in the frame τ from the corresponding track is defined as:

$$w_{j,\tau} = \Theta(q_{j,(k)} - q_{j,\tau}) \tag{15.18}$$

where Θ denotes the Heaviside function. Therefore, the expression for re-identification (15.16) can be rewritten as below:

$$\ell_{ij} = [1 - \max\{c(\mathbf{d}_{i,t}, \mathbf{d}_{j,1})w_{j,1}, \ldots, c(\mathbf{d}_{i,t}, \mathbf{d}_{j,t-1})w_{j,t-1}\}]$$

The described approach is based on the selection of image areas obtained from observations contained in the track that would be closest to the data that was utilized for training the neural network $\mathbf{h}(\mathbf{I}\,|\,\boldsymbol{\theta})$ that defines the mapping into the descriptors space. As a fixed number of the most relevant candidates is selected for re-identification, the majority of comparisons with uninformative descriptors obtained from images that do not contain desired objects are successfully excluded from the algorithm.

Various approaches to constructing a quality assessment function $q(\mathbf{I})$ are described in biometrics [12–14] and, in particular, in the field of face recognition [15, 16]. Quality in biometrics refers to the expected accuracy of object recognition from the given image, that is the value that correlates with the usefulness of the object's image for the recognition problem [17]. The indicators based on contrast, brightness, focusing, sharpness, and illumination that are approximated with Gaussian distributions are proposed for image quality assessment in [17]. The main disadvantage of this method is the utilization of the Gaussian approximation that may be not a feasible assumption in practice.

The method of quality assessment proposed in [18] does not rely on the assumptions about the indicator distribution. The problem of quality assessment is reduced to supervised learning, specifically, a regression problem where the answers are obtained as a result of the data markup by assessors, and the support vector machine is used as a machine learning algorithm [19]. Preliminarily, training set images are mapped into deep features by a convolutional neural network that is pre-trained for solving the classification problem.

The alternative approach to image quality assessment could be considered in order to avoid any additional calculations. The correspondence of an object to the data which was utilized in the descriptor extraction algorithm training correlates with the probability of object presence in the given image area. Thus, the detector's confidence can be treated as a simple estimate of the "quality" indicator of an image area containing an object. The assumption of such correlation provides an opportunity to select for re-identification objects corresponding to the areas with the highest value of the detector's confidence.

In practice, the assignment problem (15.10) with a modified cost matrix (15.15) can be efficiently solved with the Jonker–Volgenant algorithm [6]. Thresholds for components $\mathbf{C}_{i,j}$ and $\ell_{i,j}$ cost matrices (15.15) are introduced to control the correctness of assignments. Exceeding the threshold excludes the corresponding assignment. In addition, if within a specified time interval there are no new observations that are assigned to the track, the track is considered to be finished and terminates.

15.5 Computational Experiments

Two datasets of video images MOT20-01 and MOT20-02 [20] were utilized as test data to evaluate various re-identification-based tracking methods. The datasets are video sequences of pedestrians taken by a CCTV camera in a crowded place from two different angles. Most objects move toward the camera or across the frame. A description of the datasets is given in Table 15.1.

In the experimental setup, two different neural detector models that are SSD and RetinaNet were used for detection and rough localization of objects [1, 3]. Both models were pre-trained on facial images. The models differ significantly in complexity levels: the number of trainable parameters for the SSD and RetinaNet detectors is 4 and 45 million, respectively.

Two descriptors selection methods that are based on the "quality" assessment were considered for the re-identification procedure. The first quality assessment method utilized the detector confidence as a "quality" indicator, this approach is denoted as Deep-Conf-SORT. The second method corresponded to the "quality" assessment algorithm proposed in [18], this approach is denoted as Deep-QA-SORT.

The purpose of the experiments was to compare the MOT methods with re-identification and pre-filtering of objects with the original Deep-SORT approach [7] with the same hyperparameters. A set of standard quality measures was selected as criteria for the tracking system operation [21–23]: Precision $= \frac{TP}{TP+FP}$ where FP denotes the number of type I errors (false positives), TP is the number of true positives; Recall $= \frac{TP}{TP+FN}$ where FN is the number of type II errors (false negatives), TP is the number of true positives; IDS denotes the total amount of identity for all tracks; the processing rate in frames per second (Hz) that does not take into account the computational time of the "quality" assessment algorithm since it is built into the face recognition system and is available for each detection without additional calculations.

To estimate the optimal number of descriptors selected for comparison in the re-identification procedure using formula (15.18), experimental dependences of IDS on the parameter k were estimated for all considered methods and test datasets. The function of IDS for all the studied methods has a minimum in the vicinity of the point $k = 5$ (Figure 15.1). Thus, this value that is characteristic of the dataset scene geometry was used for further experiments.

Table 15.1 Datasets description.

Dataset	Frame rate (Hz)	Mean number of objects per frame	Number of frames	Number of tracks
MOT20-01	25	42.1	429	90
MOT20-02	25	72.7	2782	296

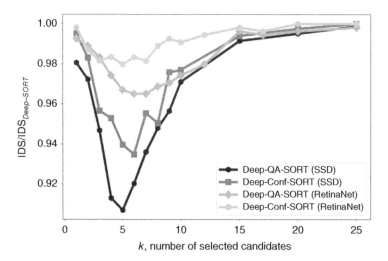

Figure 15.1 Dependence of the ratio of the IDS method to the IDS of the basic method on the number of selected candidates.

The results of computational experiments are presented in Tables 15.2 and 15.3. The results reveal that both methods for descriptor pre-selection for re-identification lead to the more computationally efficient operation of the tracking system compared to the basic approach at the same level of Precision

Table 15.2 Results of the MOT20-01 dataset.

Detector	Tracking method	Precision	Recall	IDS	Operation rate (Hz)
SSD	Deep-SORT	0.847	**0.895**	203	39.2
	Deep-Conf-SORT	**0.850**	0.890	202	**44.7**
	Deep-QA-SORT	0.848	0.892	**195**	44.7
RetinaNet	Deep-SORT	0.871	**0.922**	168	40.9
	Deep-Conf-SORT	**0.875**	0.919	166	**46.0**
	Deep-QA-SORT	0.873	0.921	**162**	46.0

Table 15.3 Results of the MOT20-02 dataset.

Detector	Tracking method	Precision	Recall	IDS	Operation rate (Hz)
SSD	Deep-SORT	0.851	**0.910**	828	22.6
	Deep-Conf-SORT	**0.860**	0.904	778	**31.8**
	Deep-QA-SORT	0.855	0.907	**751**	31.8
RetinaNet	Deep-SORT	0.887	**0.945**	543	22.4
	Deep-Conf-SORT	0.891	0.941	533	**31.4**
	Deep-QA-SORT	**0.896**	0.939	**526**	31.4

and Recall, and they also reduce the number of incorrect identity switches (IDS) between tracks. The relative gain in IDS is higher when using the SSD detector compared to RetinaNet. This observation can be explained by the fact that the SSD model has less complexity in terms of the number of parameters, it is less accurate, which leads to a larger number of false detections, which are eliminated by the proposed methods of quality assessment in the re-identification procedure.

15.6 Conclusion

To sum up, the original problem of MOT in the online mode is reduced to a set of subtasks of linear state filtering independently for each object and assigning new observations to the observed tracks with the cost matrix determined by the distance of new observations to the filtered states. In the assignment subtask, the adjustment of the cost matrix with the help of the additional term obtained on the basis of candidate re-identification by comparing with descriptors from the trajectory is considered. A generalization of the basic DeepSORT method [7] is presented. The modification includes using of the pre-filtering object descriptors in a trajectory. The pre-filtrating is based on the assessment of the "quality" indicator and it selects the most informative descriptors by previous observations. The computational experiments revealed the effectiveness of the tracking methods that include the pre-filtering procedure, an increase in the speed of the tracking system up to 1.4 times, and a decrease of identity switches in comparison with the basic approach with the same level of precision and recall.

Acknowledgments

The work is supported by the Russian Foundation of Basic Research (Grant no. 21-51-53019).

References

1 Liu, W., Anguelov, D., Erhan, D. et al. (2016). Single shot multibox detector. *Computer Vision – ECCV* 21–37.

2 Redmon, J., Divvala, S., Girshick, R., and Farhadi, A. (2016). You only look once: Unified, real-time object detection. *IEEE Conference on Computer Vision and Pattern Recognition (CVPR)* 779–788.

3 Lin, T.-Y., Goyal, P., Girshick, R. et al. (2017). Focal loss for dense object detection. *IEEE International Conference on Computer Vision (ICCV)* 2999–3007.

4 Bewley, A., Ge, Z., Ott, L. et al. (2016). Simple online and realtime tracking. *IEEE International Conference on Image Process* 3464–3468.

5 Kuhn, H.W. and Yaw, B. (1955). The Hungarian method for the assignment problem. *Naval Res. Logist. Quart* 83–97.

6 Jonker, R. and Volgenant, A. (1987). A shortest augmenting path algorithm for dense and sparse linear assignment problems. *Computing* 38: 325–340.

7 Wojke, N., Bewley, A., and Paulus, D. (2017). Simple online and realtime tracking with a deep association metric. *IEEE International Conference on Image Processing (ICIP)* 3645–3649.

8 Grigorev, A. and Gneushev, A. (2021). Re-identification with pre-filtering by image quality for multiple object tracking. *Information* 409–418. (in Russian).

9 Kalman, R.E. (1960). A new approach to linear filtering and prediction problems. *Transactions of the ASME–Journal of Basic Engineering* 82: 35–45.

10 Barker, A.L., Brown, D.E., and Martin, W.N. (1995). Bayesian estimation and the Kalman filter. *Comput. Math. Appl.* 30 (10): 55–77.

11 Wojke, N. and Bewley, A. (2018). Deep cosine metric learning for person re-identification. *IEEE Winter Conference on Applications of Computer Vision (WACV)* 748–756.

12 Novik, V., Matveev, I.A., and Litvinchev, I. (2020). Enhancing iris template matching with the optimal path method. *Wireless Networks* 26 (7): 4861–4868.

13 Matveev, I. (2010). Detection of iris in image by interrelated maxima of brightness gradient projections. *Applied and Computational Mathematics.* 9 (2): 252–257.

14 Matveev, I.A., Novik, V., and Litvinchev, I. (2018). Influence of degrading factors on the optimal spatial and spectral features of biometric templates. *Journal of Computational Science* 25: 419–424.

15 Desyatchikov, A.A., Kovkov, D.V., Lobantsov, V.V. et al. (2006). A system of algorithms for stable human recognition. *Journal of Computer and Systems Sciences International* 45 (6): 958–969.

16 Lobantsov, V.V., Matveev, I.A., and Murynin, A.B. (2011). Method of multi-modal biometric data analysis for optimal efficiency evaluation of recognition algorithms and systems. *Pattern Recognition and Image Analysis.* 21 (2): 494–497.

17 Abaza, A., Harrison, M.A., and Bourlai, T. (2012). Quality metrics for practical face recognition. *IAPR International Conference on Pattern Recognition (ICPR)* 3103–3107.

18 Best-Rowden, L. and Jain, A.K. (2018). Learning face image quality from human assessments. *IEEE Transactions on Information Forensics and Security* 13: 3064–3077.

19 Suykens, J., Van Gestel, T., De Brabanter, J. et al. (2002). Least squares support vector machines. *World Scientific* 29–70.

20 Dendorfer, P., Rezatofighi, H., Milan, A. et al. (2003). MOT20: a benchmark for multi object tracking in crowded scenes. *arXiv.09003[cs]*, 2020.

21 Wu, B. and Nevatia, R. (2006). Tracking of multiple, partially occluded humans based on static body part detection. *IEEE Computer Society Conference on Computer Vision and Pattern Recognition (CVPR'06)* 951–958.

22 Bernardin, K. and Stiefelhagen, R. (2008). Evaluating multiple object tracking performance: the clear mot metrics. *EURASIP Journal on Image and Video Processing* 2008: 1–10.

23 Ristani, E., Solera, F., Zou, R. et al. (2016). Performance measures and a data set for multi-target, multi-camera tracking. *European Conference on Computer Vision* 17–35.

Index

Artificial Intelligence in Industry 4.0 and 5G Technology, First Edition.
Edited by Pandian Vasant, Elias Munapo, J. Joshua Thomas, and Gerhard-Wilhelm Weber.
© 2022 John Wiley & Sons, Inc. Published 2022 by John Wiley & Sons, Inc.

Printed and bound by CPI Group (UK) Ltd, Croydon, CR0 4YY
27/09/2022
03150477-0001